Marine production mechanisms

THE INTERNATIONAL BIOLOGICAL PROGRAMME

The International Biological Programme was established by the International Council of Scientific Unions in 1964 as a counterpart of the International Geophysical Year. The subject of the IBP was defined as 'The Biological Basis of Productivity and Human Welfare', and the reason for its establishment was recognition that the rapidly increasing human population called for a better understanding of the environment as a basis for the rational management of natural resources. This could be achieved only on the basis of scientific knowledge, which in many fields of biology and in many parts of the world was felt to be inadequate. At the same time it was recognised that human activities were creating rapid and comprehensive changes in the environment. Thus, in terms of human welfare, the reason for the IBP lay in its promotion of basic knowledge relevant to the needs of man.

The IBP provided the first occasion on which biologists throughout the world were challenged to work together for a common cause. It involved an integrated and concerted examination of a wide range of problems. The Programme was co-ordinated through a series of seven sections representing the major subject areas of research. Four of these sections were concerned with the study of biological productivity on land, in freshwater, and in the seas, together with the processes of photosynthesis and nitrogen fixation. Three sections were concerned with adaptability of human populations, conservation of ecosystems and the use of biological resources.

After a decade of work, The Programme terminated in June 1974 and this series of volumes brings together, in the form of syntheses, the results of national and international activities.

INTERNATIONAL BIOLOGICAL PROGRAMME 20

Marine Production Mechanisms

Edited by

M. J. Dunbar

Marine Sciences Centre, McGill University, Montreal

CAMBRIDGE UNIVERSITY PRESS

CAMBRIDGE

LONDON · NEW YORK · MELBOURNE

CAMBRIDGE UNIVERSITY PRESS
Cambridge, New York, Melbourne, Madrid, Cape Town, Singapore, São Paulo, Delhi

Cambridge University Press
The Edinburgh Building, Cambridge CB2 8RU, UK

Published in the United States of America by Cambridge University Press, New York

www.cambridge.org
Information on this title: www.cambridge.org/9780521105576

First published 1979
This digitally printed version 2009

A catalogue record for this publication is available from the British Library

Library of Congress Cataloguing in Publication data
Main entry under title:
Marine production mechanisms.
(International Biological Programme; 20)
Includes index.
1. Marine productivity. 2. Marine ecology.
I. Dunbar, Maxwell John. II. Series.
QH91.8.M34M37 574.5′2636 77-88675

ISBN 978-0-521-21937-2 hardback
ISBN 978-0-521-10557-6 paperback

Contents

List of Contributors *page* xiii

1 Introduction 1
M. J. Dunbar

2 Primary production in Frobisher Bay, Arctic Canada 9
E. H. Grainger

3 Primary production in some tropical environments 31
S. Z. Qasim

4 Biological productivity of some coastal regions of Japan 71
K. Hogetsu

5 Factors determining the productivity of South African coastal waters 89
J. R. Grindley

6 The Strait of Georgia Programme 133
T. R. Parsons

7 Biological production in the Gulf of St Lawrence 151
M. J. Dunbar

8 Patterns of the vertical distribution of phytoplankton in typical biotopes of the open ocean 173
H. J. Semina

9 The Dutch Wadden Sea 197
M. van der Eijk

10 Seaweed utilization in the Philippines 229
G. T. Velasquez

11 Trophic relationships in communities and the functioning of marine ecosystems: I Studies on trophic relationships in pelagic communities of the Southern Seas of the USSR and in the tropical Pacific 233
T. S. Petipa

12 Trophic relationships in communities and the functioning of marine ecosystems: II Some results of investigations on the pelagic ecosystem in tropical regions of the ocean 251
E. A. Shushkina & M. E. Vinogradov

13 Soviet investigation of the benthos of the shelves of the marginal seas 269
A. A. Neyman

14 Studies of trophic relationships in bottom communities in the southern seas of the USSR 285
E. A. Yablonskaya

15 Studies of the pattern of biotic distribution in the upper zones of the shelf in the seas of the USSR 317
A. N. Golikov & O. A. Scarlato
Index 333

Table des matières

Liste de collaborateurs page xiii
1 Introduction 1
M. J. Dunbar
2 Production primaire dans la baie Frobisher, Arctique Canadien 9
E. H. Grainger
3 Production primaire dans quelques milieux tropicaux 31
S. Z. Quasim
4 Productivité biologique de quelques regions côtières du Japon 71
K. Hogetsu
5 Facteurs déterminants la productivité des eaux côtières Sud-Africaines 89
J. R. Grindley
6 Le programme 'Detroit de Géorgie' 133
T. R. Parsons
7 Production biologique dans le Golfe Saint Laurent 151
M. J. Dunbar
8 Types de distribution verticale du phytoplancton dans des biotopes caractéristique en mer libre 173
H. J. Semina
9 La Wadden See néerlandaise 197
M. van der Eijk
10 L'utilisation des algues aux Philippines 229
G. T. Velasquez
11 Relations trophiques dans les communautés et fonctionnement des écosystèmes: I Recherches sur les relations tropiques dans les communautés pélagiques dans les mers australes de l'URSS et dans Le Pacifique tropical 233
T. S. Petipa
12 Relations trophiques dans les communautés et fonctionnement des écosystèmes: II Quelques résultats de recherche sur les écosystèmes pélagiques dans les régions océaniques tropicales 251
E. A. Shushkina & M: E. Vinogradov
13 Recherches soviétique sur le benthos des plateaux continentals des mers bordières 269
A. A. Neyman
14 Etude des relations trophiques des communautés benthiques dans les mers australes de l'URSS 285
E. A. Yablonskaya

15 Etudes sur les types de distribution biotique dans les zones supérieures du plateau continental dans les mers de l'URSS 317
A. N. Golikov & O. A. Scarlato
Table analytique 333

Содержание

Список авторов *страница* xiii

1 Введение 1
M. J. Dunbar

2 Первичная продукция во Фробишер Бей (Арктическая Канада) 9
E. H. Grainger

3 Первичная продукция в некоторых тропических условиях 31
S. Z. Qazim

4 Биологическая продуктивность некоторых прибрежных районов Японии 71
K. Hogetsu

5 Факторы, определяющие продуктивность южноафриканских прибрежных вод 89
J. R. Grindley

6 Программа исследований пролива Джорджия 133
T. R. Parsons

7 Биологическая продуктивность в заливе Сент Лоуренс 151
M. J. Dunbar

8 Типы вертикального распределения фитопланктона в типичных биотопах открытого океана 173
H. J. Semina

9 Голландское Ваdден Зее 197
M. van der Eijk

10 Использование морских водорослей на Филиппинах 229
G. T. Velasquez

11 Трофические взаимоотношения в сообществах и функционирование морских экосистем: I Изучение трофических взаимоотношений в пелагических сообществах южных морей СССР и в тропической Пацифике 233
T. S. Petipa

12 Трофические взаимоотношения в сообществах и функционирование морских экосистем: II Некоторые результаты исследований пелагической экосистемы в тропических районах океана 251
E. A. Shushkina & M. E. Vinogradov

13 Советские исследования бентоса шельфов окраинных морей 269
A. A. Neyman

14 Исследование пищевых взаимоотношений в донных сообществах южных морей СССР 285
E. A. Yablonskaya

15 Исследования типов распределения жизни в верхних зонах шельфа морей СССР 317
A. N. Golikov & O. A. Scarlato

Указатель 333

Contenido

Colaboradores *página* xiii

1 Introducción 1
M. J. Dunbar

2 Producción primaria en Frobisher Bay, Canadá ártico 9
E. H. Grainger

3 Producción primaria en algunos ambientes tropicales 31
S. Z. Qasim

4 Productividad biológica de algunas regiones costeras del Japón 71
K. Hogetsu

5 Factores determinantes de la productividad en aguas costeras de Sudáfrica 89
J. R. Grindley

6 El Programa del Estrecho de Georgia 133
T. R. Parsons

7 Producción biológica en el Golfo del San Lorenzo 151
M. J. Dunbar

8 Modelos de distribución vertical del fitoplancton en tipicos biotopos oceánicos de alta mar 173
H. J. Semina

9 El Mar de Wadden, en Holanda 197
M. van der Eijk

10 Empleo de las algas marinas en Filipinas 229
G. T. Velasquez

11 Relaciones tróficas en las comunidades y funcionamiento de los ecosistemas marinos: I Estudios sobre las relaciones tróficas de las comunidades pelágicas de los mares meridionales de la URSS, y del Pacifico tropical 233
T. S. Petipa

12 Relaciones tróficas en las comunidades y funcionamiento de los ecosistemas marinos: II Algunos resultados de las investigaciones sobre el ecosistema pelágico en las regiones tropicales del océano 251
E. A. Shushkina & M. E. Vinogradov

13 Investigaciones soviéticas sobre el bentos de la plataforma continental en los mares marginales 269
A. A. Neyman

14 Estudios sobre las relaciones tróficas en las comunidades de fondo de los mares meridionales de la URSS 285
E. A. Yablonskaya

15 Estudios sobre el modelo de distribución biótica en las zonas superiores de la plataforma continental, en los mares de la URSS 317
A. N. Golikov & O. A. Scarlato
Indice 333

List of contributors

Dunbar, M. J., Marine Sciences Centre, McGill University, Montreal PQ, Canada

Golikov, A. N., Zoological Institute, Academy of Sciences, Leningrad, V-164, USSR

Grainger, E. H., Arctic Biological Station, PO Box 400, Ste-Anne de Bellevue, Quebec, Canada H9X 3L6

Grindley, J. R., School of Environmental Studies, University of Cape Town, Rondebosch 7700, South Africa

Hogetsu, K., Department of Biology, Metropolitan University, Tokyo, Japan*

Neyman, A. A., All-Union Research Institute of Marine Fisheries and Oceanography, VNIRO, Moscow, USSR

Parsons, T. R., Institute of Oceanography, University of British Columbia, Vancouver, British Columbia, Canada V6T 1W5

Petipa, T. S., Institute for Biology of Southern Seas, Academy of Sciences of Ukraine, Sevastopol, USSR

Qasim, S. Z., National Institute of Oceanography, NIO Post Office, Dona Paula 403 004, Goa, India

Scarlato, O. A., Zoological Institute, Academy of Sciences, Leningrad, V-164, USSR

Semina, H. J., Institute of Oceanology, Academy of Sciences of USSR, Moscow, USSR

Shushkina, E. A., Institute of Oceanology, Academy of Sciences of USSR, Moscow, USSR

van der Eijk, M., Department of Inland Fisheries, Bezuidenhoutseweg 73, The Hague, The Netherlands

Velasquez, G. T., Department of Botany, University of the Philippines, Quezan City, Philippines

Vinogradov, M. E., Institute of Oceanology, Academy of Sciences of USSR, Moscow, USSR

Yablonskaya, E. A., All-Union Research Institute of Marine Fisheries and Oceanography, VNIRO, Moscow, USSR

* Present address: Uchikoshi-cho 715–313, Hachioji, Tokyo, Japan 192.

1. Introduction

M. J. DUNBAR

The International Biological Programme served a most useful purpose in bringing together biologists, particularly ecologists, from many different countries to focus their efforts on 'the biological basis of productivity and human welfare', as part of the objective of IBP was originally expressed. It was very successful in arranging for meetings between biological teams from many countries, in stimulating the exchange of ideas and in organizing joint programmes of research. The research covered land, sea and fresh water, and took the 'biome' approach in general. The marine programme (IBP/PM) was slow in starting, and for a time it almost looked as though it was not going to survive. The reason for this was not any lack of enthusiasm for international cooperation in marine research, but rather the conviction that this cooperation already existed and was in good health, in the form of the 'Conseil Permanent' for the North Atlantic, which was founded in 1902, ICNAF, and a similar and later development in the Pacific, and the International Indian Ocean Expedition. But it turned out, as IBP developed, that there was a useful job that IBP/PM could do, namely the concentration of international effort on specific problems of production, ecosystem structure and function, aquaculture, and the effects of mankind, usually deleterious, upon the marine environment.

This volume contains a selection of results from nine countries in the basic study of productivity and ecosystem structure. The chapters fall clearly under those two headings, with a little overlap between them, as follows: basic productivity studies, chapters by Qasim, Van der Eijk, Grainger, Grindley, Dunbar, Parsons, Hogetsu, Semina, and Velasquez; and studies on ecosystem structure and trophic relations, chapters by Petipa, Shushkina & Vinogradov, Yablonskaya, Neyman, and Golikov & Scarlato.

The world map of marine biological productivity has changed somewhat in the last few decades, partly because of the work of IBP. Compare, for instance, the classic map of Sverdrup (1955), which emphasized Antarctic waters and the equatorial east Pacific, with that produced for the 1972 IBP Fifth General Assembly by Rodin, Bazilevich & Rozov (1975), which gives greater prominence to the northern Indian Ocean, the waters of Indonesia, and the northeast Pacific. There has also been considerable revision of our ideas on the relative productivity of land, fresh water, and ocean. To quote from Rodin *et al.* (1975): 'the total annual increment of the world ocean exceeds 30000 percent of its total phytomass reserve. This is all too natural, since the ocean is dominated by unicellular plants with extremely rapid reproductive potentials. And yet, the annual increment of terrestrial communities is three times greater than the world's oceans (including the increment of rivers and lakes...). It is significant that even though the ocean covers 180 times the area of rivers and lakes, its phytomass reserves are only four times as large.' This latter statement is another way of saying that in the sea it is the shallow-water shelf areas that are productive, and that there are enormous oceanic areas in the Pacific, Atlantic, Indian, and Arctic oceans which are very low indeed in biological production.

This global pattern refers to primary production, which is responsible for some 99% of the total biomass on earth (Rodin *et al.*, 1975), and it is usually expressed as weight per unit surface area, as are most of the data presented in the following chapters. Considering the great difficulties involved in transforming data expressed as weight per unit volume (mg/1, for instance), which is the form in which the measurements are made at sea or in any aquatic environment, and the difficulties involved in the measurements themselves, it is surprising that the results are almost invariably published with apparent confidence and usually without any statement of their limits of accuracy. Mathematical approximations for the conversion of volume measurements to area estimates have been devised (for instance, see IBP Handbook No. 12, edited by Vollenweider (1969)), which no doubt carry with them an aura of authority for many workers, but not for all. Beyond that source of inaccuracy, it is then necessary to add the dimension of time, to express the production per unit area per day, or per year; another hazard to accuracy.

Finally there are the pitfalls and uncertainties of the original measurements themselves. Riley (1972) wrote that 'we still know very little about marine productivity', and 'there are several different ways of measuring ^{14}C fixation, and at least three other techniques of investigating primary production'. On the subject of the comparison between the Winkler and the ^{14}C methods, Qasim, who mentions the point in his chapter on IBP work in Indian waters, points out that the oxygen and ^{14}C techniques measure the rates of different reactions, and that therefore the two methods do not always agree (Fogg, 1969). High light intensities, the presence of oxidizing or reducing substances, organic matter in the sample, including the unpredictable presence of metabolites excreted by the plant cells themselves, and a high bacterial count; all these factors can increase the discrepancy between the results of the two methods. With all these hurdles to overcome, there is little doubt that our present values for biological production per unit area, in the sea and to a lesser extent in lakes, must be accepted as only very approximate.

The chapter by Grainger, in this volume, gives in Table 2.2 a number of measurements by several authors of phytoplankton production from the Arctic Ocean to the Caribbean, expressed as grams carbon per metre squared per year. It is interesting that Grainger's own estimate of the production in Frobisher Bay, in southwest Baffin Island, is from 41 to 70 g C/m^2/year, close to that of Steemann Nielsen (1958) for three stations in west Greenland. This suggests that the placing of Frobisher Bay in the marine subarctic, or the region of mixed Arctic and non-Arctic (Atlantic) water, by Dunbar (1972 and elsewhere), has meaning in terms of basic productivity, for there is no doubt that west Greenland is sub-Arctic rather than Arctic. On this theme, it is distressing for those who like to use the same term to mean the same thing in different regions and environments, to find the continued use of the term 'sub-Arctic' to apply to the north Pacific region from the south Alaskan coast across to Japan. In the north Pacific there is no trace of Arctic water (water from the upper layer of the Arctic Ocean), and therefore no mixture of Arctic and non-Arctic water. Grainger's chapter includes measurements of diatom production in sea ice, which agree with the results of others in the far north, and also contrast with the much lower production found in the most southerly region of sea ice coverage in North America found by Dunbar in the Gulf of St Lawrence; the reasons for this difference are touched upon in Dunbar's chapter.

In impressive contrast to the Baffin Island production is the much richer growth recorded in Indian coastal waters by Qasim, in a chapter that reviews IBP and other work in that region. The average annual net production for the entire coastal waters of India to a depth of 50 m is given as 1.19 g C/m^2/day, to compare with Grainger's maximum production in Frobisher Bay at the peak of the very short season in July–August of 1 g C/m^2/day, in the best year out of three; the annual average is of course very much lower. Qasim includes results of measurements in coral reef lagoons, a net production of 3.38 g C/m^2/day, and even higher rates for mangrove swamps and sea grass beds.

The chapter by Hogetsu describes very careful work in Japan at four localities, two of which, Hiuchi Nada, and Suruga and Sagami Bays, are in the southern region influenced by the warm Kuroshio current, the third, Sendai Bay, in the intermediate zone between the Kuroshio and the colder Oyashio current, and the fourth, Akkeshi Bay, in the Oyashio influence. The winner is Akkeshi Bay, with 295 g C/m^2/year, the other three regions producing less than half that amount. Hogetsu's chapter also summarizes Japanese work on particulate and dissolved organic matter, secondary production, and the total food web in the Japanese IBP areas.

The chapter by Grindley, on South African work, is remarkable for its very detailed account of the upwelling process in southwest African waters, the region of the Benguela current. He writes: 'The extraordinary richness of the upwelling of cold nutrient-rich water from the layer of the

Benguela current on the west coast of South Africa in comparison to adjacent oceanic waters has been referred to frequently. This upwelling of cold nutrient-rich water from the layer of the South Atlantic Central Water is well known, but earlier data did not permit any quantitative treatment of the process or of the consequent organic production.' The South African IBP studies have made this quantitative treatment possible, and the results constitute something of a landmark (seamark?) in southern hemisphere marine research. Grindley's chapter gives in addition a full account of work on red tide, zooplankton distribution and standing crops, benthos and fish fauna; in fact, everything. One interesting finding is an increased copepod diversity in the northern part of the upwelling area; according to present orthodoxy upwelling and the consequent increased production should give lesser, not greater, specific diversity. But this rule has exceptions, as in the evolution of polar systems; in this present case the increased diversity can be traced to the convection into the surface waters of species from intermediate depths.

Parsons' work in the Strait of Georgia in British Columbia has covered, and continues to cover, a wide spectrum of research on the mechanisms of marine productivity, and the Georgia Strait IBP Project is reviewed here. Parsons concludes with a most useful discussion of the relation of the Georgia Strait Programme to other aquatic systems, and of the use and development of models in production studies; a subject to which I return at the end of this introduction. My chapter on the Gulf of St Lawrence Programme covers certain work done as a corollary of the IBP survey on primary phytoplankton production, such as secondary production and the study of the ice biota, mentioned above in connection with the works of Grainger on ice biota in Frobisher Bay.

Dr van der Eijk (chapter 9) has summarized the extensive Dutch IBP work on the productivity of the Dutch Wadden Sea. This is a very thorough investigation of the chemical and physical environment, and the biological production, of that shallow coastal area, the southern part of the tidal flats that extend from the Netherlands to Denmark, which have achieved a reputation for rich flora and fauna, from diatoms to birds. The reputation is well earned, but the Dutch work emphasizes that in fact the primary production is not outstandingly high when compared to other marine regions, and that the secondary production is surprisingly low. Comparison with the German and Danish work on the Waddens will be interesting. The Dutch work demonstrates the importance, in the total organic accumulation, of both living and dead material brought in from the North Sea. It also gives a measure of the present eutrophication from fresh water influx.

Semina takes a different approach to the study of primary production, that of the vertical distribution of phytoplankton populations and the layering of concentration in accordance with the stability pattern of the water column, in this case principally in the Pacific Ocean. The production is expressed as standing crop, in terms of numbers of cells per litre, which is the classical manner and is still preferred by many biologists on the grounds that at least as much can be learned about the ecosystem by measuring the degree of entrapment of energy (as matter) as by chasing the energy flow through the system; the more so in view of the inherent inaccuracies in the methods of measuring rates of production already mentioned. Semina's conclusion, that aggregations of phytoplankton are found wherever the sinking rate is reduced, is very reasonable.

The contribution by Velasquez is in a class by itself. Instead of reviewing the work on seaweed production done during the IBP in the Philippines, he decided, since the seaweed studies had been published, to produce an account of the several species of attached algae used commercially and domestically in the Philippines, together with the ways in which they are prepared for the table. His contribution makes most refreshing reading. It should sell the whole book.

The next group of five chapters, all from the USSR, deal with community structure and function, both descriptively and theoretically. Three of them concern benthic systems. The two pelagic studies form a pair by Petipa and by Shushkina & Vinogradov. Petipa distinguishes two types of interrelations in the pelagic ecosystem: '(1) prey–predator relationships, and (2) non-predatory relations based on the external metabolites excreted into the environment by one organism and consumed by the others.' Prey–predator relations, which dominate at higher trophic levels, have been given a great deal of attention; the non-predatory relations, which prevail at lower levels and which involve microbiotic elements, are only beginning to be studied. Both are described in this chapter, using material from the Black Sea and from the tropical Pacific. In the second part of this general study of pelagic systems, Shushkina & Vinogradov pay special attention to the modelling of the processes going on within systems, and they describe vividly the non-homogeneity of distribution of organisms and the rise and collapse of ecosystems in the pelagic biome. Both chapters subject the ecosystem to fine dissection and bring out the importance of the size aspects of marine organisms (compare the work of Parsons in this regard). The conceptual patterns are excellently developed.

Neyman's chapter on the benthos of the shelves of marginal seas extends this manner of analysis, to some extent, to the benthos, and does it on a global scale; it is a treatise in biogeography that treats organisms not systematically (by taxa) but by trophic group (suspension-feeders, deposit-feeders, etc.), and relates the dominance of a given group to the bottom type, sedimentation, and the influence of currents. Thus he points out that suspension-feeders are dominant where currents are strong, deposit-feeders in regions of vortices and less turbulence. Yablonskaya's study focusses the same kind of attention and method on the Azov, Caspian and Aral Seas, and does so in considerable detail. Finally, the chapter by Golikov & Scarlato, a more orthodox or classical biogeographic undertaking, describes the influence of both bottom type and water mass on the distribution of the biota, and it discusses also the settling of larvae and its relevance to aquaculture.

These fourteen chapters, representing much of the work of the IBP/PM 'theme A' group on productivity mechanisms in different climatic regions, are good examples of the modern approach to the understanding of marine ecosystems and their rational exploitation in the interests of human welfare. It cannot be said that the marine fauna and flora of the world are yet fully surveyed and catalogued; there are still many regions that stand in need of elementary faunal and floral work. But the groundwork, which goes back to the days of Aristotle, has been done. Food chains are fairly well known, and much of the specific life history work needed for exploitation and conservation has been achieved. The life histories of the species commercially used by human populations formed the basis for the elegant theories of fishing of recent decades which led to the establishment of the 'maximum sustainable yield' concept, and other products of the mathematics of fishing which have served up to now to design fishing regulations and international agreements.

There are ample signs that these methods of achieving control of marine resources, and of assuring their conservation, have not been adequate. They have indeed been superseded by more broadly based concepts of primary and secondary production, and the recent development of the modelling of energy flow through the ecosystem, which is enjoying the present fashionable sunshine. But such methods in their turn appear to be not enough, and there is now developing the realization that environmental factors, which form part of the total system, have been understated. Such matters as variation in annual influx of fresh water into the marine system and the seasonal control of the fresh water influx by large-scale hydroelectric development; small-scale and large-scale climatic change and annual meteorological fluctuations; latitudinal variation in seasonality; variation in dominance, from year to year, of different species in the same system; behavioural effects of variations of changes in the environmental ecological signals to which species are attuned; transient phenomena in water movement (advections, upwellings, etc.); the formation and decay of subsystems; the extinction of subsystems and their replacement from adjacent regions; all these affect the biological stocks in the sea which mankind uses, but they have not so far figured importantly in the mathematical equations with which fishery scientists back their recommendations for legal or international control. The contents of this volume suggest that this new approach has been launched, none too soon, and there is little doubt that IBP helped most significantly to hasten it.

I close by acknowledging with pleasure the help and support, during my time as Marine Convenor for IBP, of Mr R. S. Glover, first Marine Convenor, whose work established the pattern of development of the Productivity Marine Section (IBP/PM) and to whom much of its success must be ascribed; and of Dr Barton Worthington, Executive Director of IBP, whose intermittent needling and constant encouragement did much to ensure the successful conclusion of the marine programme.

References

Dunbar, M. J. (1972). The nature and definition of the marine subarctic, with a note on the sea-life area of the Atlantic salmon. *Transactions of the Royal Society of Canada, Series IV*, **10**, 250–7.

Fogg, G. E. (1969). Oxygen-versus ^{14}C methodology. In *A manual on methods for measuring primary production in aquatic environments. IBP handbook No. 12*, ed. R. A. Vollenweider, pp. 76–8. Oxford: Blackwell Scientific Publications.

Riley, G. A. (1972). Patterns of production in marine ecosystems. In *Ecosystem structure and function*, ed. J. A. Wiens, pp. 91–112. Oregon State University Press.

Rodin, L. E., Bazilevich, N. I. & Rozov, N. N. (1975). Productivity of the world's main ecosystems. In *Productivity of world ecosystems*, Proceedings of a Symposium, V General Assembly of SCIBP, ed. D. E. Reichle, J. F. Franklin & D. W. Goodall, pp. 13–26. Washington, DC: National Academy of Sciences.

Steemann Nielsen, E. (1958). A survey of recent Danish measurements of the organic productivity in the sea. *Rapports et procès-verbaux des réunions, Conseil international pour l'exploration de la mer*, **144**, 92–5.

Sverdrup, H. V. (1955). The place of physical oceanography in oceanographic research. *Journal of Marine Research*, **14**, 287–94.

Vollenweider, R. A. (ed.) (1969). *A Manual on Methods for Measuring Primary Production in Aquatic Environments. IBP Handbook No. 12.* Oxford: Blackwell Scientific Publications.

2. Primary production in Frobisher Bay, Arctic Canada

E. H. GRAINGER

Introduction

Nansen (1902) found little plant life in the interior of the North Polar Sea, and a succession of authors since (Braarud, 1935; Thorson, 1936; Digby, 1953; Bursa, 1961) have agreed in their observations that phytoplankton is sparse under the sea-ice at high latitudes. Lack of light has been fairly consistently put forward as the principal deterrent to plant growth beneath the ice. To quote from Nansen (1902, p. 423): 'Whenever the water of the polar currents divests itself of its ice-covering, and meets the warmer water from the south, there is a sudden and enormous development of life of all kinds'. We know now that when northern waters become ice-free in summer, whether or not they meet warmer water from the south, there can be found underway the typically short and intense phytoplankton bloom which characterizes the annual biological cycles of marine waters of high latitudes.

Marine biological observations in cold northern seas have been carried out traditionally only during the summer, open-water period. At least one Canadian Eskimo has coupled the warm-weather biologists with the migratory birds, pointing out that members of both groups customarily retreat from the arctic with the onset of severe autumn weather. As a result of this habit, our knowledge of biological conditions in winter, under and within the ice has remained meagre to this day, despite the drifts of the vessels *Fram*, *Maud* and *Sedov*, and of various ice islands, and the smaller, more biologically oriented efforts of such as Thorson and Madsen in East Greenland between 1931 and 1933 (Thorson, 1936; Ussing, 1938), Digby a little farther south in East Greenland in 1950 and 1951 (Digby, 1953, 1954), MacGinitie at Point Barrow, Alaska, between 1948 and 1950 (MacGinitie, 1955), and the present author in Foxe Basin, arctic Canada, in 1955 and 1956 (Grainger, 1959; Bursa, 1961). It was partly in response to the need for winter data that the present study was initiated in 1967, designed to provide a description of seasonal variation in several physical, chemical and biological features, an assessment of biological production, and the development of some understanding of the principal factors controlling the annual production cycles in a shallow-water arctic environment.

Frobisher Bay, at the southeast corner of Baffin Island, is approximately 1900 km north of Montreal. Most of the observations used in this paper were made at station 5, about 50 m deep, at 63° 40′ N, 68° 27′ W, 8 km south-southeast of the settlement at the head of the bay, and at station 5B, nearer shore. Additional material was acquired from adjacent parts of the bay.

Methods

Nutrients and chlorophyll *a* were analysed in the field following the methods of Strickland & Parsons (1968). Light penetration was measured with a GM submarine photometer and 'deck' cell without filters, sensitive to light from about 400 to 640 μm, and a standard secchi disc. Ice samples were collected with a 3-inch diameter SIPRE ice corer. Primary productivity measurements were made using the carbon-14 method and the services of the International Agency for ^{14}C Determination, Søborg, Denmark. Carbon uptake measurements were carried out *in situ*, using 60-ml glass-stoppered bottles innoculated with 1 ml of $NaHC^{14}O_3$ with activity of 4 μC. Three bottles were used for each sampled depth; all were suspended at the depths from which the samples originated, two of them exposed to light, the third protected from light. Incubation periods were usually 3 hours, starting at local noon.

Several approaches have been used to convert hourly to daily production rates (see Vollenweider, 1969). In this study, percentages of total daily incident light found between 12.00 and 15.00 hours were used to convert experimental data to rates per day.

Results

Data on physical and chemical matters relevant to phytoplankton production in Frobisher Bay, in particular ice and snow cover, underwater light, temperature, salinity, dissolved oxygen and nutrient cycles, have been presented (Grainger, 1971), and will be discussed only briefly here.

Annual variations in the duration of the sea-ice and snow cover depend for the most part on climatic conditions. Because incoming radiation in June and early July was not significantly less than that found farther to the south at the same time of the year (where light is not under normal circumstances a factor limiting plant production in near-surface waters at that time), the presence or absence of an ice and snow cover was obviously important, in either admitting light to the subsurface waters or preventing its entry (Fig. 2.1*a*). Later in the summer, incoming light was reduced. Adequate light for photosynthesis, found to be as deep as 20 m in July, probably rarely reached to 5 m in November. The ice cover formed in late November or early December.

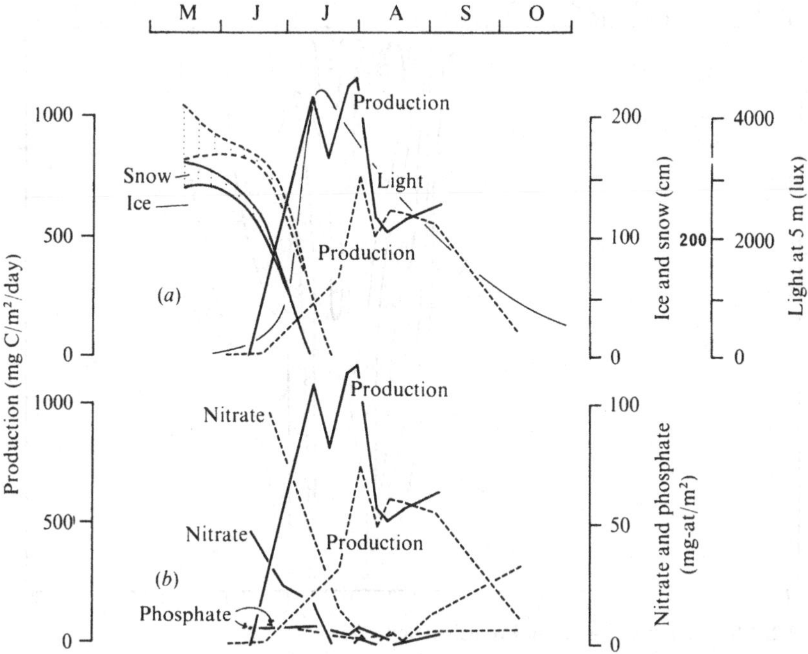

Fig. 2.1. Features of the oceanographic annual cycle in Frobisher Bay. (*a*) Ice and snow thickness and light intensity at 5 metres related to the phytoplankton production curves of 1968 and 1969. (*b*) Nitrate and phosphate related to the production curves of 1968 and 1969. ---, 1968; ——, 1969.

Annual cycles of nitrate-nitrogen and phosphate-phosphorus over 2 years (1968 and 1969) are shown in Fig. 2.1(*b*), with phytoplankton production curves for the same two years. Nitrate decline from winter levels, given in the figure as nitrate in the upper 10 m only, the level where most of the plant production occurs, was noticeably later in 1968 than in 1969. Also, more complete depletion occurred in the upper 10 m in 1969 than in 1968. Phosphate, also shown in the upper 10 m only, varied through similar cycles, but the variation was much less, and quantities remained clearly positive even at their lowest levels.

Chlorophyll *a* levels through 1969 (Fig. 2.2) were characterized by a strong vertical structure with little stratification. This is in contrast with the nutrients, especially nitrate, in which considerable stratification was developed as nutrients were consumed in waters near the surface (see above). Regrettably, chlorophyll *a* data are not available for the whole of 1968, and therefore only 1969 was completely surveyed (Fig. 2.3). Close similarities in the chlorophyll curves for 1967 and 1969 prevail for at least

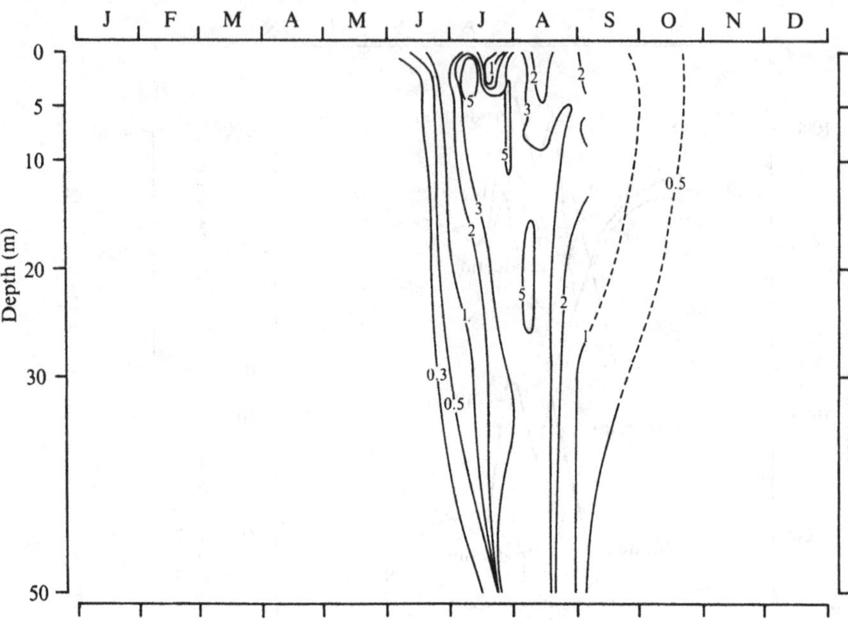

Fig. 2.2. Chlorophyll *a* (mg/m^3) in Frobisher Bay, 1969.

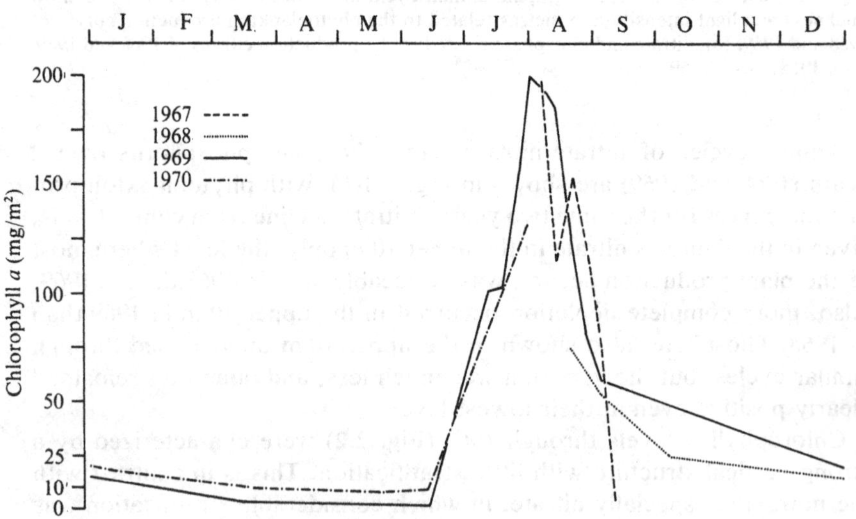

Fig. 2.3. Chlorophyll *a* in Frobisher Bay, 1967–70

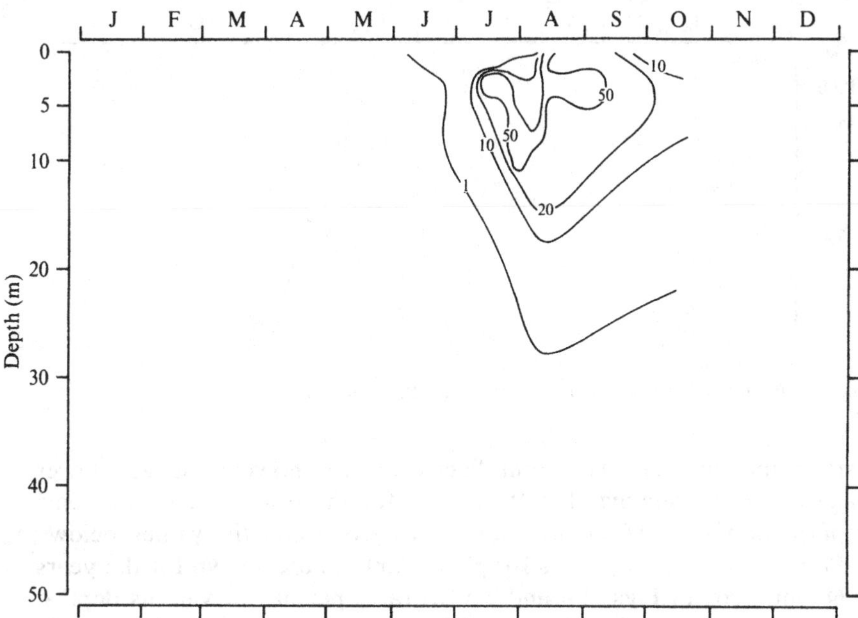

Fig. 2.4. Phytoplankton production (mg C/m³/day) in Frobisher Bay, 1968.

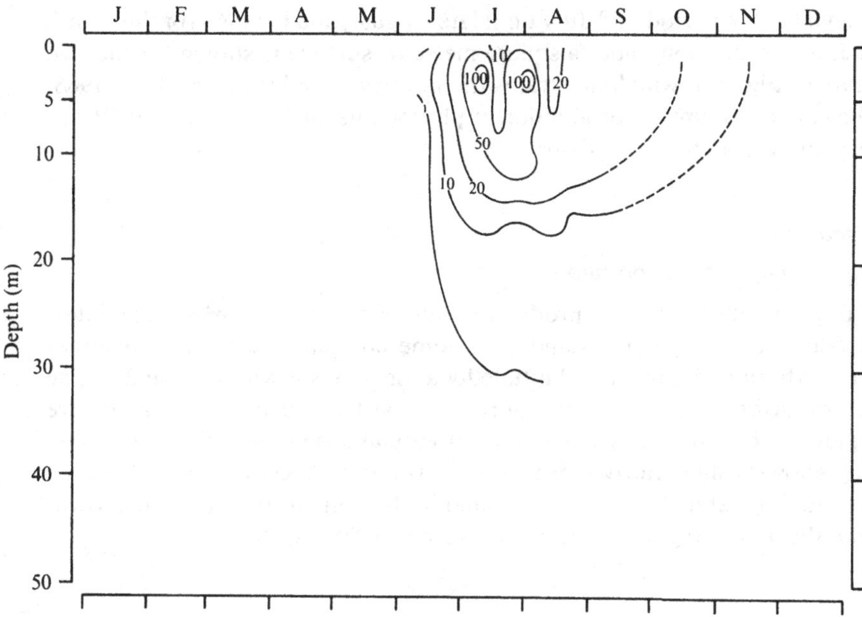

Fig. 2.5. Phytoplankton production (mg C/m³/day) in Frobisher Bay, 1969.

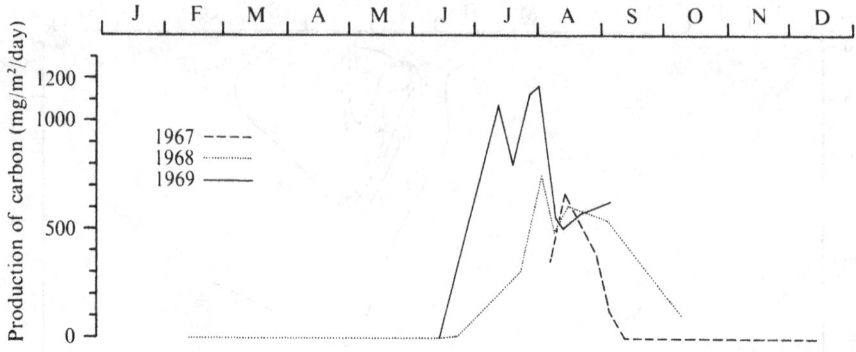

Fig. 2.6. Phytoplankton production in Frobisher Bay, 1967–9.

part of the summer. Data from August of 1968 however suggest lower values in early summer for 1968, than for the other years, and some confirmation is given for this in the discussion of productivity rates, below.

Primary productivity rates for phytoplankton are shown for the years 1968 and 1969 in Figs. 2.4 and 2.5, as rates per day at various depths. A later start in 1968 (10 mg C/m^3/day achieved only in July) than in 1969 (10 mg reached in June) may be seen. Lower maximum values too were apparent for 1968 than for 1969, with highest values in 1968 being shown for the period mid-July until late August, and for 1969 for the month of July. Production under a square metre of surface is shown for the two years in Fig. 2.6, which illustrates the later start and lower peak for 1968. The calculated annual production of phytoplankton for 1968 and 1969 was respectively 41 and 70 g C/m^2.

Discussion

The primary production rate

A phytoplankton primary production rate of 50–100 g C/m^2/yr is an intermediate value by world standards. Some comparative figures for other far northern and a few low-latitude localities are shown in Table 2.1. The central Arctic Ocean and the northernmost Canadian arctic islands are regions of exceptionally low annual phytoplankton production. Frobisher Bay shows values fairly close to those found in West Greenland, which are much greater than has been found in the high arctic, and much lower than the 100–200 g levels of waters south of 50°–60° N.

Table 2.1. *Phytoplankton production estimates from several far northern marine locations and from some comparable regions in lower latitudes*

Author	Location	Production estimate g C/m²/yr
	Far northern marine	
Apollonio (1959)	Arctic Ocean	0.6
Apollonio (1956)	Cornwallis I., Canada	15[a]
McLaren (1969)	Baffin I., land-locked fjord	12
Steemann Nielsen (1958)	West Greenland, 3 locations	29, 95, 98
Petersen (1964)	West Greenland	36
Bagge & Niemi (1971)	Gulf of Finland	30–40
	Lower latitude marine	
Platt (1971)	St Margaret's Bay, Nova Scotia	190
Parsons, LeBrasseur & Barraclough (1970)	Strait of Georgia, British Columbia	120
Ryther & Yentsch (1958)	Off New York	100–160
Steven (1971)	Near Barbados	105

[a]This figure is an extrapolation of the value 0.192 g C/m²/day during the summer.

Nutrients

The immediate sources of nutrients found in the waters of the euphotic zone of Frobisher Bay are probably several in number, including waters beneath the euphotic zone, activities within the euphotic zone itself, river run-off, and sewage from the adjacent town site. Data are inadequate for setting up a nutrient budget, but some pertinent information may be brought out; phytoplankton primary production estimates of 41 and 70 g C/m²/yr respectively for 1968 and 1969 indicate (using a C:N ratio of 6) a demand of 6.8 and 11.7 g N/m² for each of the same years. Measured consumption, from changes in quantities of nitrate and nitrite *in situ*, works out to 5.2 and 6.0 g for each of the two years, leaving a requirement for 1.6 and 5.7 g N/m²/yr for 1968 and 1969. Contributions from local sewage were not likely to have exceeded 0.1 g N in either year, and river flow in both years was almost certainly not as high as 0.1 g N (from both nitrate and nitrite). Quantities of 1.6 and 5.6 g remain, and much of this was probably introduced from deeper water. Arctic rivers like the Sylvia Grinnell, the principal freshwater contributor to the study site, are not commonly assessed as being very significant carriers of nutrients. This point of view seems to be supported here. Similarly, arctic snow may not be especially important as a source of nutrients as was found for instance in Ontario lakes (Barica & Armstrong, 1971). The few nutrient measurements made from snow as part of this study showed extremely low levels. Nutrients in sea-ice have already been considered from Frobisher Bay

Table 2.2. *Nitrate values (µg-at N/l) at station 5 related to times of 'spring' tides (ST)*

Depth (m)	ST−6 days	ST−2 days	ST+3 days	ST+5 days
0	0.0	0.0	0.4	0.2
1	0.0	0.0	0.4	0.2
3	0.0	0.0	1.2	0.7
5	0.0	0.0	1.5	0.1
7	0.0	0.0	1.7	0.5
10	0.0	0.2	1.7	1.7
20	2.6	1.2	2.8	2.0
30	3.3	2.6	2.9	3.0
50	3.6	3.3	2.7	4.3

(Grainger, 1977). Fairly large quantities of nitrate and phosphate were shown to be released to the water from melting sea-ice in early summer. It was concluded however that the nutrients had originated in the water from which the ice had formed, and that this was therefore not a factor in the nutrient budget.

There is evidence that 'spring' tides bring additional nitrate to the region of station 5; frequently an increase at all depths and usually a replenishment in the euphotic layers have been shown (Table 2.2). The mean value of nitrate measurements made during flood tide periods was 135 mg-at N/m^2. During ebb tides it was 113 mg-at. Phosphate too was higher during flood tide.

Nitrogen requirements calculated for Frobisher Bay from production figures ranged downward from a high of nearly 14 mg-at/m^2/day (the five-day period between 27 July and 1 August 1969). Average values found during the production period were similar to Harris's (1959) calculated 5–6 mg-at N/m^2 from Long Island Sound. Phosphorus requirements were not calculated. Phosphate was not exhausted in the water column, or even in the upper layers; it did not appear to be limiting in Frobisher Bay. Nitrate however may have been limiting for brief periods of time as it became entirely consumed in the upper water layers from time to time. It was fairly quickly replenished.

Ratios (by atoms) of nitrogen (nitrate+nitrite): phosphorus (phosphate) were found to range downward from 14.3, and about one half the observations were three and less, and more than one quarter of them were less than one. A time sequence was shown; highest ratios occurring in May and June (when levels of both nutrients were highest), lowest between July and October (Table 2.3). Table 2.3 gives N:P ratios in the whole column (0–50 m) and in the upper, active zone (0–20 m). Ratios rose over winter and fell through the summer. Values were similar at all depths in winter,

Table 2.3. *Annual variation in N:P ratio, station 5, Frobisher Bay, May 1969–May 1970*

	Depth (m)			Depth (m)	
Date	0–50	0–20	Date	0–50	0–20
1969			1969		
11 May	9.6	9.6	13 August	1.0	0.9
14 June	7.7	6.8	21 August	1.3	0.9
12 July	3.3	2.8	28 August	0.8	0.5
19 July	2.3	1.1	5 September	2.1	1.8
27 July	2.6	2.4	11 September	4.3	4.2
			1970		
1 August	1.6	0.9	22 March	3.7	3.8
9 August	2.0	1.4	23 May	6.0	5.0

but soon after plankton production began, nitrate fell relatively in the near-surface zone, and remained low until winter.

Chlorophyll

As may be expected, examination of the relationship between quantities of nutrients, and of chlorophyll *a*, shows that generally high values of the first occurred in company with low values of the second. There are interesting differences however between nitrate and phosphate in this regard. The highest nitrate levels (Fig. 2.7) of May and June, were found associated with chlorophyll *a* lower than 0.4 mg/m^3 (with a single exception). This was a period of ice cover, with low light intensity, preceding the phytoplankton bloom. During the following period of July–September, nitrate fell as chlorophyll increased, and during this period the greatest chlorophyll *a* quantities were found along with some of the lowest nitrates of the year. Virtually the lowest nitrate levels of the year were found as the chlorophyll *a* reached its maximum. Later, chlorophyll declined and nitrate began to climb. The period September–April brought a continuation in the increase in nitrate as chlorophyll levels slowly fell.

A less regular trend appeared with phosphate (Fig. 2.7); the periods in which the highest levels were present included some of the lowest ones as well. More importantly, the highest chlorophyll *a* occurred in company with intermediate phosphate values, rather than the lowest ones as was found with nitrate. The chlorophyll *a* cycle clearly was more dependent upon nitrate than phosphate. Nitrate was nearly exhausted at the height of the chlorophyll *a* cycle; phosphate was not.

The rate of production of phytoplankton is obviously related in part to

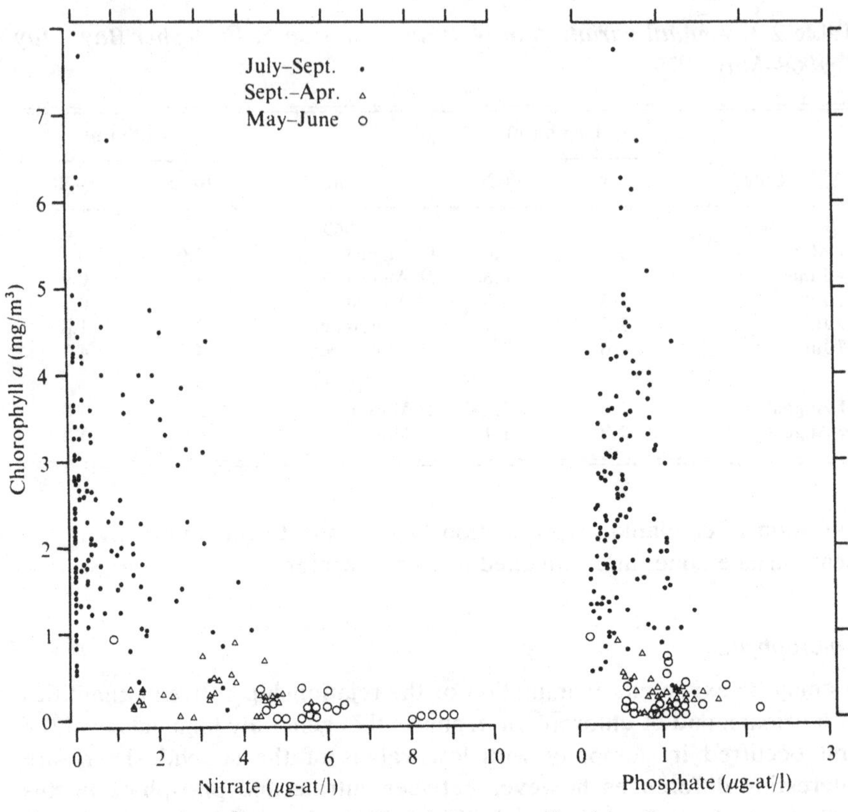

Fig. 2.7. Chlorophyll *a* related to nitrate and phosphate, Frobisher Bay.

Table 2.4. *Chlorophyll a, phytoplankton production rate and assimilation number (production per unit of chlorophyll a available) related to depth*

Depth (m)	Chlorophyll *a* (mg/m³)	Production rate (mg C/m³/h)	Assimilation number (C:Ch. *a*)
0	1.31	1.28	0.98
5	2.00	3.92	1.96
10	2.38	2.73	1.15
20	1.98	0.36	0.18
30	1.79	0	0

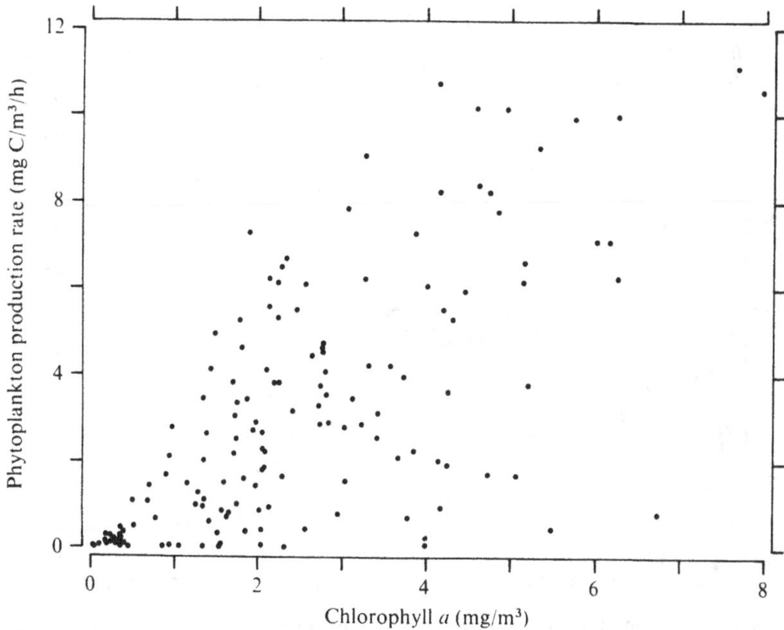

Fig. 2.8. Phytoplankton production rate related to chlorophyll *a*, Frobisher Bay.

the amount of chlorophyll *a* present (Fig. 2.8), with considerable scatter evident in the relationship. Such a relationship is expected in a given locality (Steemann Nielsen, 1963). Much of the scatter is contributed by the uneven vertical distribution of production rate values, samples from close to the surface and from 7 or 10 m downward characteristically showing relatively low production levels in relation to chlorophyll quantities present (Table 2.4). The highest production rates did not occur in fact at depths with the greatest quantities of chlorophyll *a*.

Light and production

The dependence of the photosynthetic rate on light is shown in Fig. 2.9. Seven curves represent as many sampling dates from June to September. The surface value is at the extreme right in all instances, and successive sampling depths (1, 3, 5, 7, 10, 20 m) lie along the curves towards the left. On four dates, the greatest production occurred between about 500 and 2000 lux, and on three others between about 3000 and 5000 lux. The three stations showing the highest production rates were occupied a few days following 'spring' tides, while those with lower rates were sampled during intermediate periods. The reasons for this are not clear; increase in

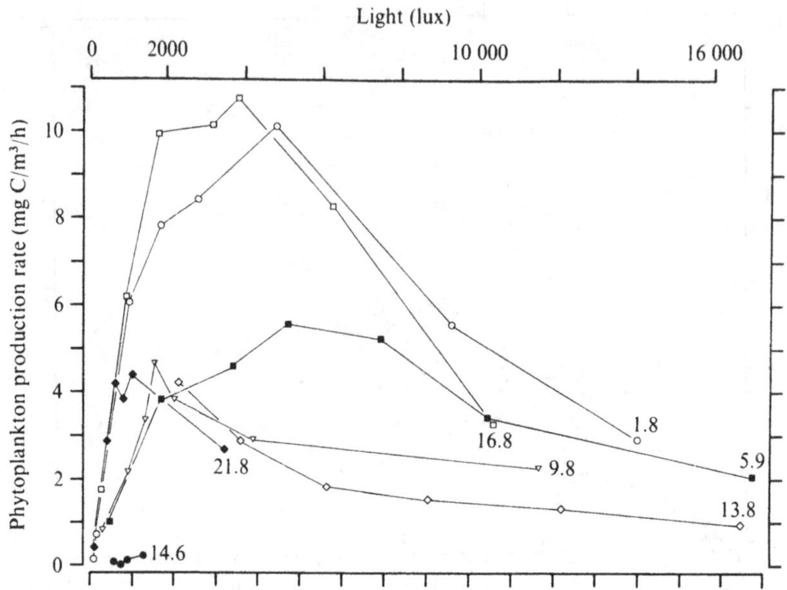

Fig. 2.9. Phytoplankton production rate related to light, Frobisher Bay. In each curve, the symbol at the right extremity represents the surface value, the next toward the left 1 metre and so on through depths of 3, 5, 7, 10 and more metres, reaching the maximum observed depth at the left extremity of the curve. Numbers on curves represent day and month of sampling.

nutrient levels or possibly change in phytoplankton composition may be factors.

The range of 1000–5000 lux appears to embrace a large proportion of the photosynthetic activity carried out in Frobisher Bay. Activity does extend on either side of that range however, and light as low as 100 lux supports photosynthesis in the ice-inhabiting plants (Grainger, 1977), and as high at least as 17000 lux allows some activity at the surface of the bay.

Ryther & Menzel (1959) showed that generally, green algae, diatoms and dinoflagellates become saturated and inhibited at progressively higher light intensities, and they referred to the three groups respectively as 'sun', 'intermediate' and 'shade' forms. In the Sargasso Sea they found in the surface 'sun' group maximum photosynthesis at about 5000 ft candles (53800 lux), in the 50-m 'intermediate' group maximum photosynthesis at about 3000 ft candles (32280 lux), and in the 100-m 'shade' group maximum photosynthesis at about 500 ft candles (5380 lux). The Frobisher Bay plants most closely resembled the 100-m 'shade' group of the Sargasso Sea in their light requirements.

Steemann Nielsen & Hansen (1959) used the terms 'sun' and 'shade' flora differently, suggesting the unlikelihood of there being separate 'sun'

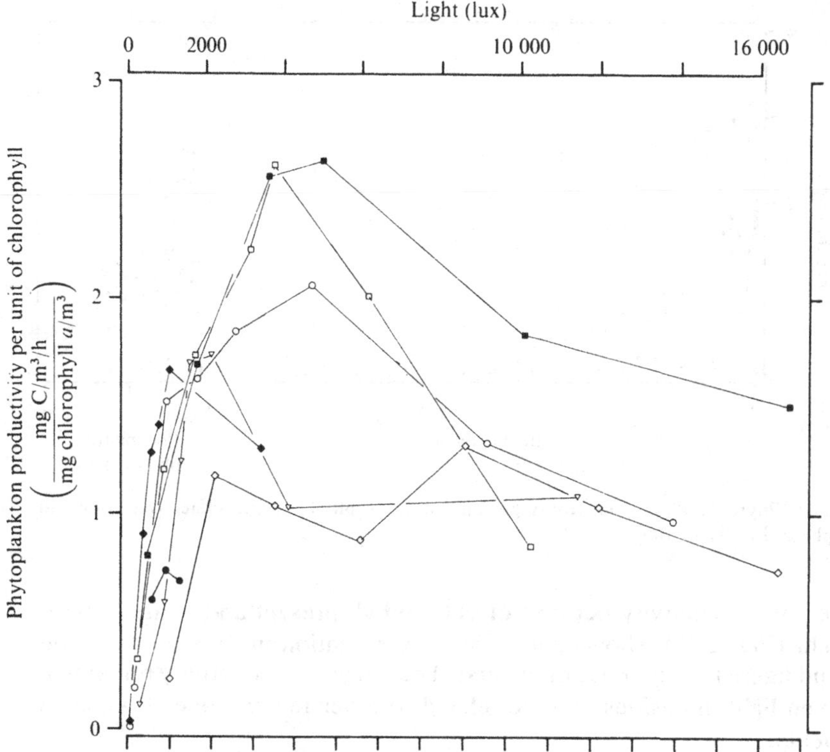

Fig. 2.10. Phytoplankton productivity per unit of chlorophyll present related to light, Frobisher Bay. Symbols as in Fig. 2.9.

and 'shade' floras in the arctic, but rather that populations of the same species may be adapted to sun and shade conditions at different times. They found a range of light intensity for maximum photosynthesis from about 25 000 lux for tropical plankton to about 6000 lux for 'arctic summer deep' plankton in Davis Strait. This end of the scale approximates the conditions met by the Frobisher Bay surface plankton. The Frobisher Bay conditions are close to Bunt's (1964) light saturation level of about 300 ft candles (about 3200 lux) found in McMurdo Sound, in the Antarctic. There, photosynthetic activity in phytoplankton was shown to be clearly inhibited by light of 1100 ft candles (11 836 lux), and the compensation point was seen to lie close to 5 ft candles (54 lux). Another example of adaptation to low light comes from the Japan Sea (Sorokin & Konovalova, 1973) where maximum photosynthesis in the phytoplankton was found under 2 to 3 klux under the mid-winter ice.

Removing differences between the curves of photosynthesis (in Fig. 2.9), which are related to variations in chlorophyll quantity, by deriving

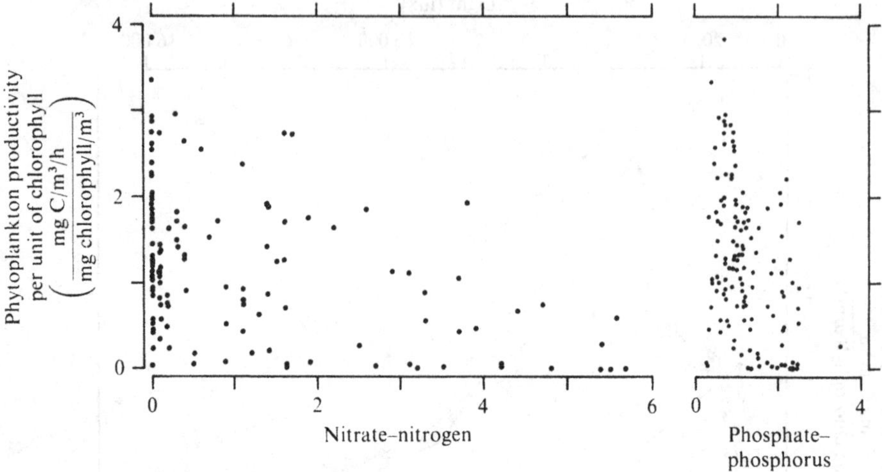

Fig. 2.11. Phytoplankton productivity per unit of chlorophyll present related to nitrate and phosphate, Frobisher Bay.

values of productivity per unit of chlorophyll present and relating these to light (Fig. 2.10), shows somewhat less variation in the curves of the second figure than in those of the first. The range of productivity remaining at given light intensities may be related to other factors, one of which is nutrients.

Nutrients and production

Nutrient supplies, specifically nitrate-nitrogen and phosphate-phosphorus, show a correlation with rate of productivity per unit of chlorophyll *a* present (Fig. 2.11). High productivity under conditions of undetectable nitrate indicates that photosynthetic activity did not cease immediately with depletion of nitrate. Low activity under conditions of high nitrate occurred at subsurface depths where high nitrate levels were retained under low illumination. In the same figure, productivity showed a similar relationship to phosphate, except that the supply of phosphate was not exhausted. As with nitrate, the highest levels were retained in light-deficient depths.

Timing of the bloom

The duration and extent of the phytoplankton blooms of 1968 and 1969 were very different (Fig. 2.1). Production began earlier in 1969, and probably most of the year's production was completed in July. In contrast, activity in July of 1968 was slow to start and to accelerate, and most of the year's production occurred in August. Thorson (1936) described the beginning of a phytoplankton bloom off northeast Greenland as taking place under the sea-ice a month before break-up. The ice at that time was free of snow and clear as glass, so that light reached the water below. Off West Greenland, the start of the production period was shown to coincide with the breaking up of the ice cover (Petersen, 1964), and maximum production coincided with maximum light in June. The fall-off in production took place while light diminished, and the end of the production season came before new sea ice formed. Certainly light has been most commonly rated as the primary factor in controlling the onset of annual plant production in the sea, in conjunction with nutrients and temperature. Occasionally, other less expected factors are introduced to explain the initiation of a bloom, for instance, the decrease in zooplankton grazing pressure has been credited with triggering the spring bloom in Narragansett Bay (Martin, 1970).

In Frobisher Bay, the events associated with the initiation of the summer phytoplankton bloom seem to be fairly well understood. In 1968, air temperatures were lower during May and June (−0.7 °C) than they were during the same period of 1969 (0.0 °C). The trend continued through July of both years, with the average temperature in 1968 (4.6 °C) well below that of 1969 (6.8 °C). This difference was great enough to have left a snow cover on the ice in 1968 which lasted about 2 weeks longer than it did in 1969, and to have contributed to a similar difference in ice break-up time. In 1968, the snow cover on the ice remained until near the middle of July; in 1969 the ice was clear by the end of June (Fig. 2.1*a*). During July 1968, light penetrated to the surface of the water only during the second half of the month, and probably only some 150 hours of sunshine reached the water during that time (Canadian Dept of Transport, 1968). In July 1969, there were 280 hours of sunshine, an exceptionally high value for the month. Occurring along with light differences were those differences in water temperature; July 1968 was colder than July 1969, in July 1968, nearly all the water deeper than 5 m was colder than −1 °C, whereas in July 1969, the −1 °C isotherm was at 10 m depth early in the month, and it descended at least to 50 m before the end of July. Nutrient potentials were similar in both seasons.

Light appears to be the decisive factor in controlling the start of the phytoplankton bloom in Frobisher Bay (Fig. 2.1*a*). Early removal of the snow–ice barrier to light (an event controlled by the weather conditions of winter and spring) permits early development of the summer bloom in water of suitable temperature and in the presence of sufficient nutrients. A late spring may delay snow-ice removal and prevent entry of light to the water until later than the normal time. This may be expected to delay the development of the bloom.

Factors controlling the level of the maximum production rate achieved during a season are not as readily explainable as those which set off the summer bloom. Light probably becomes far less important at this time, with nutrients becoming more important than earlier in the season. Sunshine was similar in late July and August in the two years 1968 and 1969. In 1968, nitrate declined more slowly and later than in 1969.

Light may be a factor in controlling production values in autumn. Normal September light would be expected to reduce the daily duration and depth range of photosynthesis found in August, other factors remaining equal. An overcast September, as was found in 1967, should have an even more profound influence, and the sharp drop between August and September values of chlorophyll (Fig. 2.6) and production rate (Fig. 2.3) in 1967 may have been largely a consequence of September weather conditions. September of 1968, in contrast, was climatically much more favourable, with many more bright days, and chlorophyll levels and productivity rates (Figs. 2.3 and 2.6) remained relatively high. In 1969, the first half of September was moderately bright; chlorophyll and production figures were relatively high then, but unfortunately there are no data on these features later in the month. Total possible light would almost certainly limit production to very low levels, regardless of other factors, by October. It is unlikely that the time of sea-ice formation in winter is relevant to the decline in seasonal photosynthetic rates, because of the late date (the end of November to early December) of sea-ice formation at Frobisher. Development of the ice cover is rather obviously related to air temperatures of autumn, for example, early ice formation (late November) in 1967 and 1969 followed mean October air temperatures of −6.2 and −5.4 °C, while late ice development (December) in 1966 and 1968 followed mean October air temperatures of −0.9 and 0.1 °C.

The sea-ice

The sea-ice at Frobisher Bay was shown to accumulate nitrate and phosphate during the course of development over winter (Grainger, 1977). At one station, the calculated quantity of phosphate-phosphorus in the ice in early January was 0.36 mg-at/m^2. This amount increased to 0.63 mg-at by late March, indicating an increase of 0.27 mg-at/m^2 during that time, then fell to 0.57 mg-at by late May, suggesting a loss by melting of 0.06 mg-at/m^2 during that period of time. The total phosphate loss by melting may have been as much as 0.57 mg-at/m^2 PO_4-P between late May and early July (Fig. 2.12). The nitrate profile (Fig. 2.13) shows a calculated loss from the ice by melting of as much as 1.0 mg-at/m^2 NO_3-N between late May and early June.

Chlorophyll *a* found in the sea-ice (Fig. 2.14) increased in quantity from 0.44 mg/m^2 in January to 1.55 mg/m^2 in March, and to 4.58 mg/m^2 in late

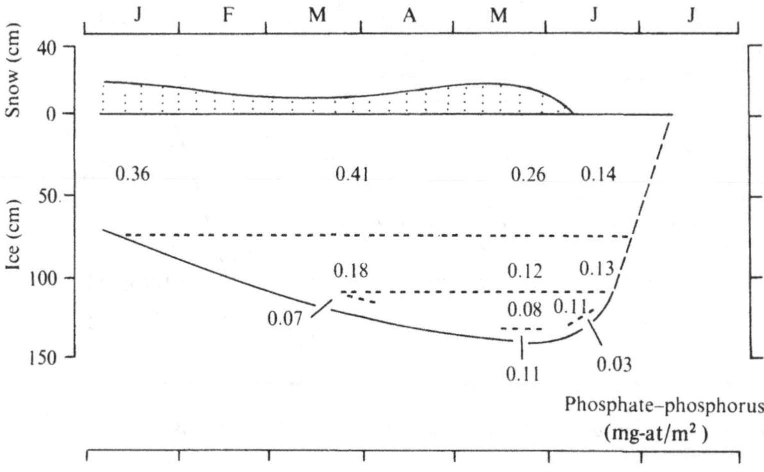

Fig. 2.12. Seasonal changes in phosphate in the sea ice, Frobisher Bay.

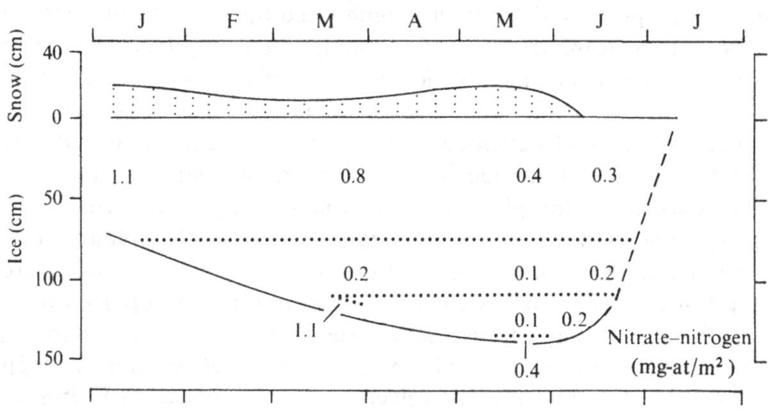

Fig. 2.13. Seasonal changes in nitrate in the sea ice, Frobisher Bay.

May. After that date, loss occurred to the sea below by melting of the ice, and probably at least 4.58 mg/m² of chlorophyll *a* were added to the sea from the melting sea ice.

Figs. 2.12–2.14 show nutrients and chlorophyll *a* as they were found at a single station in 1970 in Frobisher Bay. Essentially similar structures were found at different stations in different winters, and the principal variations which were found in these properties within the ice were related most closely to the thickness of the ice and snow cover. Especially deep snow was found above low chlorophyll *a* levels and above high

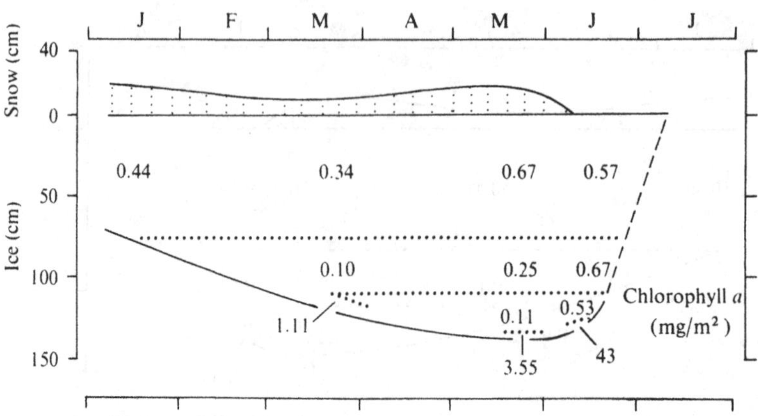

Fig. 2.14. Seasonal changes in chlorophyll *a* in the sea ice, Frobisher Bay.

nutrient levels at the start of the ice melting period. Deep snow appears to have inhibited chlorophyll *a* build-up in the underlying ice, and it seems that this in turn permitted relatively uninhibited build-up of nutrients in the ice. Nutrients in the ice probably originated entirely from the water from which the ice was formed. Chlorophyll *a* however was present only in minute quantities in the water which formed the ice cover and the increasing quantity which developed in late winter and spring must indicate plant cell division within the ice. Light and nutrients were available.

The nutrients and chlorophyll *a* were formed in largest concentrations close to the lower surface of the ice (Figs. 2.12–2.14). Phosphate, at its highest more than 8 mg-at/m^3, and nitrate more than 15 mg-at/m^3, were both found in concentrations similar to others previously reported from the lower surface of sea-ice elsewhere (Apollonio, 1958, 1961; Meguro, Ito & Fukushima, 1966; Oradovskii, 1972). Chlorophyll *a* values as high as 300 mg/m^3 were found near the lower surface of the ice in Frobisher Bay. They too are comparable to measurements made elsewhere (Meguro, 1962; Bunt, 1963; Apllonio, 1965; Meguro, Ito & Fukushima, 1966).

Considerable nutrient material apparently remained in the ice through the period of plant cell division, and was lost to the water below as the ice melted. In this way, quantities as high as 1.16 mg-at/m^2 PO_4-P and 3.3 mg-at/m^2 NO_3-N were added to the water beneath the ice. These additions were made, however, at the time of highest nutrient levels in the water, and did not exceed 5% of the total phosphate or 2% of the total nitrate available at the time in the water beneath the melting ice. The addition in the same way of chlorophyll *a* to the water beneath the ice was a more significant matter. Quantities apparently released from the melting ice were as high at least as 9 mg/m^2, representing up to twice the quantity present at the time in the water below.

Seasonal changes in plant production and nutrients in the sea water of Frobisher Bay are shown in Fig. 2.1. The period of ice melting is shown in the same figure. Clearly the time of ice melting and release of plants from the ice is very close to that of the beginning of the summer development of plants in the water beneath the ice. The two events remain however to be tied more closely together.

The beginning of the spring phytoplankton bloom, before the sea-ice disappeared, and close to the time of the start of ice melting was shown at Igloolik, in arctic Canada (Bursa, 1961). Meguro, Ito & Fukushima (1967) claimed a direct connection between the release of the flora by the ice and the immediate initiation of a sudden phytoplankton bloom under the ice. Horner & Alexander (1972) claimed however that the release of ice plants to the water was not a major contribution to the spring phytoplankton bloom at Point Barrow, north Alaska. This conclusion was based upon the finding that no Centriceae were present in the sea-ice, but that the group dominated in the water. Alexander (1974) again stated that ice does not 'seed' the spring phytoplankton bloom north of Alaska, basing her conclusion upon differences in the plant composition of the two media and the existence of a time lag between ice and water blooms.

Whether or not the sea-ice plants contribute to the later plant bloom in the water, there evidently is a significant production of organic material within the ice during the late winter and spring. The resulting flora, close to and upon the lower surface of the ice, probably forms part of a trophic series of plants and animals in, on and immediately below the ice surface (Andriashev, 1968; Horner & Alexander, 1972), which includes polychaete worms, copepods, fishes and seals, all trophically associated with the under-ice diatoms and dinoflagellates.

Nutrient supply does not appear, at least at this time, to be a prime limiting factor for plant production in the sea-ice. Probably much more important is light, which varies with the depth of the snow cover which in turn reflects variations in climatic conditions. Light measurements made through the ice in Frobisher Bay showed less than 50 lux on all observation dates in January and March, when surface light was never greater than 8300 lux and the thickness of the snow cover was between 3 and 45 cm. The light under the ice increased in late April and May to between 50 and 100 lux. Snow cover remained about the same, but surface light reached as high as 14000 lux. The snow cover on the ice disappeared in June, and light beneath the ice rose to between 500 and 800 lux. The first clear evidence of a stratified flora in the ice was found when the light beneath the ice was between 50 and 100 lux.

The compensation value in the antarctic sea-ice was found to be about 50 lux (Bunt, 1964). In temperate latitudes, Wright (1964) found autotrophic activity under the ice cover at about 60 lux. In the arctic, less than 200 lux was found at the bottom of the ice cover at the time of maximum chlorophyll development (Apollonio, 1965) and a minimum light requirement for photosynthesis in the ice was found to be 66 lux by Clasby, Horner & Alexander (1973).

Measurements of photosynthetic rate of plants in the sea-ice were not made in Frobisher Bay. Minimum chlorophyll development in the ice was found to be about 5–9 mg/m^2. Using a carbon:chlorophyll *a* ratio of between 50 and 100 to 1, the calculated minimum production in the ice was between 0.2 and 1.0 g C/m^2. Multiplying this value by a factor of 3 to 10 (ratio of maximum bloom value to total for the time period) gives a range of about 1–10 g C/m^2/year. The only available measurement (Alexander, 1974) of annual photosynthetic rate in sea-ice gave a value of up to 5 g C/m^2/year off north Alaska. From this, it seems that plant production in the sea-ice may reach higher than 10% of the annual phytoplankton production in the water.

I am indebted to many people who participated in this work, in particular to Joseph Lovrity, Gary Atkinson, John Nuyens, Gary Sleno and Douglas Fleet, and to members of the Atmospheric Environment Service at Frobisher Bay and of other federal and territorial departments there. To all these people I express my thanks.

References

Alexander, V. (1974). Primary productivity regions of the nearshore Beaufort Sea, with reference to the potential role of ice biota. In *The coast and shelf of the Beaufort Sea*, ed. J. C. Reed and J. E. Sater, pp. 609–32. Arlington, Virginia: Arctic Institute of North America.

Andriashev, A. P. (1968). The problem of the life community associated with the antarctic fast ice. In *Symposium on antarctic oceanography*, Scott Polar Research Institute, pp. 147–57. Cambridge.

Apollonio, S. (1956). Plankton productivity studies in Allen Bay. Cornwallis Island, NWT, 1956. Unpub. MS, Woods Hole Oceanographic Institution.

Apollonio, S. (1958). Hydrobiological measurements on T-3, 1957–58. Unpub. MS, Woods Hole Oceanographic Institution.

Apollonio, S. (1959). Hydrobiological measurements on IGY drifting station Bravo. *IGY Bulletin*, **27**, 16–19.

Apollonio, S. (1961). The chlorophyll content of arctic sea-ice. *Arctic*, **14**, 197–200.

Apollonio, S. (1965). Chlorophyll in arctic sea-ice. *Arctic*, **18**, 118–22.

Bagge, P. & Niemi, A. (1971). Dynamics of phytoplankton primary production and biomass in Loviisa Archipelago (Gulf of Finland). *Merentutkimuslait. Julk./Havforskningsinst. Skr.* **233**, 19–41.

Barica, J. & Armstrong, F. A. J. (1971). Contribution by snow to the nutrient budget of some small northern Ontario Lakes. *Limnology and Oceanography*, **16**, 891–9.

Braarud, T. (1935). The 'Ost' Expedition to the Denmark Strait, 1929. II. The phytoplankton and its conditions of growth. *Hvalrådets Skrifter, Norske Videnskaps-Akademi i Oslo*, **10**, 7–173.

Bunt, J. S. (1963). Diatoms of antarctic sea-ice as agents of primary production. *Nature*, **199**, 1255–7.

Bunt, J. S. (1964). Primary production under the sea ice in antarctic waters. 2.

Influence of light and other factors on photosynthetic activities of antarctic marine algae. In *Biology of the antarctic seas*, ed. M. V. Lee, Antarctic Research Series 1, pp. 27–31. Washington: American Geophysical Union.

Bursa, A. S. (1961). The annual oceanographic cycle at Igloolik in the Canadian arctic. II. The phytoplankton. *Journal of the Fisheries Research Board of Canada*, **18**, 563–615.

Canadian Dept of Transport, Meteorological Branch (1968). *Monthly record. Meteorological observations in Canada, July 1968*. Toronto, Canada.

Clasby, R. C., Horner, R. & Alexander, V. (1973). An in situ method for measuring primary productivity of arctic sea ice algae. *Journal of the Fisheries Research Board of Canada*, **30**, 835–8.

Digby, P. S. B. (1953). Plankton production in Scoresby Sound, East Greenland. *Journal of Animal Ecology*, **22**(2), 289–322.

Digby, P. S. B. (1954). The biology of the marine planktonic copepods of Scoresby Sound, East Greenland. *Journal of Animal Ecology*, **23**(2), 298–338.

Grainger, E. H. (1959). The annual oceanographic cycle at Igloolik in the Canadian arctic. 1. The zooplankton and physical and chemical observations. *Journal of the Fisheries Research Board of Canada*, **16**, 453–501.

Grainger, E. H. (1971). Biological oceanographic observations in Frobisher Bay. 1. Physical, nutrient and primary production data, 1967–1971. *Technical Report, Fisheries Research Board of Canada*, No. 265.

Grainger, E. H. (1977). The annual nutrient cycle in sea-ice. In *Polar Oceans*, ed. M. J. Dunbar, Proceedings SCOR/SCAR Polar Oceans Conference, Montreal, May 1974, pp. 285–99. Montreal: Arctic Institute of North America.

Harris, E. (1959). Oceanography of Long Island Sound. II. The nitrogen cycle in Long Island Sound. *Bulletin of the Bingham Oceanographic Collection*, **17**(1), 31–65.

Horner, R. & Alexander, V. (1972). Algal populations in arctic sea ice: an investigation of heterotrophy. *Limnology and Oceanography*, **17**, 454–8.

MacGinitie, G. E. (1955). Distribution and ecology of the marine invertebrates of Point Barrow, Alaska. *Smithsonian Miscellaneous Collections*, **128**(9).

McLaren, I. A. (1969). Primary production and nutrients in Ogac Lake, a landlocked fiord on Baffin Island. *Journal of the Fisheries Research Board of Canada*, **26**, 1561–76.

Martin, J. H. (1970). Phytoplankton–zooplankton relationships in Narragansett Bay. IV. The seasonal importance of grazing. *Limnology and Oceanography*, **15**, 413–18.

Meguro, H. (1962). Plankton ice in the Antarctic Ocean. *Antarctic Record*, **14**, 72–9.

Meguro, H., Ito, K & Fukushima, H. (1966). Diatoms and the ecological conditions of their growth in sea ice in the Arctic Ocean. *Science*, **152**, 1089–90.

Meguro, H., Ito, K. & Fukushima, H. (1967) Ice flora (bottom type): a mechanism of primary production in polar seas and the growth of diatoms in the sea ice. *Arctic*, **20**, 114–33.

Nansen, F. (1902). The oceanography of the North Polar Basin. In *The Norwegian North Polar Expedition 1893–1896. Scientific Results*, ed. F. Nansen, vol. 3.

Oradovskii, S. G. (1972). Studies on the composition of nutrients in the ice of the Barents Sea. *VNIRO Trudy*, **75**, 65–73. (In Russian.)

Parsons, T. R., LeBrasseur, R. L. & Barraclough, W. E. (1970). Levels of production in the pelagic environment of the Strait of Georgia, British Columbia: a review. *Journal of the Fisheries Research Board of Canada*, **27**, 1251–64.

Petersen, G. H. (1964). The hydrography, primary production, bathymetry, and 'Tagsâq' of Disko Bugt, West Greenland. *Meddelelser om Grønland*, **159**(10).
Platt, T. (1971). The annual production by phytoplankton in St Margaret's Bay, Nova Scotia. *Journal du Conseil international pour l'Exploration de la Mer*, **33**(3), 324–33.
Riley, G. A. (1947). A theoretical analysis of the zooplankton population of Georges Bank. *Journal of Marine Research*, **6**, 104–13.
Ryther, J. H. & Menzel, D. W. (1959). Light adaptation by marine phytoplankton. *Limnology and Oceanography*, **4**, 492–7.
Ryther, J. H. & Yentsch, C. S. (1958). Primary production of continental shelf waters off New York. *Limnology and Oceanography*, **3**, 327–35.
Sorokin, Y. I. & Konovalova, I. W. (1973). Production and decomposition of organic matter in a bay of the Japan Sea during the winter diatom bloom. *Limnology and Oceanography*, **18**, 962–7.
Steemann Nielsen, E. (1958). A survey of recent Danish measurements of the organic productivity in the sea. *Rapports et Procès-Verbaux des Réunions, Conseil international pour l'Exploration de la Mer*, **144**, 92–5.
Steemann Nielsen, E. (1963). Productivity, definition and measurement. In *The sea*, ed. M. N. Hill, pp. 129–64. New York and London: Interscience.
Steemann Nielsen, E. & Hansen, V. K. (1959). Light adaptation in marine phytoplankton populations and its interrelation with temperature. *Physiologia Plantarum*, **12**, 353–70.
Steven, D. M. (1971). Primary productivity of the tropical western Atlantic Ocean near Barbados. *Marine Biology*, **10**, 261–4.
Strickland, J. D. H. & Parsons, T. R. (1968). A practical handbook of seawater analysis. *Bulletin of the Fisheries Research Board of Canada*, **167**.
Thorson, G. (1936). The larval development, growth, and metabolism of arctic marine bottom invertebrates compared with those of other seas. *Meddelelser om Grønland*, **100**(6).
Ussing, H. H. (1938). The biology of some important plankton animals in the fjords of East Greenland. *Meddelelser om Grønland*, **100**(7).
Vollenweider, R. A. (ed.) (1969). *A manual on methods for measuring primary production in aquatic environments. IBP Handbook No. 12.* Oxford and Edinburgh: Blackwell.
Wright, R. T. (1964). Dynamics of a phytoplankton community in an ice-covered lake. *Limnology and Oceanography*, **9**, 163–178.

3. Primary production in some tropical environments

S. Z. QASIM

Introduction

In this chapter 'primary production' refers to the photosynthetic production of organic carbon in which carbon dioxide is the only source of carbon. From this definition it is easy to distinguish the chemosynthetic uptake of carbon dioxide by microbial flora where no photosynthesis is involved.

During the last decade, a considerable amount of information has accumulated on the primary production of tropical and sub-tropical regions of the world and hence to present a comprehensive review on the subject would be beyond the scope of this chapter. It was, therefore, felt

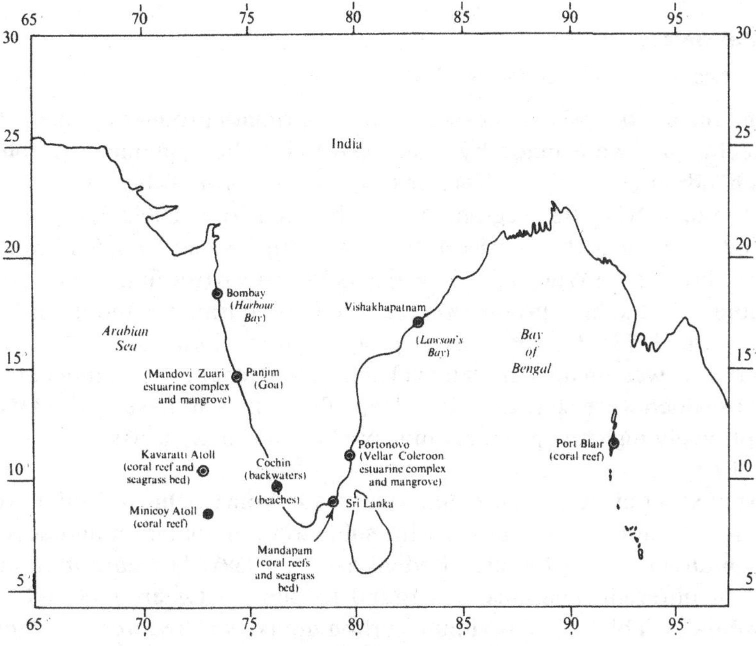

Fig. 3.1. Map of India showing the places where observations on primary production of different types of environment were made. The environments have been indicated in parentheses.

that this review should be confined to the data which have appeared on certain environments from the Indian region, largely as a part of the International Biological Programme IBP/PM. The environments dealt with here are shown in Fig. 3.1 and consist of:

1. Coastal waters
2. Backwaters and estuaries
3. Coral reefs and atolls
4. Mangrove swamps
5. Seagrass beds
6. Sandy beaches

On problems related to the productivity of the Indian Ocean, the reader may refer to some of the earlier works (Kabanova, 1968; Humphrey, 1972; Krey, 1973; Aruga, 1973; Qasim, 1977).

The methods used for measuring primary production of the six environments noted above have been diverse and these will be mentioned briefly in different sections. From the data available, attempts have been made to assign specific levels and ranges of production in each environment.

Coastal waters

West coast of India (Arabian Sea)

During the period 1965–8, measurements of primary production, using the ^{14}C technique, were made by Nair (1970) and Nair, Samuel, Joseph & Balachandran (1973) from Karwar to Cape Comorin. Tables 3.1 and 3.2 give the data from this region. As can be seen from Table 3.1, the rate of production near the coast (up to 50 m depth) ranged from 0.18 to 2.45 g C/m^2/day. At the Wadge Bank, which is 38 m deep (the first station given in Table 3.1), the high production recorded was during the month of June (Nair *et al.*, 1973). Lower rates of production within the range 0.01–0.95 g C/m^2/day, were in deeper waters (Table 3.2), except in September at two stations which were 60 m and 90 m deep. The values at these stations were exceptionally high and probably indicate bloom situations associated with upwelling.

The next set of ^{14}C measurements was made from Cochin to Quilon from August to October 1967, along with submarine illumination and several other hydrographical factors (Radhakrishna, 1969). The carbon assimilation at different locations was found to range between 0.38 and 1.11 g C/m^2/day (Table 3.3). Maximum carbon uptake was recorded at depths where the light intensity was about 50% of the incident illumination. The euphotic zone at the 13 stations ranged from 15.5 to 22 m and phosphate-phosphorus from 1.75 to 3.00 μg-at/l.

Dehadrai & Bhargava (1972*a*) made measurements of chlorophyll *a* for

Table 3.1. *Rates of primary production at 15 stations along the west coast of India (Arabian Sea) within 50 m depth*

Date	Position: Latitude N	Longitude E	Station depth (m)	Production (g C/m²/day)
1965				
5 June	8° 00′	77° 20′	38	2.09
15 December	13° 26′	75° 10′	40	0.95
16 December	Karwar Bay		7	1.39
1966				
3 February	9° 40′	76° 00′	40	0.18
6 September	9° 00′	76° 28′	25	1.24
1967				
7 August	14° 08′	74° 18′	30	0.61
6 September	9° 52′	76° 10′	18	2.37
7 September	9° 20′	76° 51′	50	1.18
7 September	8° 42′	76° 35′	35	1.26
9 September	7° 45′	77° 19′	50	0.48
9 September	7° 45′	78° 00′	47	1.43
1968				
20 July	8° 53′	76° 21′	50	1.12
21 July	10° 29′	75° 51′	37	0.89
22 July	11° 19′	75° 36′	28	1.34
24 July	12° 08′	74° 58′	37	2.45

From Nair (1970)

Table 3.2. *Rates of primary production at 53 stations along the west coast of India (Arabian Sea) beyond 50 m depth*

Date	Position: Latitude N	Longitude E	Station depth (m)	Production (g C/m²/day)
1965				
4 June	7° 30′	76° 00′	1500	0.33
6 June	8° 50′	75° 20′	1200	0.03
7 June	9° 30′	75° 10′	2000	0.13
12 October	9° 50′	75° 26′	2000	0.11
13 October	9° 20′	75° 39′	4000	0.16
14 October	8° 44′	75° 38′	350	0.05
15 October	7° 53′	77° 04′	550	0.53
16 October	8° 15′	75° 47′	1200	0.06
11 November	7° 56′	76° 55′	70	0.07
11 November	7° 52′	76° 38′	900	0.22
12 November	8° 32′	76° 00′	200	0.13
12 November	8° 32′	76° 21′	300	0.01
13 November	8° 43′	75° 26′	800	0.13
15 November			200	0.50

Table 3.2. (cont.)

Date	Position		Station depth (m)	Production (g C/m²/day)
	Latitude N	Longitude E		
1965				
24 November	11° 26′	74° 51′	82	0.11
25 November	12° 20′	74° 40′	58	0.05
25 November	12° 40′	74° 15′	86	0.27
27 November	13° 30′	73° 00′	1600	0.21
27 November	13° 30′	73° 30′	180	0.10
28 November	12° 20′	74° 21′	180	0.14
29 November	11° 15′	74° 34′	1200	0.04
14 December	11° 10′	75° 10′	60	0.57
19 December	12° 30′	74° 16′	180	0.04
1966				
6 January	14° 09′	73° 20′	160	0.25
7 January	13° 35′	72° 55′	1900	0.35
7 January	13° 06′	73° 33′	1800	0.28
8 January	12° 27′	74° 20′	120	0.25
5 February	7° 50′	77° 11′	300	0.13
7 February	9° 30′	75° 35′	1000	0.07
8 February	9° 55′	75° 09′	2000	0.39
21 April	11° 15′	74° 49′	260	0.45
22 April	11° 40′	76° 08′	1400	0.05
26 May	12° 50′	74° 05′	180	0.13
7 June	8° 12′	76° 44′	80	0.57
8 June	8° 46′	76° 10′	150	0.38
25 June	13° 30′	73° 34′	120	0.29
26 June	11° 56′	74° 11′	1700	0.22
7 September	8° 00′	77° 11′	60	4.55
7 September	8° 00′	76° 58′	90	4.55
8 November	16° 30′	73° 40′	110	0.11
8 November	16° 29′	71° 42′	300	0.12
6 December	11° 15′	74° 55′	120	0.09
1967				
9 March	9° 21′	75° 52′	188	0.05
18 April	10° 27′	72° 41′	1600	0.12
20 April	10° 43′	74° 26′	2160	0.21
8 June	10° 28′	72° 42′	1900	0.06
9 June	11° 23′	72° 46′	1900	0.04
6 August	12° 44′	74° 28′	56	0.18
31 August	11° 16′	73° 50′	2100	0.05
8 September	8° 17′	75° 44′	1400	0.40
9 September	7° 45′	76° 43′	183	0.42
10 September	7° 27′	77° 40′	117	0.95
10 September	7° 32′	76° 41′	850	0.95

From Nair (1970)

Table 3.3. *Rates of primary production at 13 stations from Cochin to Quilon along the west coast of India*

Date	Position		Station depth (m)	Production (g C/m²/day)
	Latitude N	Longitude E		
1967				
21 August	9° 58′	76° 08.5′	21	0.43
21 August	9° 28.5′	76° 13′	25	0.38
21 August	8° 55′	76° 15′	43	1.11
11 September	9° 03′	76° 19′	41	0.97
12 September	9° 31′	76° 08′	46	0.90
12 September	9° 30′	76° 18′	24	0.76
27 September	8° 51′	76° 16′	58	0.74
27 September	9° 20′	76° 24′	28	0.93
12 October	8° 32′	76° 40′	76	1.07
13 October	8° 32′	76° 32′	61	0.38
13 October	9° 08′	76° 20′	42	0.87
14 October	8° 02′	76° 44′	127	0.95
16 October	8° 03′	76° 04′	1150	1.01

Modified from Radhakrishna (1969).

a period of 9 months (September 1969 to May 1970) in coastal waters, from Goa to Bombay. During the period November–January, chlorophyll *a* was found to increase at most locations and ranged from 2.4 to 18.8 mg/m³.

In addition to the data summarized above, a series of measurements was made from the western Indian Ocean during RV *Anton Bruun* cruises of 1963 and 1964 (Ryther *et al.*, 1966). But, since these measurements were not extended up to the coast, they have not been included here.

Recently the contributions of nannoplankton and microplankton have been studied in coastal waters of Cochin. The annual contribution of nannoplankton (algae smaller than 65 μm) to the total photosynthesis was about 66.40% (Vijayaraghavan, Joseph & Balachandran, 1974).

East Coast of India (Bay of Bengal)

In the Palk Bay (near Mandapam), Prasad & Nair (1963) made a series of ^{14}C measurements. These have been given in Table 3.4. Maximum uptake was generally recorded at the surface, except in some instances when it was at a few metres below the surface. The values ranged between 0.10 and 8.68 g C/m²/day. During June and July, the values were found to be high.

In the nearshore waters, where the turbidity was high, the euphotic zone was about 6 m and the column production was relatively low. In clear

Table 3.4. *Rates of primary production at 18 stations along the east coast of India, Palk Bay (Bay of Bengal)*

Date	Place	Depth range investigated (m)	Production (g C/m²/day)
1961			
13 March	Mandapam	0–10	0.40
12 June	Mandapam	0–10	4.37
1962			
26 June	Mandapam	0–10	6.04
4 July	Mandapam	0–10	8.68
9 July	Mandapam	0–10	3.77
11 July	Thangachimadam	0–12	3.20
18 July	Uchippuli	0–12	3.22
1963			
20 February	9° 24′ N, 79° 13′ E	0–10	0.10
21 February	9° 44′ N, 79° 16′ E	0–10	0.17
11 June	Mandapam	0–10	1.07
17 June	Mandapam	0–10	2.21
1964			
26 August	Mandapam	0–8	1.40
26 August	Mandapam	0–8	1.68
2 September	Mandapam	0–8	1.11
2 September	Mandapam	0–8	1.03
16 September	Mandapam	0–8	1.02
9 October	Mandapam	0–10	1.30
13 October	Mandapam	0–7	0.74

From Nair (1970).

waters, the euphotic zone ranged from 15 to 40 m, depending upon the distance from the shore, and in these waters the daily production of the order of 3–5 g C/m²/day was often recorded (Nair, 1970).

The other relevant data on the productivity of the Bay of Bengal are from the Galathea Expedition (Steemann Nielsen & Jensen, 1957) and from the first cruise of RV *Anton Bruun*. The former are from the deeper waters and the latter from shallow regions of the coast. In the deeper parts, the values ranged from 0.12 to 0.31 g C/m²/day, while in the shelf region the range was 0.01 to 2.16 g C/m²/day (Table 3.5).

A summary of the data collected from 73 stations of both west and east coasts of India has been attempted by grouping the values into four Indian States and three depth regions (Nair *et al.*, 1973). This has been presented in Table 3.6. The average primary production for the entire coastal waters of India up to 50 m depth is 1.19 g C/m²/day, from 50 to 200 m the average value is 0.25 g C/m²/day and from depths greater than 200 m it is 0.18 g C/m²/day (Table 3.6).

Table 3.5. *Rates of primary production at 15 stations along the east coast of India (Bay of Bengal)*

Date	Position: Latitude N	Position: Longitude E	Station depth (m)	Production (g C/m²/day)
1951				
23 April	14° 20′	82° 00′	3240	0.12
24 April	17° 10′	84° 30′	2860	0.25
26 April	20° 37′	87° 33′	62	0.60
2 May	19° 53′	89° 05′	1400	0.16
4 May	13° 58′	91° 03′	3000	0.24
5 May	10° 32′	90° 59′	850	0.31
1963				
27 March	11° 49′	92° 53′	87	0.01
28 March	11° 23′	93° 31′	80	0.36
1 April	15° 08′	94° 54′	29	0.25
1 April	15° 08′	94° 04′	58	0.25
5 April	19° 41′	93° 08′	38	2.16
5 April	19° 32′	92° 52′	55	0.24
22 April	20° 35′	87° 51′	80	0.83
28 April	17° 41′	83° 19′	65	1.53
28 April	17° 35′	83° 25′	67	0.11

The first six measurements are from the Galathea Expedition and the rest from the first Cruise of RV *Anton Bruun*. (I.I.O.E Data sheets, Woods Hole Oceanographic Institution, 1964).

Table 3.6. *Average rates of primary production (g C/m²/day) obtained after pooling the data of 73 selected stations into four different states of India and three different depth regions*

States	Less than 50 m: No. of stations	Less than 50 m: Average	50 to 200 m: No. of stations	50 to 200 m: Average	Greater than 200 m: No. of stations	Greater than 200 m: Average
Tamil Nadu	3	1.33	4	0.37	6	0.18
Kerala	10	1.22	13	0.25	22	0.17
Karnataka	6	1.08	4	0.19	3	0.28
Maharashtra	—	—	2	0.12	—	—

Modified from Nair *et al.* (1973).

From the above account it is clear that very few measurements have so far been made in the coastal waters of India. The available data do not cover all seasons and hence a correct picture of the primary production of coastal waters cannot be obtained. The Arabian Sea and the Bay of Bengal are greatly influenced by the monsoon system and hence uninterrupted observations in different seasons are necessary for determining the overall productivity of shelf waters.

Backwaters and estuaries

Backwaters of Kerala

The backwaters consist of a system of brackish-water lagoons and swamps, occupying an area of several hundred square kilometres in the central part of Kerala State, south-west India (Fig. 3.2). The entire region

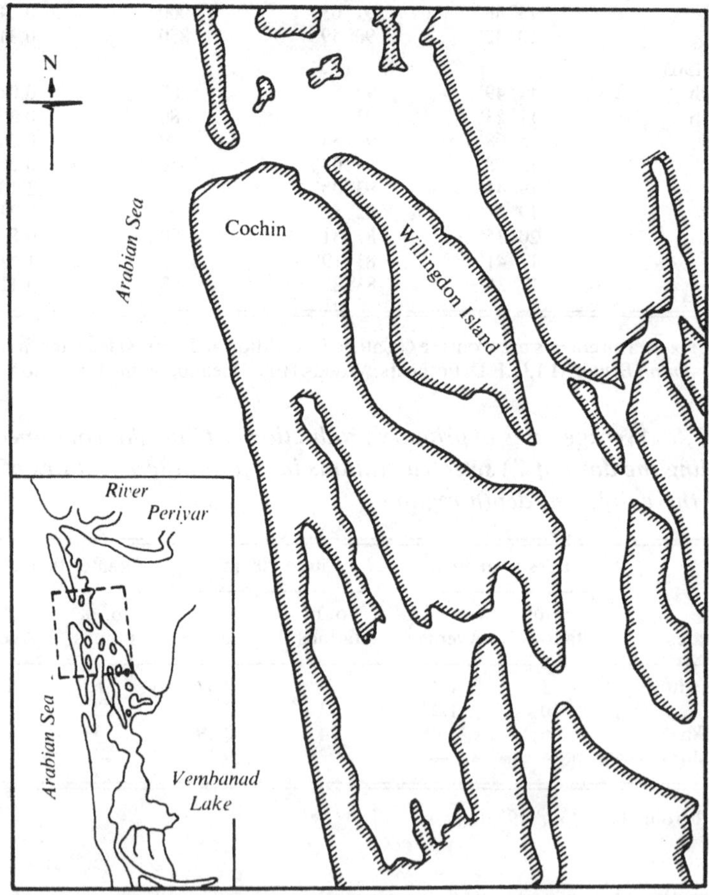

Fig. 3.2. Map showing a portion of Kerala backwaters. Inset shows the riverine and marine connections to the backwater system. The portion marked by dotted lines has been enlarged and termed Cochin Backwater. (From *The Biology of the Indian Ocean, Ecological Studies 3*, ed. B. Zeitzschel. Berlin: Springer-Verlag.)

is a centre of human activity and a number of drains, factory effluents and sewage outlets flow into the backwaters making it in certain parts intensely polluted. The northern part of the backwaters, which has been studied fairly extensively, has been termed 'Cochin Backwater(s)'. This region has also been referred to as a tropical estuary. Much of the earlier work carried out on the backwater has recently been reviewed (Qasim, 1973). However, in view of several additional investigations made during the last few years, the earlier review needs up-dating.

The hydrography of Cochin Backwater is greatly influenced by two main factors: (1) the short-term changes induced by the tides (Qasim & Gopinathan, 1969) and (2) the seasonal changes brought about by the monsoon (Sankaranarayanan & Qasim, 1969). Heavy rainfall and land runoff also affect the hydrography of coastal waters of Kerala (Derbyshire, 1967). However, the first point worth noting in the hydrography of the backwater is that a clear stratification begins to exist in the water column during the months of heavy rainfall (June to September), which makes the surface water quite distinct from the bottom water. This stratification can easily be identified from the vertical gradients of temperature, salinity and oxygen. The stratification gets broken in October–November and then the water becomes well mixed. The other point of interest is the seasonal cycle of nutrients in the backwater. All the three major nutrients (phosphate-phosphorus, nitrate-nitrogen and silicate-silicon) attain maximum concentrations during the monsoon months and their high occurrence seems to be associated with land runoff. In other words, the backwater gets maximum enrichment from May to September when the rainfall is maximum.

Light penetration in the backwater, because of high turbidity, gets considerably reduced. The euphotic zone varies from 2 to 6 m during the year with the attenuation coefficient (k) ranging from 0.60 to 3.00 (Qasim, Bhattathiri & Abidi, 1968). Sedimentation of suspended material is maximum during the pre-monsoon (January to April) and post-monsoon months (September to December). Most of the settled material gets resuspended during the monsoon months (May to August) and with the strong ebb currents gets transported into the sea (Gopinathan & Qasim, 1971).

Measurements of primary production in the backwater were made using the ^{14}C method concurrently with the light–dark bottle oxygen technique throughout the year. Experiments were conducted *in situ* using a float from which bottles were suspended at various depths of the euphotic zone. Gross and net primary production rates were determined at various depths and these have been given in Fig. 3.3 for the three seasons. Because of high turbidity, primary production was non-existent at depths greater than about 4 m.

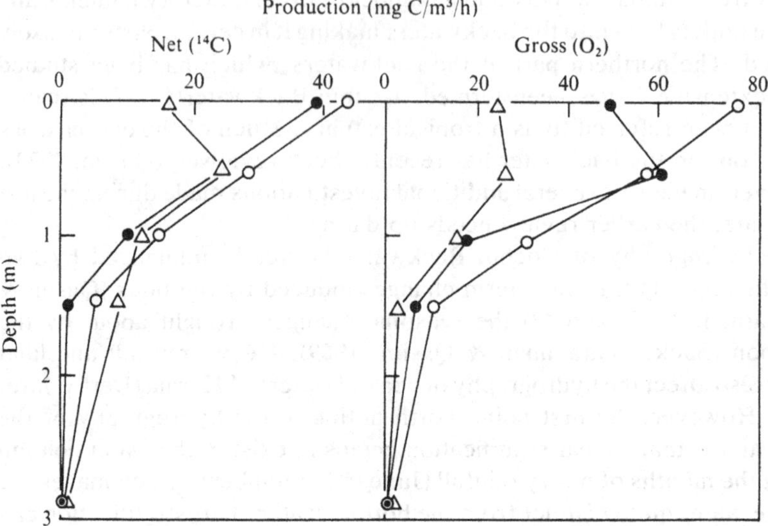

Fig. 3.3. Gross and net primary production in the backwater in relation to depth. Annual data have been pooled into three seasons each of four months. Monsoon season, closed circles; post-monsoon season, triangles; pre-monsoon season, open circles. (From *The Biology of the Indian Ocean, Ecological Studies 3*, ed. B. Zeitzschel. Berlin: Springer-Verlag.)

Monthly values of column production (gross and net) for the narrow euphotic zone have been given in Fig. 3.4. Community respiration was computed from the oxygen decrease in the dark bottle during light–dark bottle experiments. Gross production ranged from 0.35 to 1.50 g C/m²/day and net production from 0.25 to 0.88 g C/m²/day. The range in respiration was from 0.08 to 0.64 g C/m²/day (Qasim, Wellershaus, Bhattathiri & Abidi, 1969; Qasim, 1970*a*). The annual cycle of production showed three small peaks with approximately three-to-four-fold increase in April, July and October (Fig. 3.4). These peaks were in the form of brief pulses showing no regular seasonal rhythm.

In the Cochin Backwater, about 75% of the total primary production is contributed by nannoplankton and the rest (25%) by microplankton. The contributions of these two components of algae were determined by fractionating the water samples through different grades of bolting nylon and measuring the ^{14}C uptake, chlorophyll *a* concentration and cell numbers of each size group throughout the year (Qasim, Vijayaraghavan, Joseph & Balachandran, 1974).

Three main ecological factors, namely salinity, light and nutrients were found to govern the rate of primary production in the backwater. The influence of each of these factors was determined experimentally by exposing several species of unialgal cultures to varying salinities, light conditions and nutrients.

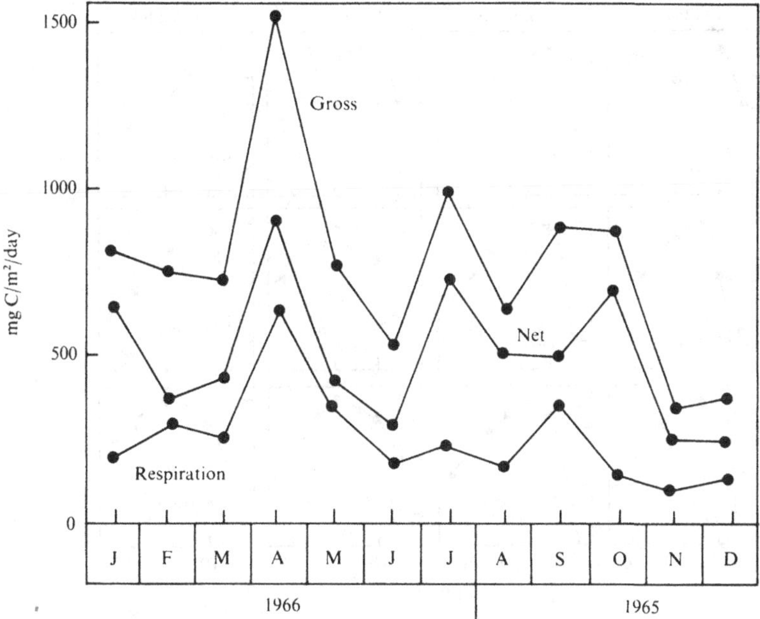

Fig. 3.4. Annual cycle of gross and net primary production in the backwater together with the seasonal changes in community respiration. The values of production and respiration refer to euphotic zone. (From *The Biology of the Indian Ocean, Ecological Studies 3*, ed. B. Zeitzschel. Berlin: Springer-Verlag.)

Fig. 3.5 gives the rates of photosynthesis of 12 organisms at different salinities. In all the organisms, maximum photosynthesis occurred at low salinities (Qasim, Bhattathiri & Devassy, 1972*a*). The ecological implications of the laboratory experiments were tested from the field data. In the backwater, maximum abundance of algae was found during a period when, because of heavy rainfall and land runoff, the salinity became low. This was confirmed once again by the total cell counts made throughout the year. Maximum numbers of algae were found in September when the salinity was low (Devassy & Bhattathiri, 1974). Many organisms were found to bloom successively at exceptionally low salinities. Such an adaptation in algae is probably to ensure that peak production occurs during a period when the nutrients in the environment are maximum. The enrichment in the estuary and in the waters of the south-west coast of India is associated with large dilutions during the monsoon months (Qasim, Bhattathiri & Devassy, 1972*a*).

Unialgal cultures of 11 different species were exposed to varying light intensities in an incubator provided with neutral density filters and their rates of photosynthesis were measured by ^{14}C uptake. The saturation point (I_k) in most of the organisms ranged between 11 and 17 kilolux. In

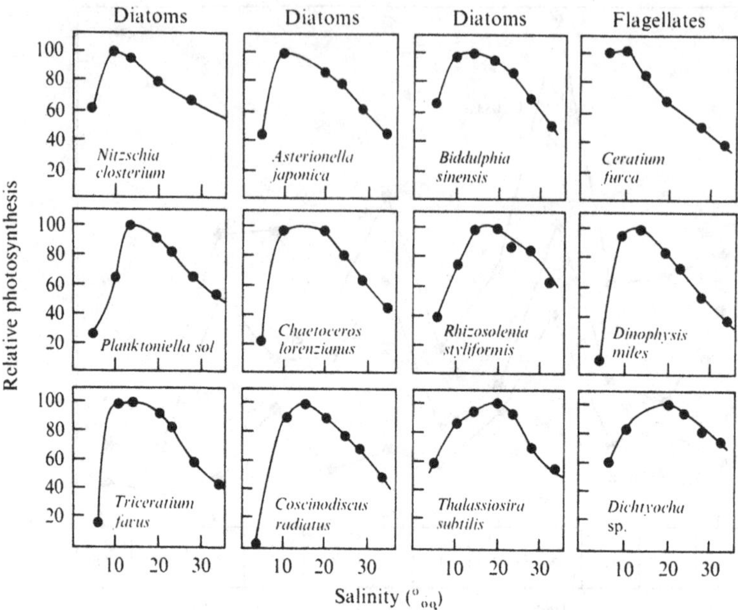

Fig. 3.5. Rates of photosynthesis (^{14}C uptake) as a percentage of maximum shown by unialgal cultures of diatoms and flagellates in relation to salinity. (From Qasim, Bhattathiri & Devassy, 1972*a*, *Marine Biology*, **12**, 201.)

addition to the intensity effect, the influence of the wavelength of light was also determined by exposing each organism to a portion of the visible spectrum, starting with the longest wavelength. Sharp cut-type glass filters were used in a specially designed incubator for cutting off the transmittance of light (Qasim, Bhattathiri & Devassy, 1972*b*). In all the organisms, no wavelength dependence of photosynthesis in saturation light was observed. All regions of the visible spectrum were found to be effective as long as there was enough light energy to stimulate photosynthesis. These observations indicate that marine planktonic algae are capable of adapting themselves to changing light conditions fairly rapidly. Such a chromatic adaptation is of a distinct advantage to the algae which have a floating existence (Qasim, Bhattathiri & Devassy, 1972*b*).

The nutrient requirements of different algae were found to be quite variable and high instantaneous concentrations of phosphate and nitrate alone did not give rise to a substantial increase in primary production. This was experimentally demonstrated by studying the growth kinetics of nutrient-depleted cells of a diatom, *Biddulphia sinensis*, and a dinoflagellate, *Ceratium furca*. The growth rate of both these organisms, as a function of phosphate and nitrate, followed the kinetics similar to that of the Michaelis–Menten equation for bacterial growth. The capacity of each organism to utilize phosphate and nitrate, when available either singly or in combination, was determined from the values of the half-saturation constant (K_s).

Both the organisms (*Biddulphia* and *Ceratium*) were found to occur in the Cochin Backwater. The former became abundant during the monsoon months when the nutrients reached their maximum concentrations, and fairly high densities of the latter were recorded during the pre-monsoon months when the nutrients were low. *Ceratium* has a low K_s and hence it is a better competitor at low concentrations of nutrients than *Biddulphia*. At high concentrations of nutrients, *Biddulphia*, which has a higher K_s, dominates over *Ceratium*. Thus the study of the growth kinetics of the two organisms gave an explanation of the observed seasonal succession of the two species (for details see Qasim, Bhattathiri & Devassy, 1973).

A simple model developed by Margalef (1967) was used to study the annual cycle of primary production and plankton succession in the Cochin Backwater. The increased flushing rate during the monsoon months seemed to govern the abundance of phytoplankton in the estuary. When flushing rate increased, the algal densities became very low. During the post-monsoon months, when the estuary remained relatively stagnant, the algal biomass was high (Wyatt & Qasim, 1973).

Besides phytoplankton, the other important component in the estuary is organic detritus. It plays a very significant role in the food chain of the estuary and leads to several additional pathways between primary production and animal nutrition (Qasim, 1970*a*, 1972*a*). The importance of detritus, its rate of sedimentation, its chemical composition and nutritive value have been described by Qasim & Sankaranarayanan (1972). The total quantity of settled detritus ranged from 67 to 1013 g/m^2/day. Detritus is perhaps the most readily available food material in the estuary. It occurs at the bottom in coarsely particulate form and is derived from the dead material of plants and animals. Inorganic matter such as fine silt and sand particles form a substrate around which organic material adheres and forms larger aggregates of detritus. Caloric content of detritus ranges from 200 to 500 cal/g dry wt.

To determine the nutritive value of detritus, laboratory experiments were conducted on the energy conversion of a penaeid prawn, *Metapenaeus monoceros*, by feeding it on estuarine detritus. The shrimp readily consumed detritus and its energy budget was worked out after fourteen days of experiments. Measurements were made of energy utilized for growth, *P*; intake of food or consumption, *C*; and assimilation, *A*. Gross growth efficiency (P/C or K_1) ranged from 10.5 to 35.2% (average 22%) and net growth efficiency (P/A or K_2) ranged from 11.7 to 36.4% (average 24%). Food efficiency was inversely related to growth efficiency and the assimilation efficiency was of the order of 93% (Qasim & Easterson, 1974).

Seasonal changes in the zooplankton abundance in the estuary have been studied by several authors in recent years. Copepods have been found to constitute the predominant component throughout the year (Tranter & Abraham, 1971; Pillai, 1971). Among copepods, the population of omnivores remains large in all the seasons. When the copepod biomass in the estuary was low, their population was heterogenous, i.e. the species diversity was greater (Pillai, Qasim & Nair, 1973).

Mandovi–Zuari estuarine complex

Two large estuaries, demarcated by the rivers Mandovi and Zuari, are found in Goa on the northern and southern sides of Panaji Island. About 14 and 11 km from their respective openings into the Arabian Sea, the two estuaries are connected by a natural canal, the Cumbarjua canal (Fig. 3.6). A large number of tributaries join the river Mandovi, whereas no major tributary meets the river Zuari. Both these estuaries are fed largely by the monsoon water. The average rainfall in Goa is 2611 mm per year, of which nearly 80% occurs from June to August. The Zuari estuary is broader and has a greater marine influence than that of the Mandovi estuary. The tides in the estuaries are of mixed, semi-diurnal nature with a range of about 2 m. The tidal influence is felt up to 35 km and 41 km upstream in Mandovi and Zuari respectively. The flow of water in the estuarine complex is largely dependent upon tidal conditions. During the flood tide, the water enters through Zuari and reaches Mandovi via the Cumbarjua canal. During the ebb, the flow is reversed (Das, Murty & Varadachari, 1972). The speeds of surface and bottom currents at the mouth of Mandovi estuary are 160 and 128 cm/s during the flood and 108 and 98 cm/s during the ebb respectively. These measurements were made in March using an Ekman-type current meter. In the two estuaries, the depth of water varies from 2 to 11 m.

In recent years, several communications have appeared on the hydrography of these two estuaries (Dehadrai, 1970*a*, *b*; Dehadrai & Bhargava, 1972*b*; Murty & Das, 1972). These authors reported several findings which are summarised below.

Light penetration, as measured by the Secchi disc, was found to vary considerably during the different months of the year. The euphotic zone ranged from 1 to 4 m. The range in surface temperature during the year was from 24 to 32 °C. The lowest temperature was recorded in December–January and the highest in April–May. The authors noted above have divided the year into three seasons, each of four months. Thus, the pre-monsoon season includes February to May; monsoon, June to September; and post-monsoon, October to January. Seasonal changes in the salinity were largely dependent upon the rainfall and land runoff. During the monsoon season, surface salinity varied from 0.12 to 30.20‰ in

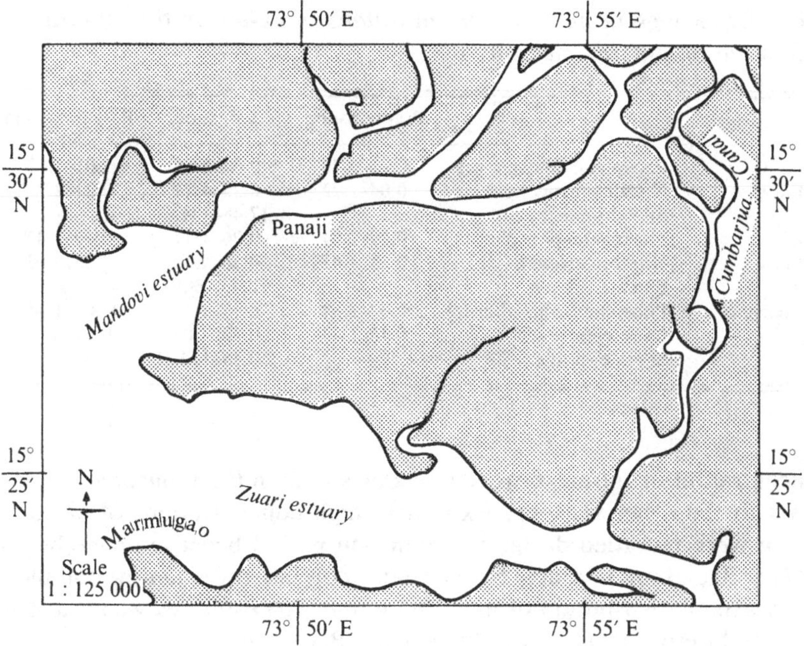

Fig. 3.6. Mandovi–Zuari estuarine complex of Goa with their marine and riverine connections.

Mandovi, from 0.21 to 32.9‰ in Zuari and from 0.12 to 15.6‰ in Cumbarjua. Salinities at the bottom were slightly higher than at the surface but no stratification in the water column was recorded. In other words, the water throughout the year remains fairly well-mixed. This feature is unlike the Cochin Backwater where there is a clear stratification in temperature, salinity and oxygen during the monsoon months (June to September), which keeps the surface and bottom waters quite distinct. In the post-monsoon season, the ranges in salinity in Mandovi, Zuari and Cumbarjua were 10.54–33.58‰, 29.07–34.95‰, and 7.74–32.89‰ respectively and in the pre-monsoon the ranges in salinity were 22.81–36.19‰, 31.51–36.25‰ and 20.39–36.26‰.

From the above data it is clear that the Zuari estuary gets the maximum marine influence and the Cumbarjua canal the least.

During the year, three major nutrients, namely phosphate–phosphorus, nitrate–nitrogen and silicate–silicon, occur in varying proportions in the estuarine complex of Goa (Table 3.7).

The nutrient ranges given in Table 3.7 indicate that there is no major difference in their concentration in the three regions. Presumably it is the same water which flows from Zuari via Cumbarjua into Mandovi. The

Table 3.7. *Range in nutrient concentrations in µg-at/l in the estuaries of Goa during different seasons*

		Pre-monsoon	Monsoon	Post-monsoon
Mandovi estuary	Phosphate–phosphorus	0.11–1.78	0.04–1.75	0.001–1.52
	Nitrate–nitrogen	0.08–1.72	0.25–4.11	0.21–2.79
	Silicate–silicon	—	17.48–134.30	—
Zuari estuary	Phosphate–phosphorus	0.29–2.04	0.26–1.17	0.10–1.42
	Nitrate–nitrogen	0.25–1.47	0.36–4.56	0.35–8.03
	Silicate–silicon	—	12.43–89.98	—
Cumbarjua canal	Phosphate–phosphorus	0.39–1.91	0.13–1.26	0.004–1.45
	Nitrate–nitrogen	0.13–1.43	0.42–3.87	0.17–5.93
	Silicate–silicon	—	20.15–114.15	—

concentration of nitrate–nitrogen becomes high in the monsoon as compared to the other seasons. Extremely high concentrations of silicate–silicon were recorded during the monsoon period because of the heavy load of silt and sand brought by the land runoff into the estuarine complex. The seasonal distribution of nutrients in the Goa estuaries was found to be very different from that in the Cochin Backwater.

For the estimation of primary production in the Goa estuaries, both oxygen and ^{14}C methods were employed (Dehadrai, 1970*a*, Dehadrai & Bhargava, 1972*b*). The values given by these authors are for the surface only. Gross production, as measured by the oxygen method, varied from 135 to 550 mg C/m^3/day in Mandovi and from 150 to 580 mg C/m^3/day in Zuari. Similarly, ^{14}C assimilation (net production) for Mandovi and Zuari estuaries ranged from 95 to 274 mg C/m^3/day and from 60 to 245 mg C/m^3/day respectively. The ranges given above pertain to high tides only. During the low tides the ranges in ^{14}C uptake were 44–122 mg C/m^3/day for Mandovi and 22–82 mg C/m^3/day for Zuari. Dehadrai (1970*a*) reports that the average net production at the surface in Mandovi ranged from 95 to 450 mg C/m^3/day. The production was lowest during the monsoon months. No determinations of primary production were made in Cumbarjua canal.

Samples of phytoplankton were collected from the mouth of Zuari estuary using a net of 0.065 mm mesh width. Cell counts were found to be low and varied from 220 to 5800 cells/l. Two peaks in cell counts were recorded during the pre-monsoon and post-monsoon periods corresponding to the maximum values of primary production. The phytoplankton crop during the monsoon months was very low.

However, recent counts of plankton organisms made from the settled samples of water collected from Mandovi, Zuari and Cumbarjua have

Table 3.8. *Cell counts (cells/l) of surface samples from Mandovi, Zuari and Cumbarjua, 1972*

	June	July	August	September
Mandovi	371 000	35 700	16 700	14 900
Zuari	387 500	7 200	3 600	18 600
Cumbarjua	430 000	3 000	89 400	4 000

failed to confirm the deductions made earlier (Bhattathiri, Devassy & Bhargava, 1976). The counts at the surface during June–September are given in Table 3.8.

Assimilation of ^{14}C in Mandovi during the four months of monsoon (June–September) ranged from 56.0 to 540.0 mg C/m^2/day. In Zuari, the range in the carbon uptake was from 108.0 to 502.0 mg C/m^2/day and in Cumbarjua from 175.0 to 440.0 mg C/m^2/day. Chlorophyll *a* values ranged from 1.60 to 12.70 mg/m^2 in Mandovi, from 1.90 to 12.40 in Zuari and from 2.20 to 12.60 in Cumbarjua (Bhattathiri, Devassy & Bhargava, 1976).

Bombay harbour bay

This bay is connected with the Arabian Sea at its southern region and with the Thana Creek at its northern region. During the south-west monsoon, a large quantity of freshwater flows into the bay.

During 1964–5, measurements of primary production were made in this bay near the jetty of the Canada–India Atomic Reactor. Water samples from the surface were incubated *in situ* with ^{14}C and the filters were counted using a liquid scintillation counter. Chlorophyll *a* concentrations were also determined from the surface samples (Krishnamoorthy & Viswanathan, 1968). Primary production was low in June because of high turbidity. The range in production during the year was from 0.015 to 4.99 g C/m^3/day with a maximum in February. Chlorophyll *a* values ranged from 1.77 to 11.98 mg/m^3. The minimum and maximum values of chlorophyll *a* were in August and December respectively (Krishnamoorthy & Viswanathan, 1968).

Vellar–Coleroon estuarine system

This estuarine system is located on the east coast of India (Bay of Bengal) and connects the Porto Novo backwaters with the adjoining mangrove forests and swamps (Fig. 3.7). Several rivulets, canals and gullies join the forest and the backwater system. The vegetation in the area is thick with

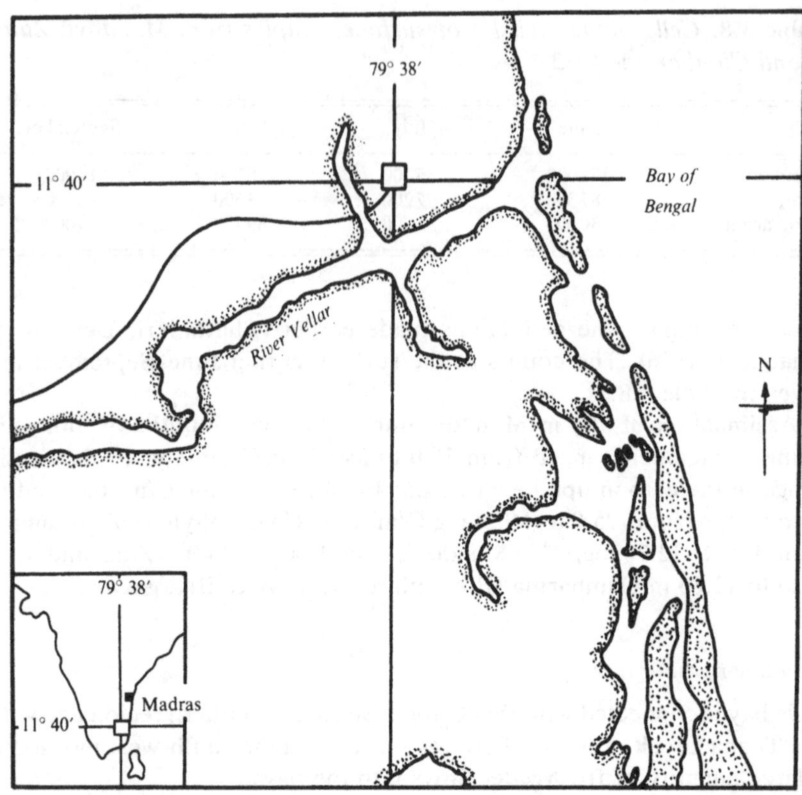

Fig. 3.7. Vellar–Coleroon estuarine system with its main connections with the Bay of Bengal. Inset shows its location on the east coast of India. (After Purushothaman and Venugopalan, 1972.)

mangrove trees and shrubs. The bottom sediment varies from fine silt, clay, coarse mud to fine mud. It becomes more sandy towards the sea. The tides in the region are of semi-diurnal type and vary from 0.5 to 1.0 m in range (Krishnamurthy, 1971; Krishnamurthy & Sundararaj, 1973).

In recent years, a series of papers has appeared on this estuarine system describing primary production, plant pigments, dissolved silicon, environmental features and trace elements in the particulate matter (Venugopalan, 1969; Krishnamurthy, 1971; Bhatnagar, 1971; Purushothaman & Venugopalan, 1972; Jegatheesan & Venugopalan, 1973; Krishnamurthy & Sundararaj, 1973). The salient features of the estuarine system are recorded below.

During the year the temperature of water in the estuary varies from 24.8 to 32.6 °C, whereas the variations in salinity are very large, of the order of 0.5–28‰ or 1.00–34.30‰ depending upon the region under investigation. Maximum salinity values are recorded from May to July and minimum, because of the influence of the north-east monsoon, from November to December. The different nutrients have been reported to range as follows (μg-at/l).

Total phosphorus 0.72–3.34
Inorganic phosphate 0.19–1.59
Dissolved organic phosphorus 0.16–1.75
Ammonia and amino acids 0.34–3.60
Nitrite 0.11–0.25
Nitrate 2.85–6.94
Silicate 20.00–140

Similarly, the ranges in plant pigments were: chlorophyll *a*, 2.05–21.56 μg/l; chlorophyll *b*, 0.30–13.40 μg/l; chlorophyll *c*, 2.70–26.50 μg/l; carotenoids, nil–5.25 MSP/m^3.

Dissolved silicon in the Vellar river was found to be 17×10^3 μg/l and that in the sea water with salinity 35‰ it was 178 μg/l. A linear inverse relationship was found between dissolved silicon and salinity.

Trace elements such as iron, copper, manganese, molybdenum, vanadium and cobalt were measured from the Vellar estuary from March 1971 to July 1972 and the seasonal cycle of each of these has been reported by Jegatheesan & Venugopalan (1973).

Measurements of primary production were carried out using the light–dark bottle oxygen method at five stations in Vellar estuary in February, March and April 1966. Incubations were made *in situ* both at the surface and 1.5 m depth. Gross and net production at the surface ranged from 11.77 to 36.99 mgC/m^3/h and 6.72 to 26.22 mg C/m^3/h respectively. At 1.5 m depth the range in the production was much higher, gross production was 21.88–58.83 mg C/m^3/h and net production was 18.48–47.07 mg C/m^3/h.

In February 1964, the ^{14}C method was also used at four stations. The two marine stations gave the production values at the surface as 11.93 and 20.17 mg C/m^3/h. The two estuarine stations, on the other hand, gave the primary production as 35.50 and 45.99 mg C/m^3/h (Venugopalan, 1969).

Bhatnagar (1971) using the light–dark bottle oxygen method, reported very high values of primary production from the Vellar estuary (Kille Backwaters). According to him, maximum gross (251.20 mg C/m^3/h) and net photosynthesis (225.44 mg C/m^3/h) based on monthly mean values, were found in August 1968 and the minimum value of gross production was 47.62 and net production was 38.09 mg C/m^3/h in February 1969. Chlorophyll *a* values during the period of investigation were 12.70 mg/m^3 (maximum) in July 1968 and 1.55 mg/m^3 (minimum) in November 1968.

A. Purushothaman & R. Natarajan (personal communication) have also estimated the primary production (using the oxygen method) and chlorophyll *a* of the estuarine system. The annual production rates (gross and net) at the mouth of Vellar estuary for the year 1969 were 598.30 and 379.70 g C/m^2 respectively. The chlorophyll *a* at the surface varied from 1.30 to 19.20 μg/l.

Lawson's Bay

Lawson's Bay is an embayment on the east coast of India (Bay of Bengal). It is located along latitude 17° 44′ N and longitude 82° 23′ E, about 7.5 km away from Visakhapatnam harbour.

Primary production in the bay was measured using light–dark bottles by Subba Rao (1973). The samples were incubated under natural light in a trough for about 6 h. Gross production ranged from 9.20 to 150.05 mg C/m^3/h. During blooms, gross production increased three- to thirteen-fold. The ratios of photosynthesis to respiration (*P*/*R*) ranged from 5.1 to 10.5. High *P*/*R* ratios coincided with the increase in production. Common organisms forming blooms were *Thalassiosira subtilis* and *Chaetoceros curvisetus*.

Coral reefs and atolls

Coral reefs are known to be among the most productive communities in the sea. The reefs of the tropical Indian Ocean include (*a*) fringing and barrier reefs, (*b*) sea-level atolls and (*c*) elevated reefs (Stoddart, 1972).

Reefs are either absent or poorly developed along the coasts of Somalia, India and Malaysia. They are, however, well developed in the Persian Gulf, Red Sea, the coasts of Kenya, Tanzania, West Malagasy and parts of Sumatra and north-west Australia. No barrier reefs are found on the continental shores except for a discontinuous submerged barrier in north-west Malagasy. Around the granitic Seychelles, the volcanic Comoros and Mascarenes, and the sedimentary Andamans and Nicobars, fringing reefs are well developed. Atolls having a dominant type of reef occur only next to fringing reefs in the Indian Ocean. One of the largest atolls in the world is Suvadiva Atoll, Maldive (70 km×35 km). Maldive atolls have formations of faros or small atoll-shaped reefs around their margins. In the eastern Indian Ocean, elevated reefs are represented by Christmas Island and Ramanathapuram, and in the western Indian Ocean by raised atolls (Stoddart, 1972; Stoddart & Pillai, 1972).

Little work has been done on the primary productivity of coral reefs of the Indian Ocean. The available information is on fringing reefs located in the Gulf of Mannar, Andaman Sea, and on atoll reefs of Minicoy and Kavaratti in the Laccadives (Nair & Pillai, 1972; Qasim, Bhattathiri & Reddy, 1972).

The flow respirometry technique of Sargent & Austin (1949, 1954) has

Table 3.9. *Estimates of gross primary production of different coral reefs from the Indian region*

Location and date	Gross production (g C/m²/day)	Authors
Manauli Reef, Gulf of Mannar 5 March 1968	7.3	Nair & Pillai, 1972
Minicoy reef 2 April 1968	9.1	Nair & Pillai, 1972
Andaman reef 24 April 1968	3.9	Nair & Pillai, 1972
Kavaratti reef 6 November 1968	6.15[a]	Qasim, Bhattathiri & Reddy, 1972
Kathuvalimuni reef, Palk Bay April 1973	6.3	Balasubramanian & Wafar, 1975

[a] Uncorrected for diffusion.

Table 3.10. *Number of genera and species of reef-building corals (hermatypic) and non-reef-building corals (ahermatypic) recorded from the seas around India*

Region	Hermatypic		Ahermatypic	
	Genera	Species	Genera	Species
Maldives	47	124	19	42
Laccadives	26	69	3	4
Ceylon	27	70	12	20
Mandapam area	25	110	7	7
Tuticorin	14	16	5	5
Andamans	23	57	8	11
Mergui Archipelago	26	51	5	14
Other regions	2	3	13	26
Total[a]	49	253	27	89

[a] From Pillai (1972). The values give the total number of genera and species recorded in all the regions put together, of which many were common to each region. The total number of genera was 76, and of species, 342.

been used in all the investigations to determine the primary productivity of the reef. The method involves the measurement of oxygen changes in water during its transport over the reef.

Table 3.9 gives the values of primary production in different reefs. All the reefs, except Andaman Reef, were found to be autotrophic. Pillai

Table 3.11. *Range in the density of algae in three different coral reefs of the Bay of Bengal*

Locality	Range in density of algae (wet weight, kg/m^2)
Gulf of Mannar reefs	0.43–1.16
Palk Bay reefs	0.08–0.57
Pamban Island reefs	0.42–0.96

From Rao (1972).

(1972) gives the total number of genera and species of corals found in some of the reefs of India (Table 3.10). The mean densities of algae found in three fringing reefs were determined by Rao (1972). These are given in Table 3.11.

The descriptions of the reefs whose productivity has been measured are given below.

Manauli Reef

This fringing reef is situated along a small island in the Gulf of Mannar, 6 km away from the south-east coast of India. The width of the reef is about 100 m. The coral fauna recorded from the reef includes: *Echinopora lamellosa, Montipora foliosa, Favites abdita, Favia pallida, Porites* spp., *Acropora indica, A. surculosa, A. erythraea, A. hyacinthus* (Nair & Pillai, 1972).

Kathuvalimuni Reef

This reef is located in the Palk Bay near Mandapam and lies in an east–westerly direction. It extends about 4 km in length and runs parallel to the shore. The average width of the reef is about 300 m and the average depth over the reef is approximately 2 m. The major corals occurring in the reef are: *Acropora* sp., *Favites* sp., *Favia* sp., *Porites* sp., *Goniopora* sp., *Goniastrea* sp., and *Pocillopora* sp. (Balasubramanian & Wafar, 1974).

Minicoy Reef

This is an oval-shaped reef and forms a part of Minicoy Atoll, the southernmost atoll of the Laccadive Archipelago. The length of the atoll is 8 km and it has a maximum width of 4.5 km. The depth of the lagoon varies from 1 to 3 m. The corals forming the reef include: *Goniastrea retiformis*, *Platygyra lamellina*, *Favia pallida*, *Porites* spp., *Diploastrea heliopora*, *Lobophyllia corymbosa*, *Acropora* spp. and *Pocillopora damicornis* (Nair & Pillai, 1972).

Andaman Reef

This reef is of the fringing type and lies at 5 to 7 m depth near Port Blair. The dominant species of corals are: *Acropora assimilis*, *A. syringodes*, *A. squamosa*, *Pocillopora damicornis*, *Favites* spp. and *Porites* spp. The Andaman and Nicobar group of islands generally have luxuriant fringing reefs (Nair & Pillai, 1972).

Kavaratti Reef

This reef forms a part of Kavaratti Atoll of the Laccadive Archipelago and is located at latitude 10° 33′ N and longitude 72° 36′ E. The atoll, on its eastern side, has an island of 3.45 km^2 in area, a shallow lagoon about 4500 m long and 1200 m wide in the middle and a reef about 300 m wide bordering the western margin of the lagoon (Fig. 3.8). The productivity of this atoll has been studied more than those of the other reefs described above.

The beach slope of the lagoon has a luxuriant growth of turtle grass, *Thalassia hemprichii* and manatee grass, *Cymodocea isoetifolia*. Some of the algae found in the lagoon are:

Green algae: *Ulva lactuca*, *Enteromorpha prolifera*, *Chaetomorpha littorea*, *Cladophora fascicularis*, *Dictyosphaeria cavernosa*, *Boergesenia forbesii*, *Halimeda incrassata*

Brown algae: *Colpomenia sinuosa*

Red algae: *Fosliella lejolisii*, *Hypnea cervicornis*, *Spyridia filamentosa*, *Laurencia papillosa*, *Acanthophora spicifera*, *Leveillea jungermannioides*

The dominant species of corals in the lagoon are: *Acropora indica*, *A. conferta*, *Pocillopora damicornis*, *P. verrucosa*, *Porites* spp., *Fungia* spp. and *Goniastrea* spp. (Qasim, Bhattathiri & Reddy, 1972).

Blooms of the blue-green alga *Trichodesmium erythraeum* seem to occur around Kavaratti Atoll practically every year during the pre-monsoon season. In April 1968 a massive bloom of *Trichodesmium* was recorded and in 1969 also a similar bloom occurred during the same period (Qasim, 1970*b*, 1972*b*).

Primary production of the lagoon was estimated from a study of diurnal changes in the oxygen concentration at two stations which were about 1200 m apart. The formula of Kohn & Helfrich (1957) was used to deter-

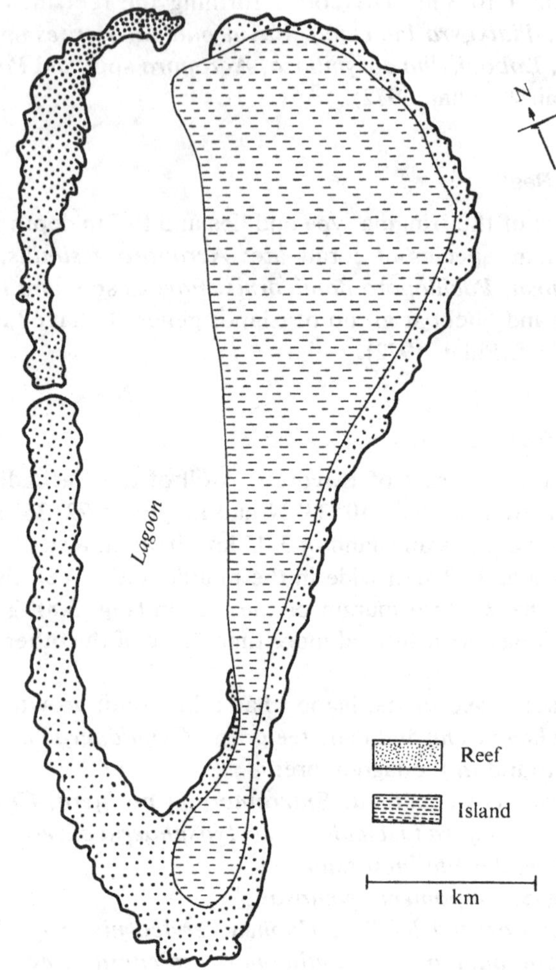

Fig. 3.8. Kavaratti Atoll of the Laccadive Archipelago showing the relative position of reef, lagoon and island. (From Qasim & Sankaranarayanan, 1970, *Limnology and Oceanography*, **15**, 574–78.)

mine the rates of changes of oxygen concentration in ml/cm strip/s between the two stations. Gross and net primary production in the lagoon were found to be 12.92 g C/m²/day and 3.38 g C/m²/day (Qasim, Bhattathiri & Reddy, 1972). Photosynthesis and respiration of some of the individual plants from the lagoon were also determined by oxygen changes, and in most species, photosynthesis for 12 hours exceeded their respiration for 24 hours. Similar experiments on three species of corals showed that their photosynthesis for 12 hours was much greater than their respiration

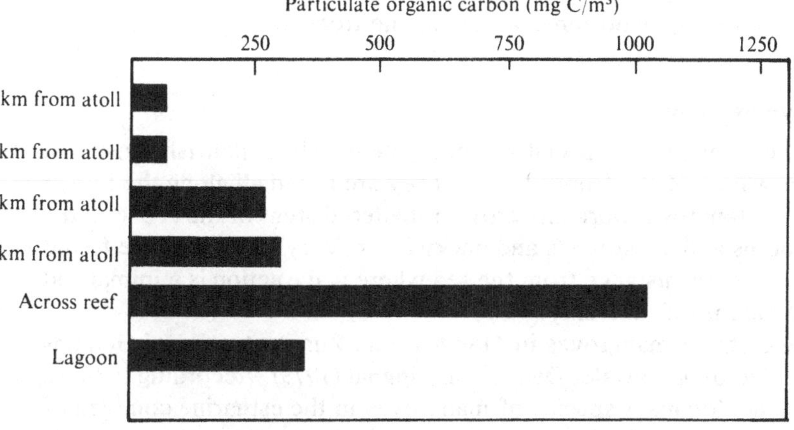

Fig. 3.9. Concentrations of particulate organic carbon at different sites. (From Qasim & Sankaranarayanan, 1970, *Limnology and Oceanography*, **15**, 574–78.)

for 24 hours. The symbiotic zooxanthellae and the boring and attached algae occurring in the corals contribute largely to the high primary productivity of the reef.

Measurements of ^{14}C uptake and chlorophyll *a* were made in the ocean water around Kavaratti Atoll. These gave extremely low values indicating that the water surrounding the reef community is low in its phytoplankton content. Particulate organic carbon, on the other hand, in the vicinity of the atoll was found to be high (Fig. 3.9). This indicates that a fairly large quantity of organic matter observed in the form of organic aggregates and microscopic particles is produced by the reef (Qasim & Sankaranarayanan, 1970). The total fractions of particulate matter added to the water as a result of its transport over the reef during day and night were also determined by the flow method (see Qasim & Sankaranarayanan, 1970). Production of particulate carbon was found to be 0.101 g $C/m^2/h$ for the day and 0.089 g $C/m^2/h$ for the night (average = 0.095 g $C/m^2/h$). This amounted to 20% of the gross primary production of the reef and 95% of the coral respiration (Qasim & Sankaranarayanan, 1970).

Zooplankton biomass recorded at Kavaratti during a period of four months varied over a wide range. In October 1968, the nocturnal abundance of zooplankton in the sea and lagoon was very high (Tranter & George, 1972). From the difference between the biomass values of sea and lagoon, Tranter & George (1972) concluded that zooplankton organisms alone were unable to provide the total energy required by the reef. The extracellular products secreted by the zooxanthellae are of possible nutritional value to the host (Muscatine, Pool & Cernichiari, 1972) and presumably these provide the extra source of energy required by the reef for its survival in oligotrophic waters of the tropics.

Mangrove swamps

Mangrove swamps are specialized ecosystems which flourish mostly in brackish waters of the tropical zone. They are found all along the Indian coastline. Mangroves normally grow in sheltered areas of the coast, at the river mouths and in estuaries and backwaters. Very often they are found at a considerable distance from the sea where wave action is minimal and the substratum soft and silty.

The ecology of mangroves in Mandovi and Zuari estuaries of Goa has been studied by Untawale, Dwivedi & Singbal (1973). According to these authors, the dominant species of mangroves in the estuarine complex of Goa are: *Avicennia officinalis*, *Rhizophora mucronata*, *Sonneratia apetala* and *Bruguiera* sp. Although these four species are found all along the estuarine complex, the growth and density of plants differ markedly from place to place. In Cumbarjua canal, the height of the plants and their abundance were found to be much greater than anywhere else in the estuarine complex. The mangroves seem to appear in succession in the swamp. The red mangrove, *Rhizophora mucronata*, establishes itself first and begins to give firmness to the shifting mud with its supporting root system. It forms the first row at the water front. Soon afterwards, the black mangrove, *Avicennia officinalis*, appears and it begins to give out a special type of aerial roots – the pneumatophores. These roots quickly occupy a large area and help in trapping silt between them. This is followed by the development of a shoot system in the black mangrove. The root systems in these two mangroves give support to the plants and prevent erosion.

The mangrove swamps of Goa undergo well marked seasonal changes. Some of the environmental features such as temperature (water and soil), salinity, oxygen, sediment load and nutrients (phosphate and nitrate) and phytoplankton counts were studied by Untawale & Parulekar (1976) and these are shown in Fig. 3.10. Temperature differences during the year, both in water and soil, were of the order of 5 to 6 °C, but the seasonal variation in salinity was very large – from almost freshwater to sea water conditions (1–34.5‰). The range in oxygen was from 2.2 to 5.4 ml/l. The swamp is characterized by a very heavy sediment load and during the monsoon period the water has a thick suspension of sediment. The variation in the phosphate–phosphorus concentration during the year is somewhat greater than that of nitrate–nitrogen (Fig. 3.10). However, the swamp does not seem to get much enrichment from external sources. During the monsoon months, when the freshwater influence is maximum, the phosphorus and nitrogen values both at the surface and bottom remain relatively low.

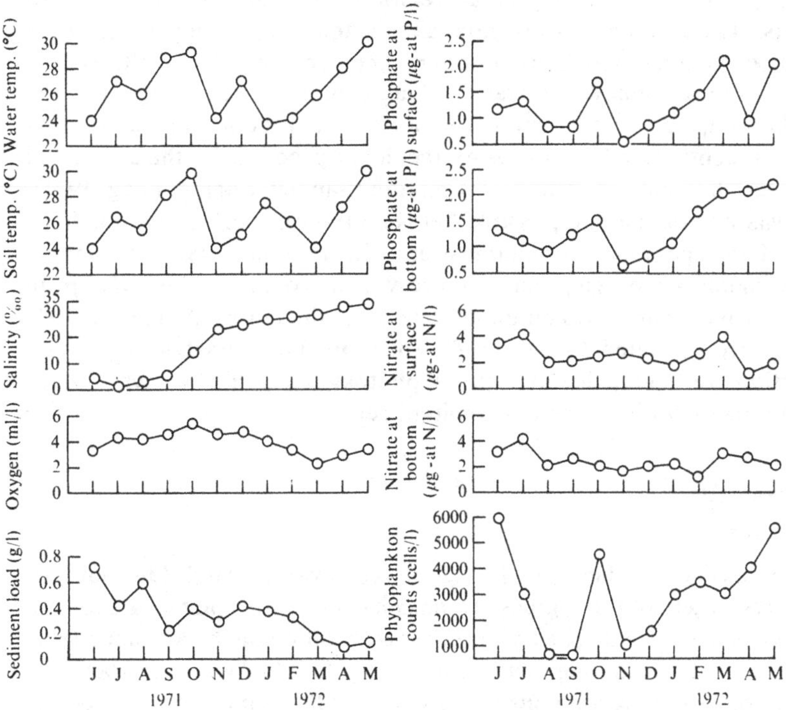

Fig. 3.10. Seasonal changes in the environmental features of a mangrove swamp at Goa. (From Untawale & Parulekar, 1976, *Mahasagar–Bulletin of the National Institute of Oceanography*, **9**, 57–62.)

Similarly, during March–May, when the swamp becomes salt water dominated, there is no significant change in the nitrogen value although phosphorus values do show an increase (Fig. 3.10).

The swamp is generally characterized by a low phytoplankton crop (50–6000 cells/l). Maximum cell numbers are found in May and June and minimum in August and September (Fig. 3.10).

Krishnamurthy & Sundararaj (1973) report a high rate of production, by what they call a 'direct estimate of primary production', in the Pichchavaram mangrove (Porto Novo) made during March–April 1972. They give the average gross primary production in the mangrove region as 7.56 g C/m^3/day and net production as 6.29 g C/m^3/day. The chlorophyll *a* concentration reported by them varied from 4.10 to 39.80 mg/m^3. It is, however, not clear what method these authors applied for measuring primary production of the mangrove swamp.

Recently Dr A. Pant (personal communication) measured the primary production in the water of a fringing mangrove in Goa using both oxygen and ^{14}C methods. Practically no correlation was found between the two methods. The oxygen method gave almost ten times higher values than the radiocarbon method for production. The average value of ^{14}C assimilation in the swamp water was 37.8 mg C/m^3/day. Assuming that ^{14}C assimilation gave an underestimation of the total production due to liberation of recently fixed carbon as extracellular products by the algae, and even after allowing a 20% underestimation (Samuel, Shah & Fogg, 1971), there was no comparison possible between the oxygen and ^{14}C methods (Dr A. Pant, personal communication). In February 1974, while the measurements were being made, there was a bloom of *Navicula* sp. in the mangrove swamp. Some experiments conducted by A. Pant in 1977 have clearly indicated that the differences in the estimates of primary production are largely due to bacterial photosynthetic activity which fixes carbon dioxide without the concomitant release of oxygen.

Seagrass beds

Kavaratti bed

The seagrass bed studied is in the lagoon on Kavaratti Atoll (Laccadives). It consists largely of turtle grass, *Thalassia hemprichii*, and manatee grass, *Cymodocea isoetifolia*. The two species grow along the beach slope from about low water neap tide down to about 100 m into the lagoon floor. The substratum in the lagoon consists of coarse white sand, coral rocks and debris. At Kavaratti, the maximum tidal range is about 1.7 m. At low tide, the grass bed is partially exposed. The bed is a luxuriant community growing in oceanic conditions. Within the intertidal zone of the lagoon, there is a rich flora of algae (Qasim & Bhattathiri, 1971).

Primary production of the seagrass bed was studied from the diurnal changes in dissolved oxygen over the bed. Experiments were also conducted on some of the isolated plants to determine their rates of oxygen production and consumption. Gross production of the seagrass community was equivalent to 11.97 g C/m^2/day and respiration 6.16 g C/m^2/day. The daily net production of the community was 5.81 g C/m^2/day with a *P*/*R* ratio of 1.94 (Qasim & Bhattathiri, 1971).

The seagrass bed is autotrophic. Its presence is ecologically important to the atoll communities, for it offers shelter to a very large number of animals and serves as food either directly, or through detritus, to many animals.

Mandapam bed

This seagrass bed is found in the Gulf of Mannar (near Mandapam) and covers a distance of about 3 km. The average width of the bed is about

200 m and the average depth over the bed is approximately 2 m. The seagrasses found in the bed are: *Cymodocea isoetifolia*, *C. serrulata*, *Halophyla ovalis*, *H. stipulacea* and *Diplanthera uninervis*. Of these, *C. isoetifolia* is the dominant species. The total quantity of seagrasses in the Mandapam bed has been estimated as 558 tonnes/km^2 (Rao, 1973).

Gross production of the bed, as determined by the oxygen changes, was 8.04 g C/m^2/day and respiration 3.25 g C/m^2/day. The *P*/*R* for July was 2.47 (Balasubramanian & Wafar, 1975).

Sandy beaches

Under the International Biological Programme, a joint project between the Marine Laboratory, Aberdeen, Scotland, and the Regional Centre of the National Institute of Oceanography, Cochin, India, was started on the 'Production process and community interrelationship of two sandy beaches'. Investigations began in 1968 and two British scientists visited Cochin every year up to 1971 for about 2–3 months. The British side of the project was funded by the Royal Society of London, and the Indian side by the Council of Scientific and Industrial Research (CSIR). The beaches selected for the study were (1) at Cochin, situated just south of the entrance of Cochin Backwater where salinity values differ over a wide range during the year and (2) at Shertallai, an open sea beach approximately 32 km away from Cochin.

The monsoon brings about conspicuous changes in the profile of both the beaches resulting in active erosion followed by gradual accretion during the post-monsoon period and restoration of the stable profile of the pre-monsoon months. The composition of the macrofauna in both the beaches varies considerably. Perhaps the increase in exposure of the beach and the seasonal fluctuation in salinity have resulted in a less diverse and often a severely restricted fauna at Cochin. At Shertallai, five species of animals are frequently represented in the samples: the bivalve *Donax incarnatus*, the gastropod *Bullia melanoides*, crustaceans *Emerita holthuisi* and *Eurydice* sp. and the polychaete *Glycera alba*; although many more, like *Oliva gibbosa*, *Mactra olorina*, *Timoclea imbricata*, appear on the beach during certain seasons. The only species recorded from Cochin throughout the year is the *Eurydice* sp. However, *Donax incarnatus* and *D. spiculum* were recorded on the beach almost throughout the pre-monsoon period. *Donax incarnatus* collected at Shertallai attains a greater size than it does at Cochin. This seems to be directly related to the greater concentration of organic carbon and chlorophyll *a* at Shertallai than at Cochin.

The studies of Trevallion *et al.* (1970) and Ansell *et al.* (1972*a*) have made it possible to understand the population density and biomass of individual species composing the beach fauna. The ghost crab, *Ocypode ceratophthalma*, present almost throughout the year at Shertallai, was found at a density of 1–2/m^2 (based on the abundance of their burrow entrances). McIntyre (1968) observed a maximum of 9/m^2 of *O. macrocera* from a beach in south-east India. *Bullia melanoides* was recorded only at Shertallai, and the greatest numbers were in January and October, when 330–340 individuals/m transect appeared on the beaches representing up to 53 g wet body wt/m transect.

Emerita holthuisi and the two species of *Donax* were studied in detail (Ansell, Sivadas, Narayanan & Trevallion, 1972*b*, *c*). The numbers of *E. holthuisi* recorded at Shertallai reached 500/m transect and the wet biomass 0.23 g/m transect. One collection made in April at Cochin produced about 4000/m transect with a biomass of 0.35 g/m transect. The bivalves *Donax incarnatus* and *D. spiculum* were present in significant numbers on the beaches and formed a major constituent of the macrofauna. *D. spiculum* appeared at Shertallai from the beginning of the monsoon, but it was found in large numbers only during the period from August to November; numbers reaching 9000/m transect were recorded, with a biomass of over 300 g wet wt/m transect. At Cochin, the species occurred from January to May; the maximum numbers recorded were 6000/m transect, with a biomass of 200 g wet body wt/m transect. For *D. incarnatus*, the data were split into two parts, one dealing with the population recruited to the beach in 1967 and the other recruited in 1968. At Shertallai, two transects were sampled, and at the beginning of the year only individuals from 1967 settlements were present. In January, the transect A at Shertallai recorded 240–250 bivalves/m transect, and transect B 660/m transect. The first indication of a 1968 settlement occurred in late January, although the major settlement came in only just prior to monsoon when in May maximum numbers of over 4000 bivalves/m transect were recorded from transect A. A detailed study of the length–weight relationship, growth, changes in body weight and reproduction of both *D. incarnatus* and *D. spiculum* has been made for Shertallai and Cochin samples (Ansell *et al.*, 1972*c*).

The low chlorophyll *a* content of the sand and the negligible ^{14}C uptake (the latter was estimated by the method given by Steele & Baird, 1968) no doubt showed that the interstitial or attached macro-organisms of the beaches at Shertallai and Cochin have essentially no primary production. Therefore the suspension feeders of the macrofauna of the intertidal sandy beach (*Emerita holthuisi* and the two species of *Donax*) may well have to depend upon the food resource of the surf water available mainly through the breakdown of plankton or through the degradation of animals higher in the food chain. For scavengers like *Bullia melanoides*, *Ocypode ceratophthalma* and *Eurydice* sp., the large quantity of fish offal and the

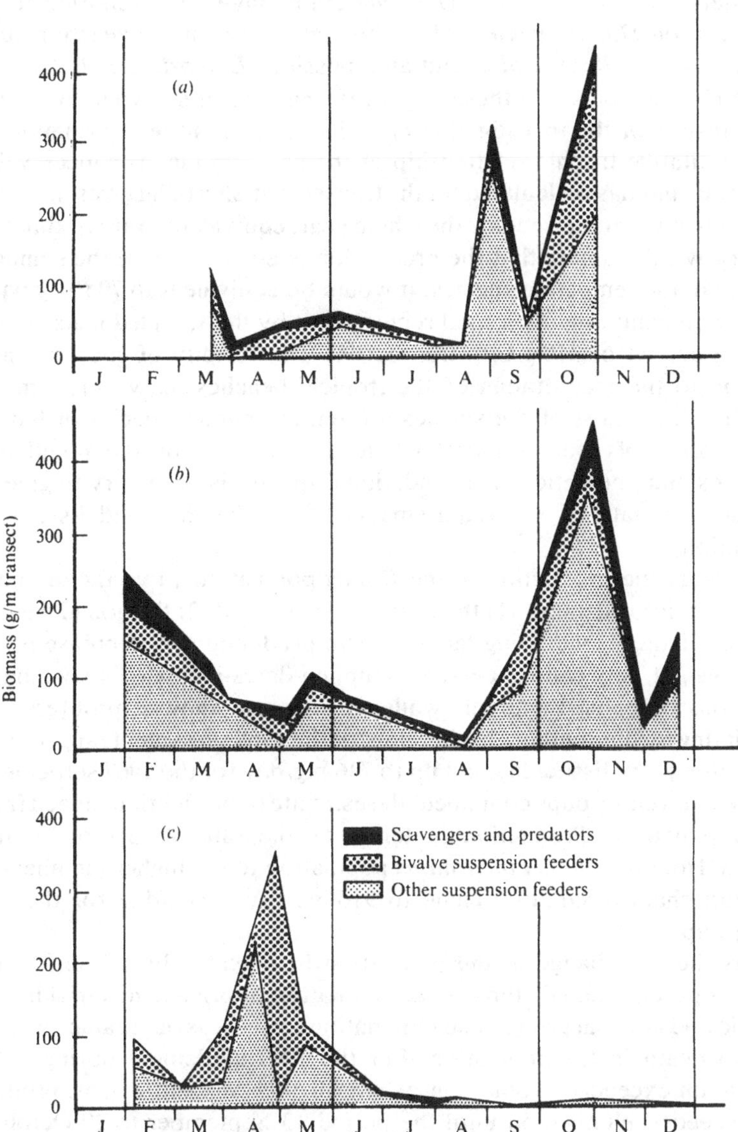

Fig. 3.11. Seasonal changes in three feeding types contributing to the total biomass of sandy beaches. (*a*) Shertallai beach, transect B. (*b*) Shertallai beach, transect A. (*c*) Cochin beach, during 1968. Deposit feeders present were too small in quantities to be shown. (From Ansell, Sivadas, Narayanan, Sankaranarayanan & Trevallion, 1972*a*, *Marine Biology*, **17**, 57.)

bodies of the neglected part of the catch after the operation of the beach seine nets offer a good diet. *Ocypode* crabs have been reported to be a predator on *Donax* species. The other carnivores in the beach include the polychaete *Glycera alba* and also possibly *Lumbriconereis latreilli*. The combined biomass of these major feeding types is shown in Fig. 3.11.

The results of the investigation provide relatively little information on the quantitative trophic relationship of the beach fauna. The mean value of the total biomass calculated for the transects at Shertallai gives the value of 140 g wet wt/m transect for the whole year, equivalent to approximately 35 g dry wt. Assuming that the production is equal to twice the standing crop, as in the temperate species, it would be equivalent to 70 g dry wt/m transect annually, giving a total requirement by the sampled macrofauna of 700 g dry wt or 350 g C/m transect/yr. The validity of such a generalization to the macrofauna of the tropical beaches, however, remains doubtful since most of the species present are annual species or have a shorter span of life. A more detailed assessment of the population dynamics and energetics of the individual species is necessary to give an accurate estimate of the requirements of the benthos and its annual production.

To assess the production of the *Donax* populations, two factors have been taken into account: (1) the somatic weight and (2) the gonad weight, the sum of these two being taken to give production. A decrease in the gonad weight between successive sampling-dates indicated spawning or other loss of gonad material, while an increase showed proliferation. Calculating by this method, production for the 1968 settlement ranged from 44 to 418 mg dry tissue/day and up to 736 mg/day for the 1967 settlement. For the two year-groups combined, the estimate of production ranged from 4 to 736 mg/day. Production in terms of the rate of loss of organic material from the *Donax incarnatus* population to the higher members of the food chain (predators) came to 916 mg dry tissue/day for the two year-groups.

Thus the net change in the population is indicated by the difference between production (*P*) through accumulation of organic material by the somatic and gonadal growth and elimination by the loss of organic material through death in the population. For the 1967 settlement during 1968, elimination exceeded production, while for the 1968 settlement, production exceeded elimination until the period 13 September to 29 October. Thereafter, production failed to keep pace with elimination and the biomass declined. For the total population, there was a net loss of material throughout the year.

Normally, production and elimination would balance over the year. In this study of *Donax incarnatus*, the 1967 settlement had been apparently greater than the 1968 settlement. This resulted in higher elimination, since the 1967 settlement obviously died during 1968, while the recruitment of a new population was not sufficient in 1968 to maintain the balance of production and elimination. This can be taken only as a peculiarity of the year.

Looking at the relationship between production (P), elimination (E) and biomass ($\bar{B}$) and calculating the ratios $P/\bar{B}$ and $E/\bar{B}$, the above model is confirmed, since production thus calculated is 6.2 g dry wt tissue/g biomass and elimination is 7.5 g dry wt tissue/g biomass. But in years when more abundant settlement occurs, the ratio $P/\bar{B}$ would be greater than $E/\bar{B}$.

The values of the ratios $P/\bar{B}$ and $E/\bar{B}$ show two important aspects of the trophic ecology of the beach – one being an indication of the food requirement of the macrobenthos population and the second an indication of the contribution by the macrobenthos to the next trophic level, assuming that the bivalve population contributes mainly to the macrofauna. Taking the annual production as 6.5 times the biomass, the total production is 227.5 g/m transect/year (average biomass is approximately 35 g dry tissue/m transect for the year). Assuming that 50% of the organic material is carbon, the production will be equivalent to 114 g C/m transect/yr. With an ecological efficiency of 10%, the annual requirement by the population would be 1140 g C/m transect, and the quantity of food available to the predators seems to be of the order of 228 g/m transect/yr.

Summary

Primary production of six different environments from the Indian region, namely; coastal waters, backwaters and estuaries, coral reefs and atolls, mangrove swamps, seagrass beds, and sandy beaches, has been described. Along both the coasts of India, the rate of production near the shores is greater than in regions away from the coast. High instantaneous rates of production are generally recorded in upwelling areas during the monsoon period.

The backwater system of Kerala is a relatively well studied estuarine system. Primary production in this turbid and polluted environment is high and is largely contributed by nannoplankton. Three main ecological factors seem to govern the rate of primary production in the backwater. These are salinity, light and nutrients. The influence of each of these factors was studied experimentally on several unialgal cultures. Maximum photosynthesis in all the organisms was recorded at low salinities. Such a dependence of algae on low salinity seems to be an adaptation to utilize the enrichment of water, associated with large dilutions in the estuary during the monsoon months, to a maximum degree. The saturation points (I_k) of several algae were determined by exposing them to varying light intensities. When the organisms were exposed to different wavelengths of light, no wavelength dependence of photosynthesis in saturation light was observed.

The capacity of two organisms to utilize phosphate and nitrate either singly or in combination was determined from the values of the half-saturation constant (K_s). High concentrations of nutrients seem to favour the abundance of diatoms in the backwaters, while at low concentrations of nutrients dinoflagellates become predominant. The annual cycle of phytoplankton production and succession was found to be related to the increased flushing of the estuary caused by the monsoon.

The other important food constituent in the backwater is organic detritus, which leads to several pathways in the food chain. The energy budget of a penaeid prawn *Metaperaeus monoceros* was worked out by feeding it on estuarine detritus.

On the west coast of India, the other estuarine system on which some information has been gathered is the Mandovi–Zuari complex of Goa. These estuaries are also greatly influenced by the monsoon cycle. The Zuari estuary gets more marine influence than the other. Primary production in these estuaries is lower than that of the backwater at Cochin. Similarly, the average primary production of the Bombay harbour bay has also been found to be low.

On the east coast of India, the Vellar–Coleroon estuarine system has been studied fairly well, and from the available data on primary production, largely based on oxygen technique, it seems that the system is very productive. The nutrient and chlorophyll concentrations in this estuary remain high practically throughout the year.

The coral reefs studied are of two types, the fringing reefs and the atoll reefs. Both these types of reefs have been found to be highly productive communities. Primary production in all the reefs, except the Andaman reef, has been found to be much greater than their respiration. The water surrounding Kavaratti Atoll (Laccadives) is low in phytoplankton content but the particulate organic carbon in the vicinity of the atoll is high, indicating that the reef produces large quantities of organic matter which appear in the form of microscopic particles.

Observations on the mangrove swamps of the Mandovi–Zuari estuaries indicate that the mangrove trees along the waterfront appear in succession. The swamps undergo marked seasonal changes induced by the monsoon and are generally characterized by a low phytoplankton production.

The two seagrass beds studied form highly productive systems. Both the beds are luxuriant communities. One grows on an atoll in oceanic conditions and the other in association with a fringing reef in coastal waters.

Primary production in the sandy beaches of the south-west coast of India has been found to be extremely low. Marked changes in the profile of the beaches occur during the monsoon, resulting in active erosion followed by accretion during the post-monsoon period and stable conditions during the pre-monsoon period. The biomass of several animals inhabiting the beaches was determined and a quantitative trophic relationship of the beach fauna has been attempted. From the relationship between production and biomass and from the estimate of total carbon, it became possible to determine the total food requirement of the beach macrofauna during the year.

I thank Mr P. M. A. Bhattathiri and Dr P. Sivadas for their help in the preparation of the manuscript.

References

Ansell, A. D., Sivadas, P., Narayanan, B., Sankaranarayanan, V. N. & Trevallion, A. (1972*a*). The ecology of two sandy beaches in southwest India. I. Seasonal changes in physical and chemical factors, and in the macrofauna. *Marine Biology*, **17**, 38–62.

Ansell, A. D., Sivadas, P., Narayanan, B. & Trevallion, A. (1972*b*). The ecology of two sandy beaches in southwest India. II. Notes on *Emerita holthuisi*. *Marine Biology*, **17**, 311–17.

Ansell, A. D., Sivadas, P., Narayanan, B. & Trevallion, A. (1972*c*). The ecology of two sandy beaches in southwest India. III. Observations on the population of *Donax incarnatus* and *D. spiculum*. *Marine Biology*, **17**, 318–32.

Aruga, Y. (1973). Primary production in the Indian Ocean. II. In *The Biology of the Indian Ocean, ecological studies 3*, ed. B. Zeitzschel, pp. 117–30. Berlin: Springer-Verlag.

Balasubramanian, T. & Wafar, M. V. M. (1974). Primary productivity of some fringing reefs of south-east India. *Mahasagar–Bulletin of the National Institute of Oceanography*, **7**, 157–64.

Balasubramanian, T. & Wafar, M. V. M. (1975). Primary productivity of some seagrass beds in the Gulf of Mannar. *Mahasagar–Bulletin of the National Institute of Oceanography*, **8**, 87–91.

Bhatnagar, G. P. (1971). Primary organic production and chlorophyll in Kille backwaters, Porto Novo (S. India). In *Proceedings of the symposium on 'tropical ecology with an emphasis on organic productivity'*, ed. P. M. Golley & F. B. Golley, pp. 351–62. New Delhi. (Mimeo.)

Bhattathiri, P. M. A., Devassy, V. P. & Bhargava, R. M. S. (1976). Production at different trophic levels in the estuarine system of Goa. *Indian Journal of Marine Sciences*, **5**, 83–6.

Das, P. K., Murty, C. S. & Varadachari, V. V. R. (1972). Flow characteristics of Cumbarjua canal connecting Mandovi and Zuari estuaries. *Indian Journal of Marine Sciences*, **1**, 95–102.

Dehadrai, P. V. (1970*a*). Observations on certain environmental features at the Dona Paula point in Marmugao Bay, Goa. *Proceedings of the Indian Academy of Sciences*, **B72**, 56–67.

Dehadrai, P. V. (1970*b*). Changes in the environmental features of the Zuari and Mandovi estuaries in relation to tides. *Proceedings of the Indian Academy of Sciences*, **B72**, 68–80.

Dehadrai, P. V. & Bhargava, R. M. S. (1972*a*). Distribution of chlorophyll, carotenoids and phytoplankton in relation to certain environmental factors along the central west coast of India. *Marine Biology*, **17**, 30–7.

Dehadrai, P. V. & Bhargava, R. M. S. (1972*b*). Seasonal organic production in relation to environmental features in Mandovi & Zuari estuaries, Goa. *Indian Journal of Marine Sciences*, **1**, 52–6.

Derbyshire, M. (1967). The surface waters off the coast of Kerala, south-west India. *Deep-Sea Research*, **14**, 295–320.

Devassy, V. P. & Bhattathiri, P. M. A. (1974). Phytoplankton ecology of the Cochin backwater. *Indian Journal of Marine Sciences*, **3**, 46–50.

Gopinathan, C. K. & Qasim, S. Z. (1971). Silting in navigational channels of the Cochin harbour area. *Journal of the Marine Biological Association of India*, **13**, 14–26.

Humphrey, G. F. (1972). The biology of the Indian Ocean. In *Bruun memorial lectures*, Technical Series 10, pp. 7–22. Intergovernmental Oceanographic Commission, UNESCO, Paris.

Jegatheesan, G. & Venugopalan, V. K. (1973). Trace elements in the particulate matter of Porto Novo waters. (11° 29′ N–79° 49′ E). *Indian Journal of Marine Sciences*, **2**, 1–5.

Kabanova, J. G. (1968). Primary production of the northern part of the Indian Ocean. *Oceanology*, **8**, 214–25.

Kohn, A. J. & Helfrich, P. (1957). Primary organic productivity of a Hawaiian coral reef. *Limnology and Oceanography*, **2**, 242–51.

Krey, J. (1973). Primary production in the Indian Ocean. I. In *The Biology of the Indian Ocean, Ecological studies 3*, ed. B. Zeitzschel, pp. 115–26. Berlin: Springer-Verlag.

Krishnamoorthy, T. M. & Viswanathan, R. (1968). Primary productivity studies in Bombay harbour bay using ^{14}C. *Indian Journal of Experimental Biology*, **6**, 115–16.

Krishnamurthy, K. (1971). Phytoplankton pigments in Porto Novo waters (India). *Internationale Revue der gesamten Hydrobiologie*, **56**, 273–82.

Krishnamurthy, K. & Sundararaj, V. (1973). A survey of environmental features in a section of the Vellar–Coleroon estuarine system, South India. *Marine Biology*, **23**, 229–37.

McIntyre, A. D. (1968). The microfauna and macrofauna of some tropical beaches. *Journal of Zoology*, **156**, 377–92.

Margalef, R. (1967). Laboratory analogues of estuarine plankton systems. In *Estuaries*, ed. G. H. Lauff, pp. 515–21. Publication No. 83. Published by the American Association for the Advancement of Science, Washington DC.

Murty, C. S. & Das, P. K. (1972). Premonsoon tidal flow characteristics of Mandovi, estuary, Goa. *Indian Journal of Marine Sciences*, **1**, 148–51.

Muscatine, L., Pool, R. R. & Cernichiari, E. (1972). Some factors influencing selective release of soluble organic material by zooxanthellae from reef corals. *Marine Biology*, **13**, 298–308.

Nair, P. V. R. (1970). Primary productivity in the Indian Seas. In *Bulletin of the Central Marine Fisheries Research Institute, No. 22*. Mandapam Camp, India. (Mimeo.), 56 pp.

Nair, P. V. R. & Pillai, C. S. G. (1972). Primary productivity of some coral reefs in the Indian seas. In *Proceedings of the symposium on 'corals and coral reefs'*, 1969, ed. C. Mukundan & C. S. G. Pillai, pp. 33–42. Cochin: Marine Biological Association of India.

Nair, P. V. R., Samuel, S., Joseph, K. J. & Balachandran, V. K. (1973). Primary production and potential fishery resources in the seas around India. In *Proceedings of the symposium on 'living resources of the seas around India'*, 1968, pp. 184–98. Special publication. Cochin: Central Marine Fisheries Research Institute.

Pillai, C. S. G. (1972). Stony corals of the seas around India. In *Proceedings of the symposium on 'corals and coral reefs'*, 1969, ed. C. Mukundan & C. S. G. Pillai, pp. 191–216. Cochin: Marine Biological Association of India.

Pillai, P. P. (1971). Studies on the estuarine copepods of India. *Journal of the Marine Biological Association of India*, **13**, 167–72.

Pillai, P. P., Qasim, S. Z. & Nair, A. K. K. (1973). Copepod component of zooplankton in a tropical estuary. *Indian Journal of Marine Sciences*, **2**, 38–46.

Prasad, R. R. & Nair, P. V. R. (1963). Studies on organic production. I. Gulf of Mannar. *Journal of the Marine Biological Association of India*, **5**, 1–26.

Purushothaman, A. & Venugopalan, V. K. (1972). Distribution of dissolved silicon in the Vellar estuary. *Indian Journal of Marine Sciences*, **1**, 103–5.

Qasim, S. Z. (1970*a*). Some problems related to the food chain in a tropical estuary. In *Marine food chains*, ed. J. H. Steele, pp. 45–51. Edinburgh: Oliver & Boyd.

Qasim, S. Z. (1970*b*). Some characteristics of a *Trichodesmium* bloom in the Laccadives. *Deep-Sea Research*, **7**, 655–60.

Qasim, S. Z. (1972*a*). The dynamics of food and feeding habits of some marine fishes. *Indian Journal of Fisheries*, **19**, 11–28.

Qasim, S. Z. (1972*b*). Some observations on *Trichodesmium* blooms. In *Taxonomy and biology of blue-green algae*, pp. 433–8. Madras: University of Madras.

Qasim, S. Z. (1973). Productivity of backwaters and estuaries. In *The biology of the Indian Ocean, ecological studies 3*, ed. B. Zeitzschel, pp. 143–54. Berlin: Springer-Verlag.

Qasim, S. Z. (1977). Biological productivity of the Indian Ocean. *Indian Journal of Marine Sciences*, **6**, 122–37.

Qasim, S. Z. & Bhattathiri, P. M. A. (1971). Primary production of a seagrass bed on Kavaratti Atoll (Laccadives). *Hydrobiologia*, **38**, 29–38.

Qasim, S. Z., Bhattathiri, P. M. A. & Abidi, S. A. H. (1968). Solar radiation and its penetration in a tropical estuary. *Journal of Experimental Marine Biology and Ecology*, **2**, 87–103.

Qasim, S. Z., Bhattathiri, P. M. A. & Devassy, V. P. (1972*a*). The influence of salinity on the rate of photosynthesis and abundance of some tropical phytoplankton. *Marine Biology*, **12**, 200–6.

Qasim, S. Z., Bhattathiri, P. M. A. & Devassy, V. P. (1972*b*). The effect of intensity and quality of illumination on the photosynthesis of some tropical phytoplankton. *Marine Biology*, **16**, 22–7.

Qasim, S. Z., Bhattathiri, P. M. A. & Devassy, V. P. (1973). Growth kinetics and nutrient requirements of two tropical marine phytoplankters. *Marine Biology*, **21**, 299–304.

Qasim, S. Z., Bhattathiri, P. M. A. & Reddy, C. V. G. (1972). Primary production of an atoll in the Laccadives. *Internationale Revue der gesamten Hydrobiologie*, **57**, 207–25.

Qasim, S. Z. & Easterson, D. C. V. (1974). Energy conversion in the shrimp, *Metapenaeus monoceros* (Fabricius), fed on detritus. *Indian Journal of Marine Sciences*, **3**, 131–4.

Qasim, S. Z. & Gopinathan, C. K. (1969). Tidal cycle and the environmental features of Cochin Backwater (a tropical estuary). *Proceedings of the Indian Academy of Sciences*, **B69**, 336–48.

Qasim, S. Z. & Sankaranarayanan, V. N. (1970). Production of particulate organic matter by the reef on Kavaratti Atoll (Laccadives). *Limnology and Oceanography*, **15**, 574–78.

Qasim, S. Z. & Sankaranarayanan, V. N. (1972). Organic detritus of a tropical estuary. *Marine Biology*, **15**, 193–99.

Qasim, S. Z., Vijayaraghavan, S., Joseph, K. J. & Balachandran, V. K. (1974). Contribution of microplankton and nannoplankton in the waters of tropical estuary. *Indian Journal of Marine Sciences*, **3**, 146–9.

Qasim, S. Z., Wellershaus, S., Bhattathiri, P. M. A. & Abidi, S. A. H. (1969). Organic production in a tropical estuary. *Proceedings of the Indian Academy of Sciences*, **B69**, 51–94.

Radhakrishna, K. (1969). Primary productivity studies in the shelf waters off Alleppey, south-west India, during the post-monsoon, 1967. *Marine Biology*, **4**, 174–81.

Rao, M. U. (1972). Coral reef flora of the Gulf of Mannar and Palk Bay. In *Proceedings of the symposium on 'corals and coral reefs'*, 1969, ed. C. Mukundan & C. S. G. Pillai, pp. 217–30. Cochin: Marine Biological Association of India.

Rao, M. U. (1973). The seaweed potential of the seas around India. In *Proceedings of the symposium on 'living resources of the seas around India'*, 1968, pp. 687–92. Special publication. Cochin: Central Marine Fisheries Research Institute.

Ryther, J. H., Hall, J. R., Pease, A. K., Bakun, A. & Jones, M. M. (1966). Primary organic production in relation to the chemistry and hydrography of the western Indian Ocean. *Limnology and Oceanography*, **11**, 371–80.

Samuel, C. T., Shah, N. M. & Fogg, G. E. (1971). Liberation of extracellular products of photosynthesis by tropical phytoplankton. *Journal of the Marine Biological Association, UK*, **51**, 793–8.

Sankaranarayanan, V. N. & Qasim, S. Z. (1969). Nutrients of the Cochin Backwater in relation to environmental characteristics. *Marine Biology*, **2**, 236–47.

Sargent, M. C. & Austin, T. S. (1949). Organic productivity of an atoll. *Transactions of the American Geophysical Union*, **30**, 245–9.

Sargent, M. C. & Austin, T. S. (1954). Biologic economy of coral reefs. Bikini and nearby atolls, II. Oceanography (biologic). *Professional Papers of the United States Geological Survey*, 260-E, 293–300.

Steele, J. H. & Baird, I. E. (1968). Production ecology of a sandy beach. *Limnology and Oceanography*, **13**, 14–25.

Steemann Nielsen, E. & Jensen, E. A. (1957). Primary organic production – the autotrophic production of organic matter in the oceans. *Galathea Reports*, **1**, 49–136.

Stoddart, D. R. (1972). Regional variation in Indian Ocean coral reefs. In *Proceedings of the symposium on 'corals and coral reefs'*, 1969, ed. C. Mukundan & C. S. G. Pillai, pp. 155–74. Cochin: Marine Biological Association of India.

Stoddart, D. R. & Pillai, C. S. G. (1972). Raised reefs of Ramanathapuram, south India. *Transactions, Institute of British Geographers*, Publ. No. **56**, 111–25.

Subba Rao, D. V. (1973). Effects of environmental perturbations on short-term phytoplankton production off Lawson's Bay, a tropical coastal embayment. *Hydrobiologia*, **43**, 77–91.

Tranter, D. J. & Abraham, S. (1971). Coexistence of species of *Acartiidae* (Copepoda) in the Cochin Backwater, a monsoonal estuarine lagoon. *Marine Biology*, **11**, 222–41.

Tranter, D. J. & George, J. (1972). Zooplankton abundance at Kavaratti and Kalpeni Atolls. In *Proceedings of the symposium on 'corals and coral reefs'*, 1969, ed. C. Mukundan & C. S. G. Pillai, pp. 239–56. Cochin: Marine Biological Association of India.

Trevallion, A., Ansell, A. D., Sivadas, P. & Narayanan, B. (1970). A preliminary account of two sandy beaches in southwest India. *Marine Biology*, **6**, 268–79.

Untawale, A. G., Dwivedi, S. N. & Singbal, S. Y. S. (1973). Ecology of mangroves in Mandovi and Zuari estuaries and the interconnecting Cumbarjua canal, Goa. *Indian Journal of Marine Sciences*, **2**, 47–53.

Untawale, A. G. & Parulekar, A. H. (1976). Some observations on the ecology of an estuarine mangrove of Goa. *Mahasagar–Bulletin of the National Institute of Oceanography*, **9**, 57–62.

Venugopalan, V. K. (1969). Primary production in the estuarine and inshore waters of Porto Novo (11° 29′ N–79° 49′ E). *Bulletin of the National Institute of Science, India*, **38**, Part II, 743–6.

Vijayaraghavan, S., Joseph, K. J. & Balachandran, V. K. (1974). Preliminary studies on nannoplankton productivity. *Mahasagar – Bulletin of the National Institute of Oceanography*, **7**, 125–9.

Wyatt, T. & Qasim, S. Z. (1973). Application of a model to an estuarine ecosystem. *Limnology and Oceanography*, **18**, 301–6.

4. Biological productivity of some coastal regions of Japan

K. HOGETSU

Discussions on the proposals of the PM Section of IBP led to oceanographical as well as biological research into the features of the coastal regions of Japan, four areas were determined for the research area of JIBP/PM, namely: Suruga and Sagami Bays in the Kuroshio Current (warm current) area, Akkeshi Bay in the Oyashio Current (cold current) area, Sendai Bay in the intermediate region between the two currents, and Hiuchi Nada in the Seto Inland Sea. One hundred and forty-five scientists from various universities and institutes were organized into four research teams, and they engaged in research from 1967 to 1973.

Besides this research, six cruises were carried out for the study of the productivity of biocoenoses in the north Pacific and equatorial Pacific, Kuroshio, Oyashio, Japan Sea, east China Sea and south China Sea, as part of the JPM activities.

Suruga Bay and Sagami Bay

As is generally known, the warm Kuroshio Current flows along the southern coast of the Japanese islands, and leaves the main island to the east of Tokyo at the latitude of approximately 36° N. From this locality northward, the marine climate is under the influence of the cold Oyashio Current, which comes down from the Okhotsk and Bering Seas and flows along the north-eastern coast of Hokkaido.

Suruga and Sagami Bays are situated side by side on the north-eastern Pacific coast of the Japanese main island. They are separated by the Izu Peninsula. Both bays are of similar dimensions, being roughly 60 km in width at their entrances and about 60 km in length. They are also similar in depth, being about 1500 m deep in their central parts and about 2000 m at their entrances. The widely open mouths of these bays, together with their great depths, allow the oceanic water, or branches of the warm Kuroshio Current, to enter freely into the bays (Fig. 4.1*c*).

In the euphotic zone of Sagami Bay, the water temperature showed its maximum (28.2 °C) in August and the minimum (12.4 °C) in April. Salinity ranged from 31.26‰ to 36.34‰. Phosphate–phosphorus concentration varied in the range of 0.01–1.65 μg-at/l in the surface layer, showing low values in July and August, and high values in February and March. These values tended to increase in the layers below the euphotic zone. The

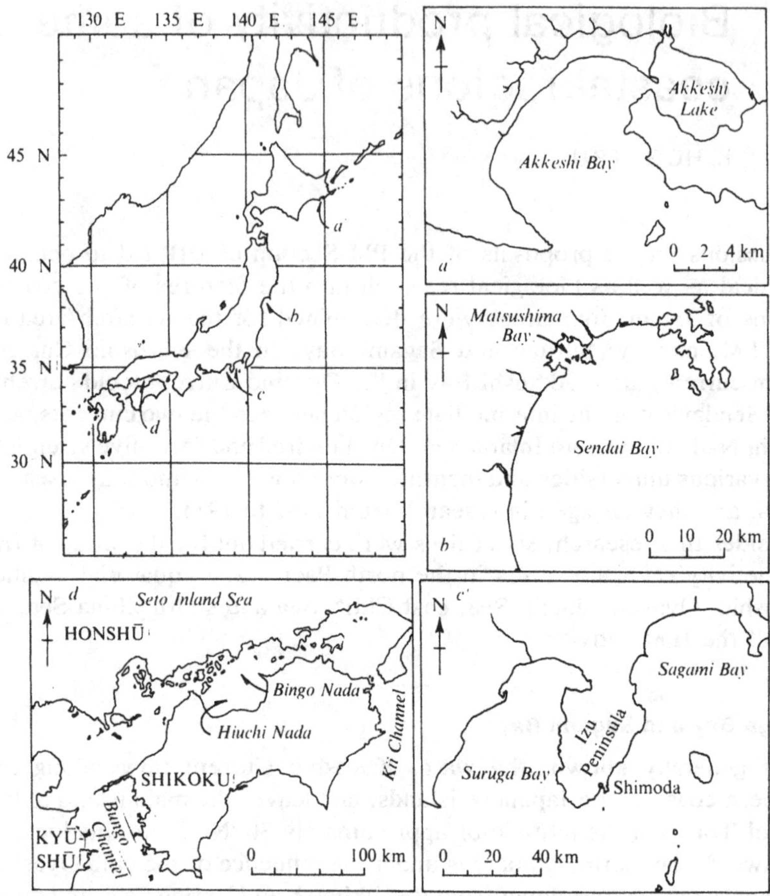

Fig. 4.1. Map showing the position of areas where JIBP/PM surveys were carried out. (*a*) Akkeshi Bay. (*b*) Sendai Bay, (*c*) Suruga Bay and Sagami Bay. (*d*) Seto Inland Sea.

concentrations of inorganic nitrogen compounds (nitrite-, nitrate- and ammonia-nitrogen) showed high values (8.0 μg N/l) in winter and early spring, and low values (0.2 μg N/l) in summer, showing remarkable increases with depth in (nitrite, nitrate)-nitrogen concentration. In these features Sagami Bay resembles Suruga Bay fairly closely.

In the *Sargassum* community at the low water mark and the *Ecklonia* community at 5 m in Shimoda Bay of Izu Peninsula, the maximum biomass (4.6 and 17.0 kg wet weight/m^2) appeared in winter and the minimum (3.4 and 7.3 kg wet weight/m^2) in summer. Based on the data of the numbers of bladelets produced and biomass measured several times a year, the annual net production of the *Ecklonia* community was estimated to be 30 kg wet weight/m^2 in Shimoda Bay.

Table 4.1. *Photosynthesis activity and net production of algal populations under different climatic conditions*

Area	mg C/ mg chl.*a*/h	g C/m²/h	g C/m²/day	g C/m²/yr	References
Suruga Bay	0.2–4.5	2.4–7.8	0.17–0.43	90	Aruga (1977)
Sagami Bay	—	—	0.15–0.5	90	Saijo *et al.* (1973)
Shimoda Bay	0.2–5.8	—	0.2–1.5	—	Shimura & Ichimura (1972)
Sendai Bay	—	—	0.19–0.72	100	Nishizawa (1977)
Akkeshi Bay	0.7–6.9	—	0.07–2.05	295	Nishizawa (1977)
Hiuchi Nada	1.0–20.0	—	0.12–0.57	127	Endo & Okaichi (1977)
Kuroshio	0.05–3.0	—	0.1–0.3	—	Saijo & Ichimura (1960) Data for summer season
Oyashio	1.0–5.0	—	0.6–0.8	—	Aruga & Ichimura (1968) Data for summer season
Pacific					
(155° W) 50° N	2.0	0.76	—	—	Takahashi *et al.* (1972)
45° N	1.71	0.76	—	—	Measured in the period
40° N	1.97	0.53	—	—	from 23 August to
25° N	1.37	0.32	—	—	5 October 1969, with
10° N	5.20	0.44	—	—	the samples from
5° N	1.07	0.17	—	—	10–30 m
0°	2.50	0.71	—	—	
10° S	3.00	0.42	—	—	

The vertical distribution of chlorophyll *a* measured by the method recommended by SCOR–UNESCO (1966), or the fluorescence method, showed the maximum at the surface or in the subsurface layer, decreasing in concentration with increase in depth below 50–80 m. Percentage degradation of chlorophylls to phaeophytins in the total pigments was relatively small in the euphotic zone (10–50%) and increased gradually to 60–90% at depth. The mean concentration of chlorophyll *a* in the euphotic zone was in the range of 0.3–1.8 mg/m³ in Suruga Bay, and 0.1–1.5 mg/m³ in Sagami Bay. Chlorophyll *a* amounts per unit area from surface to 50 m depth ranged from 15 to 90 mg/m² in Sagami Bay. The biomass of phytoplankton was estimated as 3–18 g/m², assuming 0.5% chlorophyll content.

The light-saturated rate of photosynthesis was 0.2–4.5 mg C/mg chl. *a*/h, with relatively high values in the samples from 30 m, and extremely low values in the samples from depths greater than 50 m, in Suruga Bay. The primary production of samples from both bays, measured *in situ*, ranged from 0.17 to 0.43 g C/m²/day in Suruga Bay, and from 0.15 to 0.5 g C/m²/day in Sagami Bay. Assuming an average rate of daily net primary production of 0.25 g C/m²/day, the annual net production in both bays will

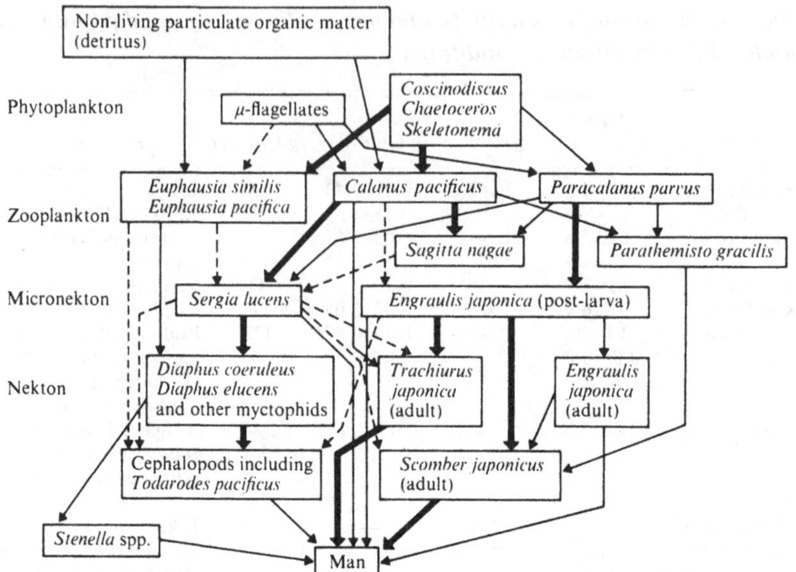

Fig. 4.2. Tentative food web in the pelagic surface community in Suruga Bay. Thicknesses of pointers are proportional to daily specific rate of food consumption. Modified from the original figure by M. Omori.

be approximately 90 g C/m²/yr. As is shown in Table 4.1, the productivity of both bays is higher than in oceanic waters and lower than in Akkeshi and Sendai Bays.

In Suruga and Sagami Bays, copepods, chaetognaths, euphausiids, macruran shrimps and some micronektonic fishes were dominant in the biomass (Fig. 4.2). Copepods were the most dominant species, found mainly in the upper 200 m, but micronektonic fishes and macruran shrimps increased in biomass in the mesopelagic layers below 500 m. The average biomass of zooplankton in the water column from the surface to 1000 m depth under 1 m² was estimated as 44.5 g in wet weight or 7.65 g in dry weight, corresponding to 3.73 g C, 0.54 g N, and 1.05 g ash.

Among copepods and euphausiids, surface feeders took phytoplankton and large quantities of chlorophyll *a*, and degraded phaeopigments were found in their stomachs and faecal pellets. The mean chlorophyll *a* and phaeopigment contents in the stomach of ten species of euphausiids were in the range of 2×10^{-3}–16×10^{-3} μg/mg dry weight. From the results obtained in rearing experiments of *Euphausia similis* in the laboratory, it was confirmed that about 75×10^{-3} μg pigments were excreted by an individual animal in one hour. If this rate is consistent, this indicates that *E. similis* excreted full stomach and gut contents within 10 hours (Nemoto,

1968). The average respiration rate of *E. nana* was 1.21 μl O_2/mg dry weight/h at 10 °C and 1.67 at 20 °C. The annual production/biomass coefficient of the commonest euphausiids was about 3.

The comparison of these amounts of pigments with the standing value of pigments in seawater showed that *E. similis* obtains the pigments by filtering 150–300 ml water in the upper layers between the surface and 100 m, and about 1000 ml water at 500 m depth, at the bottom of its vertical distribution.

Benthic animals which are retained by a sieve of 1 mm mesh and are lighter than 1 g in wet weight were defined as the 'smaller macrobenthos'. The biomasses of the smaller macrobenthos on the shelf in Sagami Bay, Suruga Bay and the Sea of Enshu were compared. The median of biomasses measured in these areas was estimated as 19.3–45 g wet weight/m^2. In Sagami Bay, the megalobenthos biomass was estimated in the range from 1 to 10 g/m^2, while the macrobenthos biomass taken in the grab was from 10 to 1000 g/m^2.

The metabolism of the intertidal benthic community at Shimoda Bay was measured by the method of the 'community as a whole'. It was found that the rates of respiration (400–500 mg O_2/m^2/h in the calcareous algal zone) and the photosynthesis (400–700 mg O_2) of the biotic community as a whole were roughly of the same magnitude, although the former exceeded the latter slightly in all cases except in the uppermost part of the intertidal zone.

The average annual commercial catch in Sagami Bay from 1956 through 1967 was about 1.5×10^4 tonnes, of which 24% was anchovy. The other major species were jack mackerel, mackerel, common squid, yellowtail and anchovy larvae. Anchovy, *Engraulis japonica*, occupies a prominent ecological niche on account of its dominance in numbers and biomass as a consumer of plankton as well as the prey of large-sized nekton in both bays (Fig. 4.3). From examination of stomach contents of anchovy and of net plankton in various seasons, it was clear that *Paracalanus*, exclusively *P. parvus*, predominate in number in the digestive tracts of the anchovy of all growth stages, except the adults fished after summer. The anchovy consumes about 0.5% of body weight per day, and the daily food take is 16 times the stomach content in the larval stage, 9–10 times in immature stages and 4–6 times in adult stage.

The first-year jack mackerel of about 50 mm body length (2 g body weight) which appears in the coastal waters in mid-June, grows to a length of 133 mm (48 g) by December, consuming about 260 g of anchovy larvae. Mackerel of 92 mm (9 g) consume 1030 g of anchovy larvae and grow to 170 g during the same period. The consumption of anchovy larvae by jack mackerel in the whole area of Sagami Bay was estimated at 4×10^{11} individuals (2×10^4 tonnes) in the year, and 1.5×10^{11} individuals (4.7×10^4

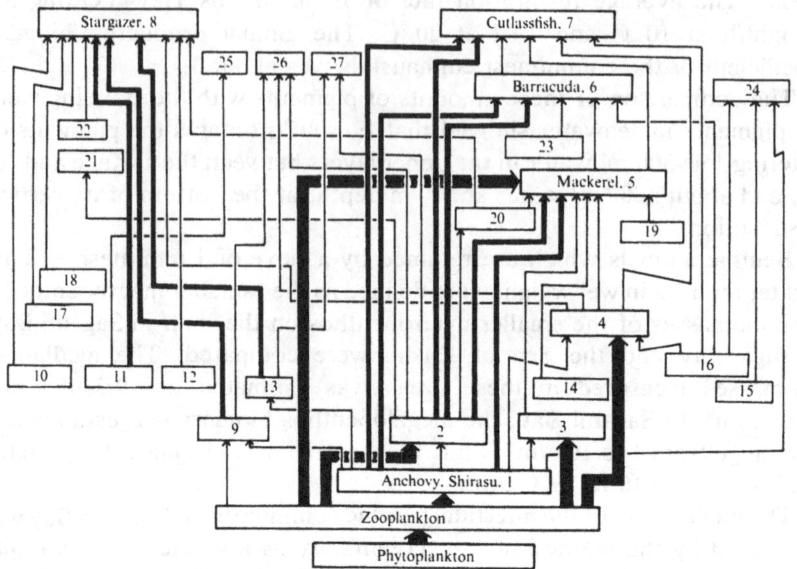

Fig. 4.3. Schematic food web based on stomach analysis of fishes collected in Suruga Bay. The thickest arrows indicate that prey–predator relationships are very close among fishes quantitatively. Medium arrows indicate a weaker relationship. Thin arrows indicate that prey–predator relationships are weak.
1, Anchovy, larvae (*Engraulis japonica*); 2, anchovy, adult (*Engraulis japonica*); 3, round herring (*Etrumeus micropus*); 4, horse mackerel (*Trachurus japonicus*); 5, Japanese mackerel (*Scomber japonicus*); 6, Japanese barracuda (*Sphyraena japonica*); 7, cutlassfish (*Trichiurus lepturus*); 8, Japanese stargazer (*Uranoscopus japonicus*); 9, leiognathids; 10, gobioids; 11, callionymids; 12, Japanese dragonet (*Calliurichthys japonicus*); 13, silver whiting (*Sillago sihama*); 14, leptocephalus; 15, Japanese codler (*Bregmaceros japonicus*); 16, lanternfishes; 17, triglids; 18, paralepidids; 19, serranids; 20, bothids; 21, sea conger (*Anago anago*); 22, Japanese eel (*Anguilla japonica*); 23, epinephelids; 24, Japanese bluefish (*Scombrops boops*); 25, spotted-tail grinner (*Saurida undosquamis*); 26, shortfin lizardfish (*Saurida elongata*); 27, snakefish (*Trachinocephalus myops*).

tonnes) by mackerel, taking into account only the commercial catches of these predators. The consumption by these predators considerably exceeds the mean commercial catch of anchovy (0.3×10^4 tonnes) in this area. As the catches of predatory fishes are only a part of the total populations, the actual consumption of anchovy larvae must be far greater than these figures.

The mean numbers of viable heterotrophic bacteria, given in terms of number per 10 ml as indicated by the plate counts on Millipore filters, were 100–250 in the upper layer (0–200 m) and 40–60 in the mid and deep layers (below 300 m) in Sagami Bay. In the heterotrophic bacterial flora, the protein-decomposers predominated throughout the year over the carbohydrate-decomposers. The biomasses of bacterial groups relevant to

Table 4.2. *Comparison of particulate organic carbon (POC) found in seawaters with different climates*

Area	POC (μg C/l)	References
Sagami Bay	40–220	Handa (1977)
Akkeshi Bay	100–800[a]	Nishizawa (1977)
Tokyo Bay	1000–3000	Handa (1973)
Kuroshio	13.0–27.5	Handa (1967)
Hiuchi Nada	300–2000	Okaichi (1971)

[a] These values correspond to 4.7–14.2 g C/m^2.

Table 4.3. *Dissolved organic carbon (DOC) measured in seas in different climatic zones*

Area	DOC (mg C/l)	References
Sagami Bay	0.75–1.97	Ogura (1977)
Hiuchi Nada	1.3–6.7	Okaichi (1971)
Tokyo Bay	0.94–5.32	Ogura (1970)
Kuroshio	0.28–0.42	Handa (1967)
Pacific (155° W, 50° N–15° S)	0.81–1.46	Ogura (1970)
East China Sea	0.96–1.42	Ogura (1970)

the nitrogen cycle were determined, and it was found that the biomasses of denitrifiers and nitrifiers were somewhat smaller than those of the other groups. The distribution of nitrogen-fixing bacteria in seawater and bottom sediment was examined; viable counts of these bacteria were 10–10^3/100 ml in seawater and 10^3–10^4/g wet wt in sediment.

The organic matter collected on the filter with a pore size of 1 μm is defined as the particulate organic matter (POM), and the matter passing through the filter as the dissolved organic matter (DOM). Particulate organic carbon (POC) ranged from 50 to 100 μg C/l in the layers from the surface to 100 m depth in Sagami Bay. The values tended to decrease with depth to 200 m, while almost uniform values of about 30 μg C/l were recorded in the layers below 200 m. High POC values were observed in May and October when chlorophyll *a* was found in greater amounts than at the other seasons. Organic nitrogen of POM was recorded as 3.2–16.0 μg N/l in the euphotic zone and 0.9–4.0 μg N/l in the layers below the euphotic zone. C/N values of POM were found to be 5–6 in the euphotic zone, and the values tended to increase with depth to over 10, suggesting faster release of carbon than of nitrogen. POM was composed of carbohydrate, protein and lipid carbons in ranges from 19.5 to 25.0%, from 30.5 to 45.0% and from 9.8 to 15.6% respectively, in the euphotic zone. In the layers from the surface to 1000 m depth, DOM was found to range from 0.5 to 1.7 mg C/l, values far higher than those found for the waters of the open sea (see Tables 4.2 and 4.3).

Both in bay water and in ocean water, a considerable amount of urea was detected. In the euphotic zone, the amount of urea was found to be roughly equal to that of the total inorganic nitrogen, and to show an almost even vertical distribution down to a considerable depth. It was confirmed that the urea is a decomposition product of organic matter, and is taken up actively by phytoplankton during photosynthesis.

Sendai Bay

Sendai Bay forms a large embayment opened widely toward the south-east (Fig. 4.1*b*). The waters of this bay are approximately 2000 km^2 in area and 75×10^9 m^3 in volume. The tidal range is rather small, only 80 and 90 cm in the mean sea level; the spring rise is 110–140 cm over the lowest low water.

Fresh water from the Kitakami, the Abukuma and many other rivers flows into the bay, forming low salinity areas along the coast. The total river water influx amounts to 114×10^8 m^3 per year. The seawater off Sendai Bay is composed of a complex of mixed waters of Kuroshio, Oyashio and Tsugaru Warm Current. The frequent variations in water system have an important influence on the hydrographic conditions of the bay. Large cities lying along the coast are Sendai, Shiogama and Ishinomaki, but there is as yet little water pollution, except in the northern part and in Matsushima Bay.

Surveys of environmental factors and of the productivity of biocoenoses were performed at 14 stations in the bay. Oceanographic observations showed that a tongue of water extending from the south-eastern region occupies the central part of the bay and divides the coastal waters into two parts, the northern and the southern, at the mouth of Matsushima Bay.

The monthly measurements of water temperature from 1967 to 1972 gave the maximum value of about 22–24 °C in August, and the minimum of about 7–10 °C in April. One of the outstanding features of the nutrient concentrations in the bay is the extremely low levels of phosphate and nitrate-nitrogen, the concentrations being barely detectable. Chlorophyll *a* values are in the general range of 0.1–9.5 μg/l, the higher values being found in winter and spring; summer levels were 0.1–1.0 μg/l. Daily averages of primary production were 0.34 g C/m^2 in May, 0.19 g C/m^2 in July, and 0.72 g C/m^2 in November. The annual net primary production of the bay would be of the order of 100 g C/m^2 or slightly more (Table

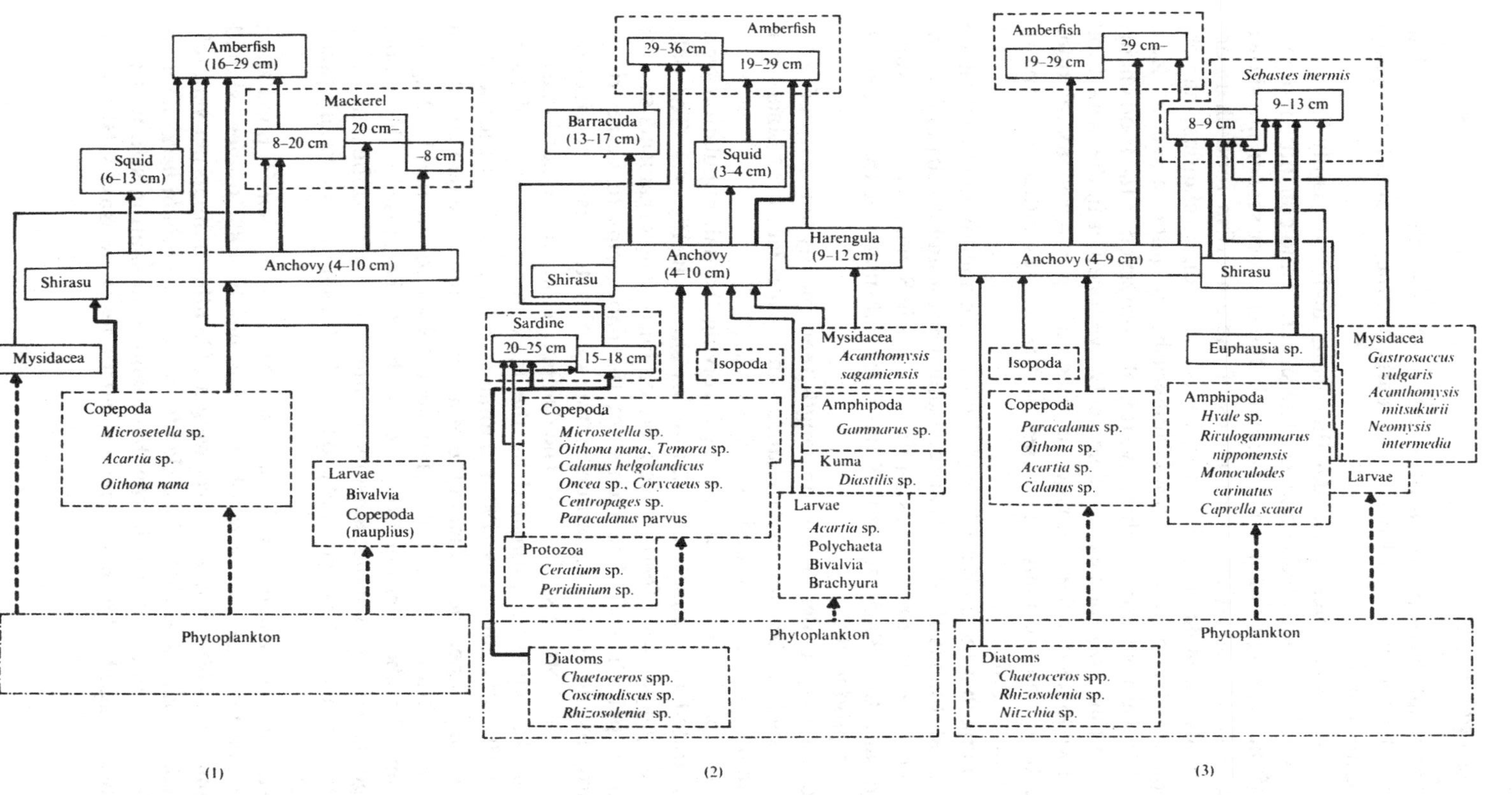

Fig. 4.4. Structures of the grazing food chain for the young amberfish community in the south-west area of Sendai Bay, 1968.
(1) 5 July–20 August; (2) 28 August–29 October; (3) 2 November–26 November.

4.1). The living phytoplankton constitutes 5–40% of the total particulate carbon. The net increase of particulate carbon in the water column would be 60 g/m^2 in a year.

Ecosystem dynamics in Sendai Bay were analysed mainly from the viewpoint of the food chain concept. The structure and constituent organisms of the grazing food chain surveyed in the summer season are shown in Fig. 4.4. The constituent organisms of the amberfish food chain are occasionally variable, but the main trunk of feeding relations, namely phytoplankton–copepoda–anchovy–amberfish, is quite stable at least during the warm seasons in this bay. It is also observed in the food chains that more than two species could scarcely live together in the same food niche. Two kinds of fishes intimately connected by prey–predator relations in the food chain usually share the food resources either in terms of species or of time.

Assuming that the amberfish feeds only on the anchovy, it is estimated that 216–300 tonnes of anchovy are consumed by the amberfish population, whose size is about 1000 tonnes, during 15 days in Sendai Bay in the warm season, and that 30–51 tonnes of the body substance of the amberfish would be produced.

The anchovy under 60 mm in length feed mostly on selected zooplankton such as *Oncaea* sp., Calanidae and cypris larvae of Macrura at the rate of about 10% of body weight per day.

Some typical detritus food chains involving the dab population were demonstrated in summer and autumn as shown in Fig. 4.5. The dab, *Limanda yokohamae*, feeds selectively on burrowing organisms such as sedentary polychaetes. The habitat of the dab seems to be restricted primarily by the density of certain benthic species, to a lesser and indirect extent by the bottom sediments. The dab population in the north-eastern area of the bay was estimated by the mark–recapture method to be 30×10^4–50×10^4 individuals, 100–170 tonnes, in May 1970. The amount of food consumed per individual in the main age group in the dab population has already been reported by Hatanaka (1966) to be nearly 2 kg/yr. Assuming the above population size in May to be close to the annual average, these dabs should consume benthic organisms to the amount of 600–1000 tonnes wet weight, and produce 47–75 tonnes of body substance per year.

It was also estimated that the starfish (*Asterias amurensis*) population in the coastal area, the size of which is nearly 1900 tonnes, should consume about 8000 tonnes in wet weight of benthic animals per year (cf. Hatanaka & Kosaka, 1959). The slender halibut, *Limanda herzenteini*, is another important flat fish in this bay, whose population size was previously estimated to be approximately 380 tonnes (Kawasaki & Hatanaka, 1951).

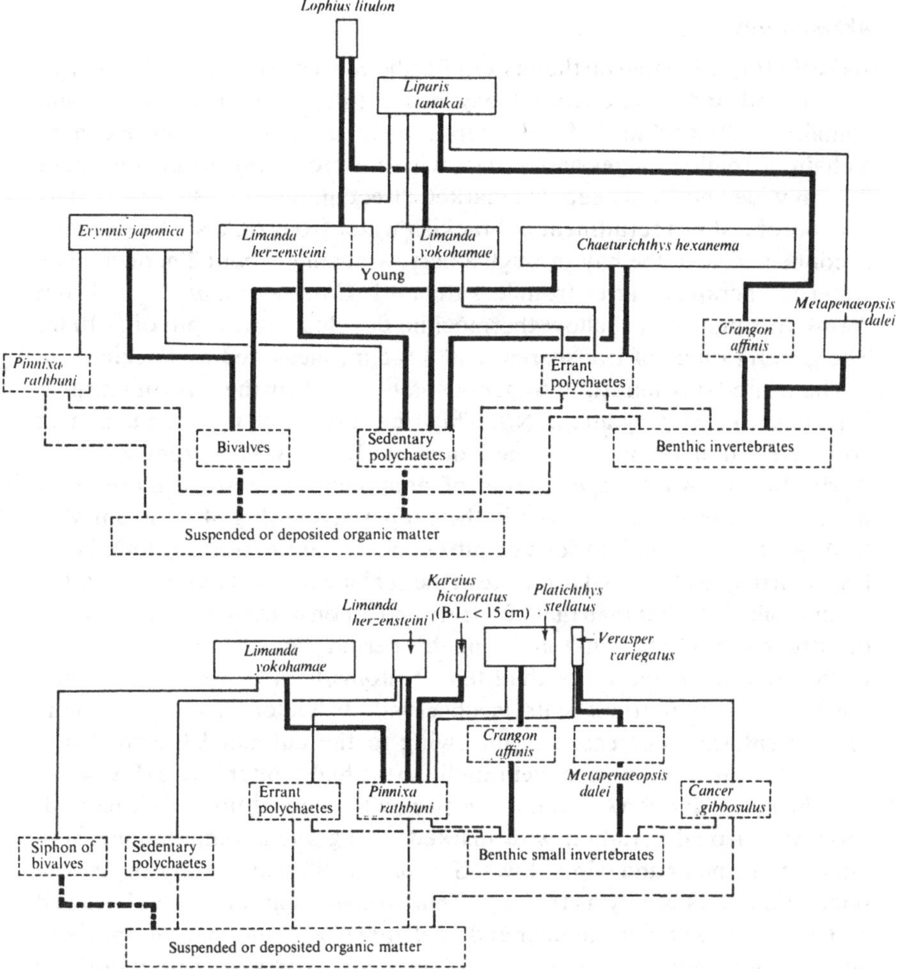

Fig. 4.5. Feeding relationships of bottom fishes in Sendai Bay. Upper, in autumn; lower, in summer. The width of box for each fish indicates relative abundance in catches by coastal trawl net. Dotted lines: not investigated in this study.

The supply of organic detritus was estimated at about 60 g C/m^2/yr in mean value, as food for benthic animals. The estimated net annual production of the crab, *Pinnixa rathbuni*, in the north head of the bay, was 40–80 tonnes wet weight, sedentary polychaetes 8500 tonnes, and other invertebrates which feed on detritus about 9000 tonnes. Therefore the net annual production of all these benthic animals amounts to nearly 18×10^3 tonnes in total. Thus the supply of organic detritus (355×10^3 tonnes) is about 20 times the net production of invertebrates using it.

Akkeshi Bay

Akkeshi Bay is located on the eastern Pacific coast of Hokkaido (Fig. 4.1*a*). This round and semi-enclosed bay has a total area of 128.4 km^2 and contains 1.421×10^6 km^3 of seawater, separated from the outer ocean by a chain of shallow ridges and islands with a narrow central opening about 2.4 km wide and 30 m deep. No marked direct inflow of fresh water exists except for a slow recruitment of brackish water from Akkeshi Lake which is connected with the bay through a narrow channel about 2 m deep. The water temperature varies from less than −1 °C to more than 15 °C. From spring to summer, a shallow thermocline develops at a depth of 5–10 m. The general range of transparency is 1.5–5 m, measured by Secchi disc.

The winter maximum of nitrogenous salts, mainly in the form of nitrates, is as high as 10–15 μg atoms N/l. This rich reserve of nitrogen is almost completely utilized by the spring bloom of phytoplankton from March to April. In late April, regeneration of ammonia and a slight increase in nitrate and nitrite are observed in the deeper layers (Fig. 4.6). From May to August, the concentrations of nitrogen salts are extremely low, being less than 0.5 μg atoms N/l in nitrate in the surface water. The concentration of ammonia is higher than that of nitrate and is considered the major source of nitrogen for phytoplankton during this period. Phosphate is ample even in this season, being more than 0.4 μg atoms/l. In September, marked regeneration of nutrient salts is observed. In October, concentrations of nutrient salts decreased again owing to the autumnal bloom. From November on, nutrients recover rapidly to the high winter levels (Fig. 4.6).

Primary production, measured *in situ* by the conventional ^{14}C method, showed seasonal variation with marked spring and autumn blooms separated by summer and winter lows. Seasonal succession of phytoplankton populations was briefly as follows: *Thalassiosira* spp. in spring, followed by *Chaetoceros* spp. in summer and *Skeletonema costatum* in autumn, and later *Asterionella japonica*, *Eucampia zoodiacus* and *Thalassionema nitzschioides* in winter. The highest daily net production was 310 mg C/m^3 observed in the surface layer in March, and the lowest was 10 mg C/m^3 in January in the ice-covered surface layer (Table 4.1).

The total annual net primary production was estimated as 295 g C/m^2, accompanied by respiration loss of 341 g C/m^2/yr. This means that the total annual gross production in this bay was more than 600 g C/m^2/yr, and more than one half of it was consumed by phytoplankton itself.

Measurements were made of zooplankton respiration rates, excretion rates, productivity and filtering, and also the estimation of the biomasses of dominant zooplankton. The biomasses of zooplankton collected by net sampling were determined on a dry weight basis, and converted to the equivalent amounts of nitrogen, phosphorus, etc. by using the conversion

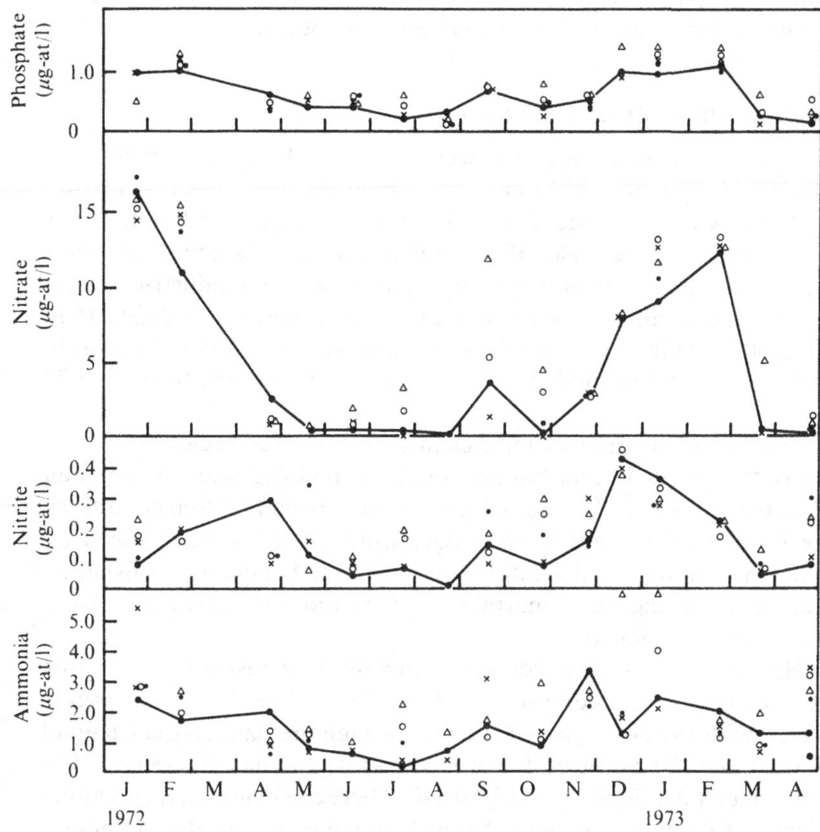

Fig. 4.6. Seasonal variations in concentrations of phosphate, nitrate, nitrite, and ammonia at the central station of Akkeshi Bay; 0 m, (–●–); 5 m, (×); 10 m, (·); 15 m, (○); 20 m, (Δ).

factors prepared previously. The measurements of zooplankton activities were carried out with specimens collected by gentle vertical hauls with a conical net and acclimatized to the dark condition for 8 hours in a vessel containing the surface water filtered through a net of 0.33 mm mesh aperture.

Based on the results obtained, it was established that only 9.5% of the net primary production was consumed by net zooplankton larger than 0.1 mm, and the consumption due to benthic animals, namely *Tecticeps japonicus*, *Diastylopsis dawsoni* and *Protomedeia kryer*, was less than 1%. The remainder of about 90% was attributed to consumption by micro-zooplankton and bacteria. The calculated secondary production in this bay was only 8.1–10.0 g C/m^2/yr. Thus, the ecological efficiency between primary production and secondary production was 3–4%, showing a poor zooplankton population inhabiting the bay. This efficiency is significantly lower than so far reported for various seas and oceans.

Hiuchi–Bingo Nada (Seto Inland Sea)

Seto Inland Sea is enclosed by the western part of Honshu, Shikoku and Kyushu, and has an area of 22400 km^2. A branch current of the Kuroshio enters from Kii and Bingo Channels. There are several hundreds of islands in the sea. Because of the relatively small amount of water exchange and the large pollutant loads from sewage and industrial waste, environmental conditions vary remarkably. Transparency exceeds 15 m in spring at the entrance channels, but in the inside waters it frequently becomes less than 2 m, and the water temperature ranges from 7 to 28 °C.

Construction of industrial factories in several coastal regions has been carried out since 1960, and huge amounts of polluted water have been introduced into the sea. Accordingly, water quality and bottom conditions change for the worse year by year. Occurrences of red tide caused by rapid increase of lower unicellular algae, lowering the quality of fisheries products and causing mass mortality to fishes and shellfish, etc., have gradually become common.

The Hiuchi–Bingo Nada water area, being the main research area of this study, is located in the central part of the Seto Inland Sea. This area is connected with the outer part of the sea through the Kurushima Channel in the west and the Strait of Bisan in the east, and has an area of 2500 km^2 and a mean depth of 23 m (Fig. 4.1*d*). There are many factories along the coast and the waste effluents from them have an important effect on the seawater and bottom sediments in some regions. As the water of this area is strongly eutrophicated, frequent occurrences of red tide mainly composed of *Skeletonema costatum*, *Eutreptiella* sp., *Gymnodinium mikimotoii* are becoming a serious problem.

Phytoplankton communities in this area were composed mainly of diatoms. The cell numbers were usually estimated to range from 10^4 to 10^5 per litre, but are sometimes as high as 10^6 due to proliferation of *Skeletonema costatum*.

Carbon and nitrogen contents of POM collected at 13 stations in different seasons were determined with an automatic micro CHN analyser. The results showed $6.7\times10^3-13.3\times10^3$ tonnes carbon (POC) in the whole area, and $1.05\times10^3-2.3\times10^3$ tonnes nitrogen (PON) (Table 4.2). Based on the ratio of POC to chlorophyll *a* content obtained for living planktonic algae, chlorophyll *a* content in waters, carbon content of 33.5% in dry weight of phytoplankton, and 80% water content for living algae, the biomasses of the phytoplankton communities were calculated as 64×10^3 tonnes in

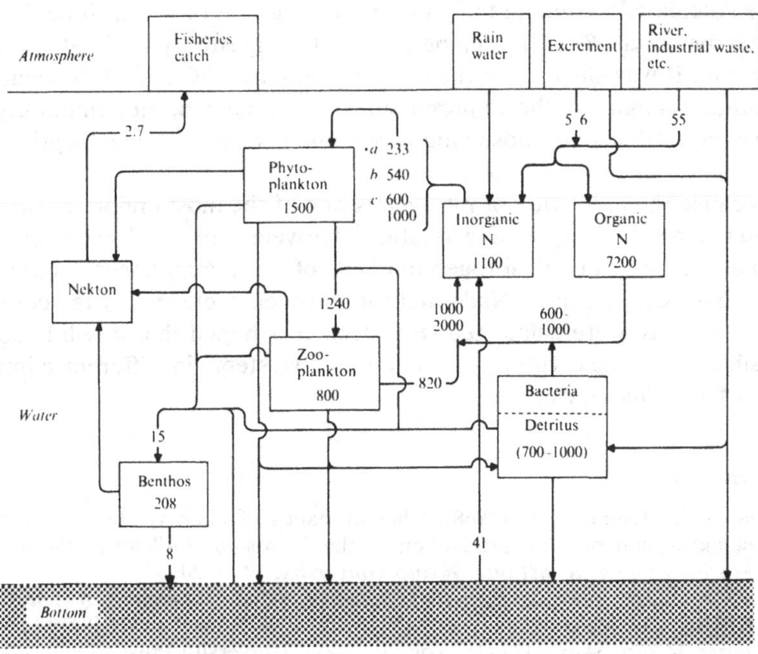

Fig. 4.7. Nitrogen flow chart in Hiuchi-Bingo Nada. Numbers in boxes are standing crop expressed in tons of nitrogen. Numbers along lines with arrow heads show the quantity of nitrogen in tons/day transported along the line. Nitrogen assimilated by phytoplankton from inorganic nitrogen is estimated in three ways; (*a*) obtained by ^{14}C-tank method; (*b*) calculated from production rate of amino-acids in phytoplankton; (*c*) calculated from the decrease of nitrogen concentration in the water during a day. Nitrogen of fisheries catch is calculated assuming protein content is 12% of catch and nitrogen content is 16% of the protein.

February, 123×10^3 tonnes in June and 90.3×10^3 tonnes in September. Primary net production measured *in situ* ranged from 0.12 to 0.57 g C/m²/day (Table 4.1). If the average net production in the high water temperature period (8 months) is assumed to be 0.42, and in the low water temperature period (4 months) to be 0.22 g C/m²/day, the annual net production is estimated at about 130 g C/m²/yr (Table 4.1).

Most of the zooplankton species found in this area were the common species found in the inlets along the coast of Japan. Among them, *Noctiluca scintillans* was very abundant, and Copepoda frequently predominated. The biomasses of zooplankton in June, September and December were calculated by multiplying the numbers and weights of individuals of respective main species. The biomasses thus obtained ranged from 3.22 to 5.38 g/m², including 3.00–5.25 g/m² of herbivorous and omnivorous zooplankters. The quantities of carbon and nitrogen which would be accumulated in the biomass of the zooplankton community as growth were calculated assuming the daily increasing rate of 7% in June, 8% in September and 5% in December, and 70% digestivity of food by zooplankton. It was shown that the standing crop of POC and PON seems to be quite enough for the requirement of the zooplankton community in every case other than those where *Noctiluca scintillans* is exceptionally abundant.

Because the utilization of nitrogen is one of the most important factors in analysing the quantitative relations between prey and predator, the biomasses, in terms of nitrogen amount, of each component composing the ecosystem in Hiuchi Nada and the nitrogen cycle among the components in a day were calculated (Fig. 4.7). It is hoped that it will become possible in the near future to compare ecosystems in different climatic zones from this viewpoint.

References

Aruga, Y. & Ichimura, S. (1968). Characteristics of photosynthesis of phytoplankton and primary production in the Kuroshio. *Bulletin of the Misaki Marine Biological Institute, Kyoto University*, **12**, 3–20.

Aruga, Y. (1977). In *Productivities of biocenoses in coastal regions of Japan*, ed. K. Hogetsu, M. Hatanaka, Y. Kawamura & T. Hanaoka, JIBP publication series 14, pp. 54–62, 127–33. Tokyo: Tokyo University Press.

Endo, T. & Okaichi, T. (1977). In *Productivities of biocenoses in coastal regions of Japan*, ed. K. Hogetsu, M. Hatanaka, Y. Kawamura & T. Hanaoka. JIBP publication series 14, pp. 318–20. Tokyo: Tokyo University Press.

Handa, N. (1967). The distribution of the dissolved and the particulate carbohydrates in the Kuroshio and its adjacent areas. *Journal of the Oceanographic Society of Japan*, **23**, 115–23.

Handa, N. (1973). In *Data book for studies on the productivity of biocenoses and circulation of bioelements in coastal regions of the Kuroshio Current area*, ed. K. Hogetsu.

Handa, N. (1977), In *Productivities of biocenoses in coastal regions of Japan*, ed. K. Hogetsu, M. Hatanaka, Y. Kawamura & T. Hanaoka. JIBP publication series 14, pp. 107–119. Tokyo: Tokyo University Press.

Hatanaka, M. & Kosaka, M. (1959). Biological studies on the population of the starfish, *Asterias amurensis*, in Sendai Bay. *Tohoku Journal of Agricultural Research*, **4**, 159–78.

Hatanaka, M. (1966). Coaction in marine fishes. In *Coaction in Biotic Communities*, ed. M. Kato, pp. 203–39. (In Japanese.)

Hogetsu, K., Hatanaka, M., Kawamura, Y. & Hanaoka (eds.) (1977). *Productivities of biocenoses in coastal regions of Japan*. JIBP publication series 14. Tokyo: Tokyo University Press.

Kawasaki, T. & Hatanaka, M. (1951). Studies on the populations of the flatfish in Sendai Bay. I. *Limanda angustirastis* Kitahara. *Tohoku Journal of Agricultural Research*, **1**, 83–103.

Nemoto, T. (1968). Chlorophyll pigments in the stomachs of euphausiids. *Journal of the Oceanographic Society of Japan*, **24**, 253–60.

Nishizawa, S. (1977). In *Productivities of biocenoses in coastal regions of Japan*, ed. K. Hogetsu, M. Hatanaka, Y. Kawamura & T. Hanaoka. JIBP publication series 14, pp. 235–49. Tokyo: Tokyo University Press.

Ogura, N. (1970). Dissolved organic carbon in the equatorial region of the central Pacific. *Nature*, **227**, 1335–6.

Ogura, N. (1977). In *Productivities of biocenoses in coastal regions of Japan*, ed. K. Hogetsu, M. Hatanaka, Y. Kawamura & T. Hanaoka. JIBP publication series 14, pp. 113–14. Tokyo: Tokyo University Press.

Okaichi, T. (1971). Effects of pollutants on the occurrence of red tide in the Seto Inland Sea. '*Kagaku to Seibutsu*', **9**, 566–71. (In Japanese.)

Saijo, Y. & Ichimura, S. (1960). Primary production in the northwestern Pacific Ocean. *Journal of the Oceanographic Society of Japan*, **16**, 139–45.

Saijo, Y. *et al.* (1973). In *Data book for studies on the productivity of biocenoses and circulation of bioelements in coastal regions of the Kuroshio current area*, ed. K. Hogetsu, 180 pp.

SCOR–UNESCO. (1966). Working group 17. Determination of Photosynthetic Pigments. In *Monographs on oceanographic methodology* I, p. 69. Paris: UNESCO.

Shimura, S. & Ichimura, S. (1972). Primary production in coastal water adjacent to the Kuroshio off Shimoda. *Journal of the Oceanographic Society of Japan*, **28**, 8–17.

Takahashi, M., Satake, K. & Nakamoto, N. (1972). Chlorophyll distribution and photosynthetic activity in the north and equatorial Pacific Ocean along 155° W. *Journal of the Oceanographic Society of Japan*, **28**, 27–34.

5. Factors determining the productivity of South African coastal waters

J. R. GRINDLEY

Introduction

Marine biological research in South Africa during IBP included work on all of the main themes relating to marine productivity. Most of the work related to comparative ecology of different regions and some of the more significant results of this work are reviewed here. This work covered the coastal waters and adjacent oceans around the subcontinent as well as estuaries and lagoons along the coast. The organisms studied ranged from micro-organisms to the largest marine mammals in these regions. Consequently the results of the research are wide ranging and diverse as may be seen from the bibliography at the end of this chapter. In order to bring together some of the findings of general importance the theme of this review has been taken as 'factors determining the productivity of South African coastal waters'. This enables some of the findings of general significance to be placed in perspective but means that many interesting results of a more restricted relevance have not been considered here. It is hoped that the bibliography, which is subdivided into sections dealing with different subjects, will allow readers to trace original work of a more specialised nature not reviewed here.

When plans were being formulated for the programme of research on marine productivity for IBP, one of the basic principles was that: 'The programme should be directed towards an improvement of our understanding of the basic ecological mechanisms which control the abundance, distribution and productivity of marine organisms of all kinds, throughout the trophic chain in the sea.' It is findings relating to these aspects in the context of South African coastal waters that are considered here.

South African marine research for the IBP was co-ordinated by the South African IBP/PM Working Group. The first convener was Professor D. H. Davies but after his death in 1966 Professor J. R. Grindley became the convener. Members of the committee were Dr T. H. Barry of the South African Museum, Professor J. H. Day of the University of Cape Town, Dr B. V. D. de Jager of the Sea Fisheries Branch, Professor A. E. F. Heydorn of the Oceanographic Research Institute and Mr W. D. Oliff of the Natal Regional Laboratory for Water Research.

Upwelling

The extraordinary richness of the upwelling region of the Benguela Current on the west coast of South Africa in comparison to adjacent oceanic waters has been referred to frequently. This upwelling of cold nutrient-rich water from the layer of South Atlantic central water is well known, but earlier data did not permit any quantitative treatment of the process, or of the consequent organic production.

This was attempted in the area off the Cape Peninsula by Andrews, Cram & Visser (1970). Upwelling was found to average 21 m/day and to occur in an area of 200 km^2. The amounts of nitrates and phosphates brought up by upwelling in the area were calculated to provide an estimate of the potential productivity. Estimates for the upwelling region ranged from 6.192 to 9.029 tonnes C/day. Annual potential productivity for the area due to upwelling was estimated to range from 0.38×10^6 to 0.93×10^6 tonnes C.

Airborne radiation thermometer studies of the upwelled component of the Benguela Current system on the west coast have revealed the existence of discrete upwelling areas on the south and west coasts of southern Africa. Interactions occur between the upwelled water and South-Atlantic surface water in the south.

Bang (1971) attempted to define the major structural elements of the Benguela region and to clarify the development and character of an active 'centre of upwelling'. He identified three regions. (1) The southern region from Cape Point to St Helena Bay is dominated by active coastal upwelling influenced by the persistent south-easterly winds that prevail here in summer. (2)Upwelling in the area further north from St Helena Bay to the Orange River reveals a number of structural features (Fig. 5.1). (*a*) An inshore zone of spasmodic and discontinuous upwelling where direct and short-term wind stress is the main driving force. (*b*) An intermediate zone consisting of water derived from a previous spell of upwelling which has been warmed and mixed with adjacent water. This zone is separated from an inshore zone of upwelling by a very sharp discontinuity, usually visible as an abrupt colour change from soupy-green to greeny-blue. (*c*) An offshore divergence belt where indications of vertical water movement may be observed and natural slick belts may be present on the surface. (*d*) The oceanic region of northward-flowing water of relatively simple structure and weaker gradients sometimes referred to as the Trade Wind Drift to distinguish it from the true Benguela system. (3) The northern region between Orange River and Walvis Bay is apparently dominated by an active 'centre of upwelling' in the vicinity of Lüderitz. Upwelling occurs consistently in the Lüderitz area while north and south of this upwelling is a more sporadic phenomenon. To the north of the centre of upwelling a complex pattern of tongues and eddies of upwelled water develops.

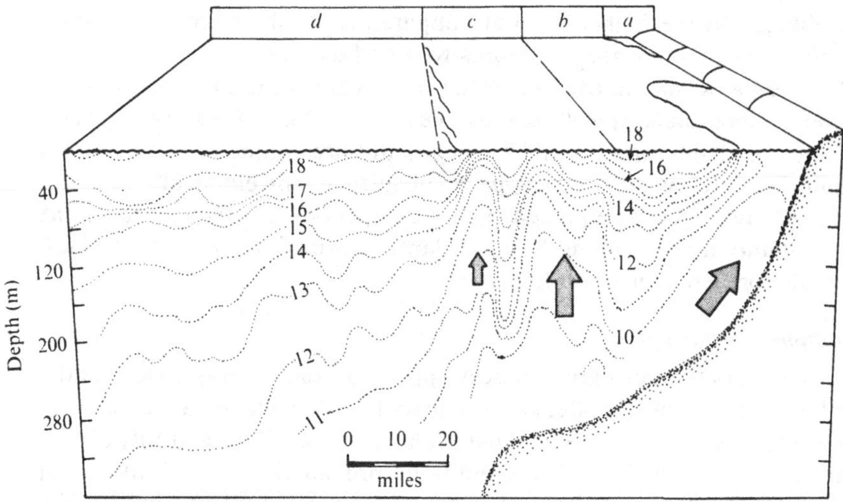

Fig. 5.1. Idealisation of main structural features of the Benguela region between 29° and 32° S. Annotations are discussed in the text. Dotted lines indicate isotherms. (From Bang, 1971.)

The term Benguela Current' needs to be defined more carefully than in the past. If it is considered to consist solely of recently upwelled water it would be discontinuous in both space and time. If it is defined in terms of northward flow of water it is not differentiated from the offshore oceanic water of the Trade Wind Drift. Bang (1971) has suggested that the Benguela Current is that area east of the offshore divergence within which oceanic processes are dominated by short-term atmospheric interactions. The shelf edge upwelling phenomenon itself appears to have an internal hydrodynamic cause as opposed to the inshore upwelling apparently induced primarily by local wind stress.

Structures affecting upwelling

Various geological features including the form of the continental margin and the presence of seamounts can influence the process of upwelling on the coast.

The Cape submarine canyon causes northward-moving deep water to be funnelled through it to upwell in the region west of Cape Columbine. A study by De Decker (1968) revealed that the Copepoda which occurred at the north (top) end of the canyon included thirteen species typical of intermediate depths not found in surface waters at the lower end of the canyon. The greater species diversity and abundance of Copepoda observed in the upwelling waters towards the northern end of the canyon is interesting, as upwelling systems are usually characterised by low diversity. The greater abundance is to be expected in waters enriched by upwelling, and the higher diversity apparently results from the addition of species from intermediate depths to a surface community.

The Vema Seamount 640 km off the west coast of South Africa appears to cause some local upwelling (Grindley, 1967; Berrisford, 1969). Temperatures and salinities around the seamount are characteristic of South Atlantic central water and suggest subsurface upwelling of Antarctic intermediate water. The presence of the copepods *Calanoides carinatus* and *Calanus tonsus* among South Atlantic warm-water forms also indicates the presence of upwelled water.

Anomalies in upwelling

In some years the upwelling process appears to fail to some extent in the northern region of the Benguela Current with widespread ecological consequences. During 1963, and particularly in May, June and July of that year, anomalous hydrological conditions prevailed off the South West African coast. Temperatures, salinities and oxygen content were usually high and various biological processes including fish spawning were affected. Stander & De Decker (1969) in their study of this oceanographic anomaly point out that an analysis of wind conditions fails to explain the reduced upwelling or inshore movement of offshore waters. It can only be concluded that local conditions were not responsible for the observed anomalies.

In 1963 several species of Copepoda appeared in the plankton of the surface waters – species previously unknown in this area. Practically all of these newcomers are known to occur further north in the tropical Atlantic and particularly in the coastal region between Angola and Senegal. It would seem that there was a southward advance of warm surface waters. However amongst the deep-water Copepoda found only in hauls from 300 m there was some replacement by species known from regions further south. Stander & De Decker suggest that the decrease in the northward flow of the Benguela Current in 1963 may have caused a corresponding decrease of the southward compensating current at deeper levels. Failure of the normal influx of subsurface water from the north could explain the decline of the northern component of the copepod fauna and the appearance of southern forms at depths below 100 m. However doubt has been cast on the existence of a compensating counter-current for it appears that there is a north-westerly drift of the oxygen minimum layer from the regions of decay off South West Africa. It has been suggested (Grindley, 1971) that a decrease in the upwelling process off the coast of South West Africa caused a reduction in productivity and consequently a reduction in the benthic decay process during 1963. Such more favourable conditions might have caused a change in the species composition of the copepod fauna at those depths. Species from further south drifting in the Benguela Current might perhaps have survived that year because of the more favourable bottom conditions.

The periodic reduced upwelling with influxes of warm surface waters occurring off South West Africa shows an obvious relationship to the phenomenon known as El Nino occurring off the coast of Peru. However the scale of the phenomenon is smaller and the extent of the ecological consequences is far less off South West Africa.

Local wind-induced upwelling

Upwelling is not restricted to the west coast of southern Africa and periodic wind-induced upwelling occurs along the south and south-east coasts. Upwelling appears to be pronounced where topographic features aid the process and it has been noted particularly in False Bay, at Hermanus, Danger Point, Knysna, Mossel Bay and Algoa Bay. Cram (1970) and Grindley & Tayler (1970) carried out studies of wind-induced upwelling in False Bay. Grindley & Tayler demonstrated the relationship between wind and surface temperature as a measure of wind-induced upwelling. Serial correlations were computed and the lag time between wind and resultant upwelling was found to be approximately 30 h. Periodic cycles of upwelling followed by surface warming and 'maturation' of the water apparently favour the generation of red water in that area.

Seasonal nutrient availability

A pronounced seasonal cycle of phytoplankton production related to nutrient fluctuations has been observed off the west coast of South Africa. Maximum productivity occurs during the season of maximum upwelling in spring and summer while the maximum standing stock has been recorded in autumn. It appears that *Chaetoceros* spp. which are dominant during most of the year are able to increase rapidly after upwelling has occurred. Increases in phytoplankton are correlated with decreases in inorganic phosphate in surface waters. A succession of different species of diatoms prevails through the year and dinoflagellates become dominant in later summer. Plankton blooms develop and drift offshore to the west, losing connection with their centre of origin.

The warmer offshore waters of the Trade Wind Drift have a much sparser phytoplankton because of the limited nutrient supply available. The phytoplankton of the offshore waters is dominated by forms such as *Planktoniella sol* and *Thallasiothrix longissima.* The transition zone corresponding to the hydrological divergence usually contains a scanty microplankton consisting mainly of the more adaptable forms from the upwelled and offshore waters.

Atmospheric pressure and current variations

Off the coast of Natal intensive investigations (Anderson, Sharp & Oliff, 1970) over a number of years have revealed a relationship between atmospheric pressure and short-term variations in currents. The passage of an atmospheric low-pressure system is almost invariably accompanied by a reversal of inshore currents from southward to northward flow. The return to a southward flow usually accompanies the appearance of the succeeding peak in atmospheric pressure.

Local small-scale upwelling close inshore often occurs during the reversals and the passage of the low-pressure systems. This upwelling results in nutrient enrichment of surface waters and occasional plankton blooms. Weather changes result in almost instantaneous changes in temperature not only at the surface of the sea but also at depths down to 400 m and more.

Warm Agulhas Current

The phytoplankton of the south-west Indian Ocean influenced by the warm Agulhas Current is notable for the high species diversity (Taylor, 1967; Nel, 1968*a*). Diatoms are the most significant group and heavy standing crops appear in certain areas inshore of the Agulhas Current. These blooms are composed of large numbers of species unlike those of the west coast. The distributon patterns of the species are complex. Nel (1968*a*) plotted the distribution of various species of diatoms in the south-west Indian Ocean and found ten different types of distribution pattern, each broad type with many variations. The inshore region and the area south of the subtropical convergence appears to be particularly distinctive. Higher cell counts were found in the cooler sub-Antarctic surface water than in the warmer waters further north.

Thorrington-Smith (1969) found measures of similarity and diversity useful in interpreting the complex phytoplankton in the Agulhas Current region. Analyses using shared species and using a similarity index taking into account relative abundance, indicate that stations in the main stream of the Agulhas Current have greater affinity with each other than with stations in waters further inshore or offshore. Diversity indices based on the number of species present at each station and their relative abundance, show that there is a greater species diversity in the main stream of the Agulhas Current than further inshore or offshore. On the limited data available it appears that there is an inverse correlation between diversity and productivity.

Light penetration

Diatoms, because of their utilisation of light for photosynthesis, are normally found in surface layers down to perhaps 200 m. Nel (1968*b*) found live diatoms in Munro–Eckman bottle samples down to 1500 m and in net hauls down beyond 1500 m in the south-west Indian Ocean. In some cases the amount of phytoplankton in the aphotic zone exceeded the amount found in the euphotic zone which seems unlikely for various reasons. Some species occurring below 500 m were actually not present in the surface layer in the same area. The most abundant diatoms in the waters below 500 m were *Chaetoceros* spp., *Rhizosolenia* spp. and *Thalassiothrix* spp. It was concluded that they must be existing by heterotrophic means or 'dark metabolism' using organic substrates as energy sources.

Primary productivity

Many plankton studies in the seas around South Africa have included measurements of standing crop or estimates of it by techniques such as chlorophyll extraction, but few measurements of primary productivity have been made. Some studies using the ^{14}C technique have been described by Mitchell-Innes (1967) and Burchall (1968*a*, *b*) (Fig. 5.2.). In the south-west Indian Ocean production was found to range from 14 to 1080 mg C/m^2/day (Mitchell-Innes, 1967; Nel, 1968*a*). In the area north of the subtropical convergence the paucity of nutrients probably limited production. In areas where upwelling or other factors increased the availability of nutrients higher production values were obtained. Pockets of high productivity (500 mg C/m^2/day) were found in Delagoa Bay and off Algoa Bay. Solar radiation is apparently a limiting factor south of the subtropical convergence, where considerable seasonal variation in productivity was apparent, dropping to below 250 mg C/m^2/day in winter. A survey of primary organic production off Natal (Burchall, 1968*b*) produced a range of values from 17 mg C/m^2/day to 942 mg C/m^2/day. The maximum was obtained south-east of the Tugela River mouth. This area is possibly enriched by outflow of nutrients from the Tugela River. Measurements ranging from 109 to 563 mg C/m^2/day were obtained in 1961 in the south-west Indian Ocean. Although no direct correlation of productivity with cell counts was apparent, it has been suggested (Grindley, 1971) that there is an indication that maximum cell counts occur downstream of areas of high productivity.

Studies of primary productivity in Saldanha Bay and the Langebaan lagoon on the west coast have been carried out. The oxygen (light–dark bottle) method for determining primary production was used. It was found that productivity was dependent on the availability of sunlight on any one

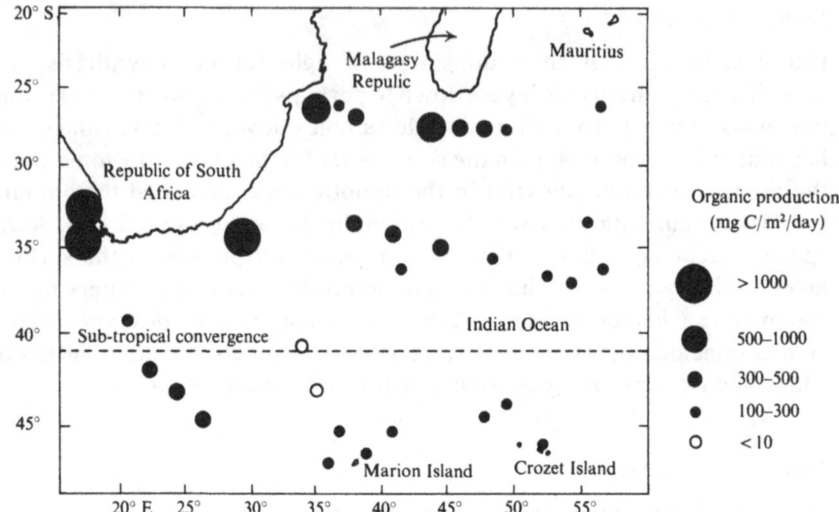

Fig. 5.2. Distribution of organic production in the south-west Indian Ocean. (From Mitchell-Innes, 1967.)

day and related to nutrient availability at different stations. At the most productive station (Salamander Bay) productivity varied from 116 to 1142 mg C/m^2/day. Andrews, Cram & Visser (1970) on the basis of their upwelling study west of the Cape Peninsula mentioned above, estimated the mean rate of primary production at 8.5 g C/m^2/day on the basis of changes in the oxygen concentration *in situ* in the euphotic zone.

More recently (Andrews, 1973) the primary production and phytoplankton standing crop have been studied west of the Cape Peninsula over a two year period. A maximum standing crop of 52 mg chlorophyll *a*/m^3 (equivalent to 3.7 g C) or 1 g chlorophyll *a*/m^2 (equivalent to 72 g C) was found in January 1973. A maximum rate of gross production of 140 mg C/m^2/h or 37 g C/m^2/day was obtained in December 1972. High oxygen concentrations *in situ* were found to accompany high production rates. Gross production rates were almost always above 1 g C/m^2/day although in June 1972 (mid-winter) they fell below this level.

Red water and marine fauna mortality

Red water is a common phenomenon round the coasts of southern Africa and has been shown to be caused by a variety of different organisms. These include various dinoflagellates (Grindley & Nel, 1968; Grindley & Sepeika, 1969; Grindley & Nel, 1970; Grindley & Heydorn, 1970; Grindley &

Taylor, 1970; Reinecke, 1967; Pieterse & van der Post, 1967). The ciliate *Cyclotrichium meunieri*, has also been recorded as a cause of red water.

In False Bay near Cape Town red water accompanied by a fauna mortality appeared on two occasions. It was caused primarily by a very high concentration (approx. 10000000 cells/l) of *Gonyaulax polygramma* associated with *Prorocentrum micans* in smaller numbers. The cause of the mortality could not be determined with certainty, but observations suggested that it was due primarily to the depletion of oxygen in the water by the decaying plankton, aggravated by the products of their putrefaction. It was shown that no toxins of the 'mussel poison' type were present but there remains the possibility that a toxin of some other type affecting the marine fauna was released. However, the marine fauna mortality only occurred when the plankton decayed as evidenced in three examples. On the basis of these and other plankton blooms some generalisations have been made about associated conditions and possible controlling factors (Grindley & Taylor, 1970).

Pieterse & van de Post (1967) described two recent fish mortalities at Walvis Bay. The first and largest was preceded by a particularly dense red water bloom with 26675000 dinoflagellate cells/l. This bloom was caused by *Peridinium triquetrum* associated with *Gymnodinium galatheanum* in smaller numbers. On the day preceding the mortality, oxygen concentration was reduced to 2.8 cc/litre. Pieterse & van der Post suggested that the mortality was caused by a combination of the reduced oxygen concentration and the mechanical clogging of gills by layers of dinoflagellate cells. It is not known whether *Gymnodinium galatheanum* produces a toxin affecting other marine fauna, but it is well known that a related species, *G. breve*, produces toxins which cause fish mortalities on the west coast of Florida.

During December 1966 red water caused by an unknown species of *Gonyaulax* appeared at Elands Bay on the west coast of South Africa (Grindley & Nel, 1968; Grindley & Nel, 1970). This bloom resulted in the death of hundreds of thousands of white mussels and black mussels and many other marine invertebrates. Toxicity tests on mussels from the affected area revealed that they were highly toxic. This suggested that more than one species might be involved as marine fauna mortalities and mussel poisoning are usually produced by different species. The species apparently responsible for the mortality was described as a new species, *Gonyaulax grindleyi*, by Reinecke (1967). It occurred in concentrations up to 6667500 cells/l. There was no evidence of a mass decay of plankton such as occurred in False Bay in 1962. The sea remained luminescent at night and the plankton was found to be alive. It was suggested that the mortality may have been caused by mechanical clogging of the gills of the affected organisms or by a toxin affecting marine invertebrates.

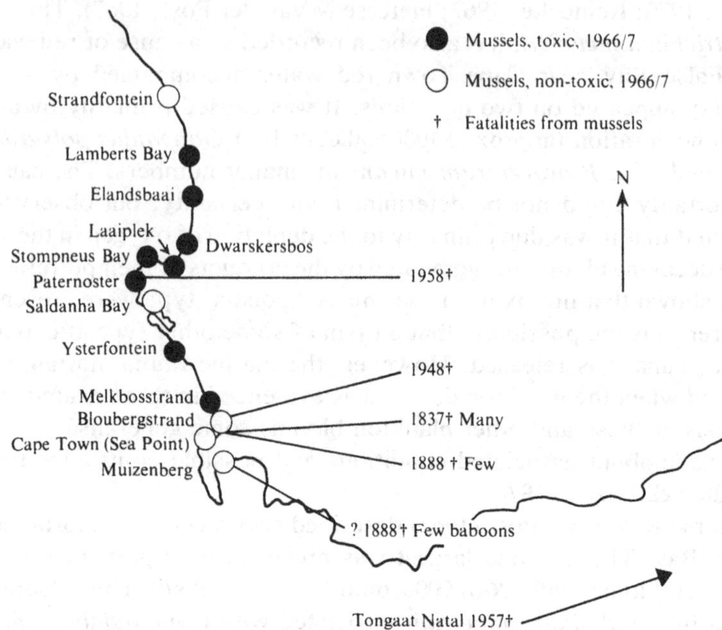

Fig. 5.3. Distribution of toxic mussels in summer and autumn 1966/7, and localities from which shellfish causing human fatalities were collected. Human fatalities from mussel poisoning in South Africa have been recorded in 1837, 1888, 1948, 1957 and 1958. In 1888 at the same time several baboons were found dead near Simonstown with clam shells in their paws. (From Grindley & Sapeika, 1969.)

Red water and mussel poisoning

In 1966–7 mussels were found to be toxic at a series of places on the west coast from north of Lambert's Bay to Melkbos only a few miles north of Cape Town (Fig. 5.3). It was not clear at that time what was the source of the toxin but the presence of a few *Gonyaulax catenella* offshore suggested that they might be responsible (Grindley & Nel, 1968).

The cause of mussel poisoning on the west coast of South Africa was finally found at Elands Bay in 1968 (Grindley & Sapeika, 1969). *G. catenella* comprised up to 40% of the phytoplankton cells in the red water with a maximum concentration of 1 140 000 cells/l. Extraction of mussel toxin directly from the phytoplankton confirmed that *G. catenella* was the source of the toxin. The toxicity of the species was found to be similar to that determined off California. The toxicity of the black mussels was found to be 42 000 mouse units/100 g (estimated lethal dose for man 36 500 mouse units). In 1967 mussels at Elands Bay were found to remain toxic for 4 months. As black mussels form an important part of the diet of the

Cape rock lobster which is widely eaten by man it was considered advisable to examine rock lobsters for toxicity. It was found that mussel poison was not taken up by the bodies of the lobsters although it appeared in the proventriculus presumably in the form of undigested portions of mussel. The tail of the lobsters, which is the part normally eaten by man, was quite unaffected.

It is remarkable that no evidence of toxicity in shellfish was found by Pieterse & van der Post (1967) in their investigation at Walvis Bay where *Gonyaulax tamerensis* occurred commonly. This species has been found to be toxic elsewhere and occurred in numbers up to 4000000/l at Walvis Bay.

Zooplankton sampling

The distribution of zooplankton is notoriously variable and patchy, so that differences in sampling methods require careful consideration. Plankton sampling in the seas around South Africa has been carried out very largely with Discovery pattern nets. The introduction of the Indian Ocean standard net necessitated a comparison between the two types (Zoutendyk, 1968). A significant difference in the estimation of standing crop by the two nets was found, and a conversion factor of 0.92 for N70 discovery net samples or 1.08 for Indian Ocean standard net samples is required to convert data for comparison. A modification of the Indian Ocean standard net incorporating a depth-flowmeter was found to be valuable particularly in areas of strong current.

A plankton pump for fine scale sampling of zooplankton was described by Hutchings, Robertson & Allen (1970). This allows sampling at discrete depths down to 50 m.

Williams (1969) described an apparatus for sampling pelagic marine bacteria employing a vacuum chamber and membrane filter operated by messenger.

Robertson (1970*a*) described an improved apparatus for determining displacement volumes of plankton based on the apparatus of Yashnov. This instrument works well except with samples containing large numbers of tunicates and coelenterates which hold water.

Seasonal variation of zooplankton

The basic seasonal cycle of plankton off the west coast of South Africa was established before the advent of the IBP. Plankton studies were concentrated in and around the St Helena Bay area in the earlier investigations. Variations of plankton volume and the succession of zooplankton had been found to be somewhat irregular. Greatest volumes of zooplankton occurred inshore in the summer. The copepods *Centropages* and *Calanoides* were found to be most abundant in late spring and summer while *Oithona* and *Metridia* were more abundant in winter than any other season.

Zooplankton: temperature distribution

In studies of zooplankton distribution interesting relationships have been found between zooplankton concentrations and water temperature. Sampling at discrete depths with a newly constructed plankton pump (Hutchings, 1973) revealed that this maximum was near the bottom, close inshore, during active upwelling, instead of at the usual depth of 20–40 m. The great importance of thermocline topography has been confirmed by pump sampling. A very marked increase of zooplankton density is always found along the thermocline and the abundance of certain species seems to depend on the presence of a thermocline. Monthly monitoring cruises provided evidence of a constant zooplankton maximum along the oceanic front and of an occasional maximum within 11 km of the west coast.

The copepoda typical of the regions of cold upwelling include *Calanoides carinatus*, *Metridia lucens* and *Centropages brachiatus*. In the warmer waters offshore there is a great diversity of species including *Eucalanus* spp., *Nannocalanus minor* and *Corycaeus* spp. (De Decker, 1973*a*). At depth of 50–200 m below the offshore waters a copepod community strongly resembling that of the surface waters nearer the coast occurs. Species such as *Metridia lucens* and *Calanoides carinatus* reappear here. The distribution of the young stages of certain important copepod species has been charted for each season. The distribution of juveniles has been found to be very localised with the youngest stages concentrated in certain small areas.

The distribution of certain species of Chaetognatha is clearly correlated with the hydrology of the areas in which they occur. *Sagitta friderici* is a neritic species typical of the cool upwelled inshore waters, while *Sagitta minima* occurs in largest numbers in the warmer waters offshore. Computations of coefficients of correlation show that the association of these two species with the two different water masses is highly significant. Several other species of Chaetognatha have proved to be good biological indicators of water movements (Stone, 1966, 1969; Masson, 1972).

Euphausia lucens and *Nyctiphanes capensis* are the dominant euphausiid species near the surface in the regions of upwelling, while in the same regions in even colder water at greater depths *Euphausia similis* var. *armata* is the dominant species (Grindley & Penrith, 1967).

Off the west coast salps and doliolids are most abundant in the warmer offshore waters beyond the regions of upwelling. Van Zyl (1960) found a direct relationship between the abundance of salps and two species of doliolids and the mean integral temperature of the sea. *Thalia democratica* is the commonest of the salps found in this area, but seven other species were recorded. The presence of *Salpa fusiformis* and *Ihlea magellanica* during the spring of 1955 suggested an influx of oceanic water from the South Atlantic to the area off the south-west Cape. The presence of *Thalia longicauda*, *Pegea confoederata* and *Cyclosalpa pinnata* in the summer of 1955–6 suggested that Agulhas Current water penetrated into this area at that time.

Agulhas Current zooplankton

De Decker (1973*a*) has stressed that the copepod communities of the south-west Indian ocean are very different from those of either inshore or offshore water of the south-east Atlantic. The species dominant on the west coast are rare in this region. The shallow Agulhas Bank area has a distinctive plankton community including copepods such as *Pseudodiaptomus nudus* and *Calanus finmarchicus*. The Copepoda of the Agulhas Current are a sharply characterised group distinguished by a high percentage of Indo-Pacific species. De Decker (1973*b*) found 274 species of Copepoda in the south-west Indian ocean. Maxima of species diversity occur in the surface waters in the north of the area and throughout the area at depths between 750 and 1500 m. The surface waters of the southern area have a relatively low species diversity and low plankton volumes.

Zooplankton biomasses in the Agulhas Current area of the south-west Indian Ocean are usually low but high biomasses may occur inshore of the main current. Studies by Zoutendyk (1970) revealed a sharp distinction between inshore stations (shallower than 200 m) and deep stations (Fig. 5.4). Much geater zooplankton biomasses occur in the inshore areas of local upwelling at stations over the continental shelf than offshore in the south-west Indian Ocean except of a few isolated stations in the far south and east of the area. It is remarkable that the biomass of zooplankton at both deep and shallow stations is similar to that from comparable areas in the west coast in the same seasons. The available data for the west coast do not appear to be exactly comparable, but nevertheless it seems that the zooplankton inshore in the south-west Indian Ocean is much richer than was previously believed. The inshore waters are particularly rich in areas of local upwelling off Durban and Port Elizabeth.

Stone (1966, 1969) examined the distribution of Chaetognatha in the Agulhas Current region using the grouping technique of Fager & McGowan. Groups of species of Chaetognatha derived on the basis of

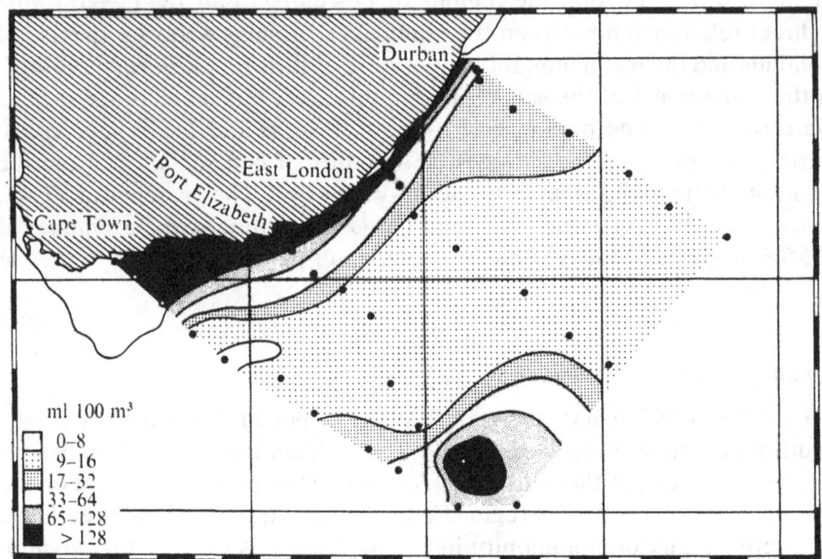

Fig. 5.4. Zooplankton standing crop in the south-west Indian Ocean in January 1963. Abundance as settled volume of plankton in millilitres per hundred cubic metres of sea water. The solid circles indicate the positions of sampling stations. (From Zoutendyk, 1970.)

co-occurrence were separated and found to correspond to the hydrologically distinct inshore and offshore water masses. Samples from the Agulhas Current off the continental shelf were very poor by comparison and an analysis of the egg production of the common chaetognath *Sagitta enflata* showed that the fecundity in neritic waters was significantly higher than in oceanic waters.

Similar distribution patterns have been found with various other groups. Talbot (1974) in an analysis of the species composition of euphausiids showed that the fauna of the neritic waters differed significantly from that of the Agulhas Current. Two euphausiids in particular, namely *Nyctiphanes capensis* and *Euphausia lucens*, were found to be good indicators of inshore waters.

Distribution of planktonic larvae

Coastal currents, and particularly eddies and return currents, play an important role in the distribution of planktonic larvae.

A study of the phyllosoma larvae of the rock lobster *Jasus lalandii* was undertaken by Lazarus (1967). The early stages are most abundant in October after the peak of breeding, and occur mainly near the coast where the adults are found. The later phyllosoma stages spread more widely and there are interesting seasonal variations in their distribution. As the larvae lead a planktonic existence in the Benguela Current for nine months, it is remarkable that a sufficient number are able to return to maintain the adult population. However the distribution of the later stages indicates that considerable return drift of later stages must occur in eddies or counter-currents. It seems probable that vertical migration plays a role in their return drift although there is not yet evidence of this.

Distributional history and phylogeny

Studies of the distribution of various plankton groups are usually made in relation to hydrological conditions. However, in zoogeography it is desirable to study not merely the environmental causes of regional distribution, but also the distributional history and phylogeny of the group which the present pattern of distribution reflects. Historical as well as ecological causes must be investigated. This has been done with the Pseudodiaptomidae (Copepoda) and a preliminary summary of the findings (Grindley, 1969) has already been published. The morphological and distributional relationships of the various groups of Pseudodiaptomidae indicate that they arose in the Ponto-Caspian region in the late Tertiary and that they escaped to the coasts of the Indian Ocean at some time between the middle Pliocene and lower Pleistocene. The separation of the modern genera, subgenera and species and their dispersion to the coasts of America, Africa, Asia and Australia apparently took place within the Quaternary.

Natural radioactive nuclides

While natural radioactive nuclides do not affect the distribution or productivity of plankton, these nuclides may find useful applications in the tracing of physical, chemical and biological processes in the sea. Papers by Shannon & Cherry (1967), Shannon (1969) and Cherry & Shannon (1974) have shown that alpha-activity due to natural nuclides accounts for most of the internal radiation dose in marine organisms. Real variations occur in the concentrations of the major alpha-emitters in marine plankton. These may to some extent be explained in terms of different currents and water masses. Biological removal is important in the removal of certain radio-nuclides from the near-surface layer of the sea. Polonium-210 is the major contributor to the alpha-radioactivity of most marine organisms including zooplankton. It is only in phytoplankton that polonium-210 loses its position as the main contributor in favour of radium-226 but even here the polonium-210 contribution is substantial. It would appear that there are many interesting problems related to the biochemical balance of the natural radioactive nuclides occurring in plankton.

Benthic fauna

Exploitation of rock lobsters

Exploitation of the west coast rock lobster *Jasus lalandii* was one of the first sections of the South African fishing industry to be established. *Jasus lalandii* occurs commonly along the rocky shores of the west coast, in depths ranging from 1–150 m but in most areas their availability has decreased in recent years. It would seem that over-exploitation is a major factor limiting the productivity of this resource today. Catches remained at about 3200 tonnes/yr from 1942 for more than 20 years due to rigidly enforced annual export quotas. Despite this there were marked decreases in availability. Analysis of available catch statistics showed that reductions in availability, associated with high fishing intensity and a decline in stocks, occurred in various areas on the west coast from 1952. A decrease in the mean size of rock lobsters caught occurred from 1958 along the whole west coast. Assessment of rock lobster stocks has revealed that the rate of exploitation on some grounds was very high. During the 1968–9 fishing season tag returns were up to 28%, indicating a very high fishing mortality in such a long-lived organism.

Studies also indicated that the minimum size limit was clearly not beneficial to the stocks. Sexual maturity is reached by males of *Jasus lalandii* at a carapace length of 6.5 cm and is reached by females at a carapace length of 7.0 cm. A highly significant increase in reproductive potential occurs between lobsters of 7.6 and 8.9 cm carapace length.

The availability of *Jasus lalandii* has improved recently (1974) as a result of stringent conservation measures applied over the past few years.

A closely related rock lobster *Jasus tristani* occurs on the Vema Seamount off the west coast of South Africa and on Tristan da Cunha and Gough Islands in the South Atlantic. Underwater observations at all three localities were made by Heydorn (1969*b*). He found that at the Vema Seamount pronounced changes in the size and sex composition of the population had occurred between 1964 when the seamount was first visited and 1966 when the seamount was again visited (Grindley, 1967). It was clear that the rapid depletion of the population could be attributed to the effects of an excessively high level of commercial exploitation.

Environmental factors affecting rock lobsters

Studies of moulting and growth of rock lobsters have revealed that moulting is fairly regular but that growth is very variable (Heydorn, 1969*c*). Differences in growth rate appear to be related to location. Comparative studies in selected areas, where mean growth rates were highest and lowest, revealed differences in the biomass of other benthic fauna. In the areas studied benthic biomass per unit area showed a decline with increasing depth and highest rock lobster growth rates occurred in shallow water. Comparisons within the same depth range also revealed that benthic biomass values were lower in areas where smaller growth increments were recorded. Between depths of 25–40 m the mussel *Aulacomya magellanica* constituted more than 90% of the standing crop. In areas of high growth increments, 7.8 kg/m^2 of mussels was recorded, whereas only 2.9 kg/m^2 was recorded for the area where growth was slow (Newman & Pollock, 1973, 1974). Feeding studies have revealed that *A. magellanica* is the major item in the rock lobster's diet. As mussels are long-lived it is likely that standing crop is closely related to productivity and that differences in rock lobster growth rates are related to differences in the availability of their benthic food organisms. This in turn is apparently related to proximity to the coastal kelp beds where high primary production takes place.

Growth rates can also be affected by temperature and there is an inverse relationship between temperature and depth. The slight decrease in temperature with depth may thus also affect growth rates in deeper water. The growth rates of juveniles have been studied by observing changes in size composition in the field, as well as by measuring the growth of animals maintained in aquaria. Good agreement was obtained between the two sets of data. Small-size rock lobsters are not found on the commercial grounds and their habitat has not yet been located, so that their biology can not yet be documented.

The populations of rock lobster within sanctuaries at Cape Town and Saldanha Bay have been assessed for comparison with exploited areas. A mass stranding of rock lobsters associated with abnormal environmental conditions at Elands Bay has been described (Newman & Pollock, 1971).

Factors affecting the reproduction of Natal lobster

The genera *Palinurus*, *Panulirus*, *Linuparus*, *Puerulus*, *Palinustus* and *Projasus* all occur in the waters off Natal. The genera *Palinurus* and *Panulirus* are of commercial and recreational importance. The life history and reproduction of *Panulirus homarus* have been studied by Berry (1970, 1971*a*). Some breeding activity was found in the summer months from November to February. *P. homarus* was found to breed repetitively through the year and females produce more broods as they increase in size. The estimated number of broods produced per year ranged from one to four. Sexual maturity is attained at the beginning of the third year after settlement but full reproductive potential (multiple broods) is only attained in the fourth year. The number of eggs carried increases in direct proportion to size and ranges from 100000 to 550000 per brood. An analysis of population structure allowed an estimate of relative fecundity to be made, taking into account the number of eggs and broods produced by each length class. It was found that the largest females (70–79 mm carapace length class) made the greatest contribution despite their relatively low numbers.

The time taken for incubation of eggs was found to be dependent on water temperature and ranged from 29 to 53 days. The eggs hatch into phyllosoma larvae of which the earliest stages occur mainly inshore of the Agulhas Current. Most of the later stages are found in the Agulhas Current, but concentrated towards the western edge of the current. The larval life apparently lasts for three to four months during which time they probably survive by making use of the eddies which circulate on the western edge of the Agulhas Current. Then metamorphosis takes place into the puerulus stage which swims actively towards the shore, where they appear between February and June.

Penaeid prawns and environments for production

Penaeid prawns represent a valuable resource in the warm coastal waters of Natal and Mozambique. Littoral prawns spend the major part of their life cycle in estuaries and are caught both in estuaries and, at the time of spawning, on trawling grounds in the open sea. The knife prawn, *Hymenopenaeus triarthrus*, on the other hand, spends its whole life cycle in the open sea at depths ranging from 320 to 540 m and is captured only by means of trawling. Aspects of the biology of *Penaeus indicus* and *Metapenaeus monoceros* and associated Penaeidae occurring off Natal on the east coast of South Africa have been investigated. *Trachypenaeus curvirostris* and *Penaeus marginatus* have been found in South African waters. Research on the deep-sea knife prawn, *H. triarthrus*, has cast much light on the distribution, size composition and biology of the heavily exploited stocks of this species in waters off Mozambique and Natal in depths of 400–440 m. These catches consist almost entirely of sub-adults. Apart from the obvious undesirable implications from stock management point of view, this situation limits the value of the biological data obtained.

Comprehensive studies on the biology of penaeid prawns with emphasis on *Penaeus indicus* have been carried out (Champion, 1970). The estuarine phase of the life cycle of these prawns extends from early juvenile stages to sexual maturity, at which stage a migration out to sea for spawning purposes takes place. Laboratory experiments have indicated that growth rates and survival are greatly affected by water temperature. Development was most rapid at 30 °C with only eight days elapsing between spawning and metamorphosis to post-larvae, but survival was greatest at 28 °C.

Population density and fecundity in intertidal Mollusca

Studies of intertidal Mollusca and particularly of *Patella* spp. have been carried out by Branch (1974*a*, *b*). On the lower shore normal population densities of *Patella cochlear* range from 100 to over 2000/m^2. At densities below 400/m^2 biomass increases in proportion to density. Above this, a critical level of about 125 g/m^2 is reached, and further increase in density does not increase the standing crop. This is because increase in population density is associated with a decrease in size. What is more important is the effect of population density on gonad output which rises to a maximum at a density of 400/m^2 and then progressively decreases. Growth rates and survival rates were also found to be lower in high-density populations. All of these factors have been experimentally measured and can be explained in terms of intraspecific competition. There is thus a natural density-dependent population control mechanism.

Factors affecting South African oysters

Studies of the reproduction of *Crassostrea margaritacea* (non-incubatory) and *Ostrea algoensis* and *O. atherstonei* (incubatory) have been carried out by Genade (1973). Spawning of *C. margaritacea* occurs after a monthly mean water temperature of 20 °C has been reached. Fertilisation of ova (50 μm diameter) occurs in the water. Veliger larvae (length 76 μm) develop within 36 hours, and mature larvae with eyes (length 280 μm) may develop within three weeks under optimal conditions. Larvae of *O. algoensis* are liberated at an average length of 147 μm while *O. atherstonei* larvae swarm at a length of 172 μm. Larvae of the latter two species mature and metamorphose within approximately two weeks of liberation. Shell growth in *C. margaritacea* occurs only above 16 °C and in the Knysna estuary marketable oysters are obtained after two growth seasons. Natural mortality during post-larval development was found to be approximately 40%. Spat settlement in South African estuaries is too irregular to provide a basis for the oyster cultivation industry, so hatchery techniques of spat production for the indigenous oyster species are being developed.

Factors affecting the distribution of shelf fauna

With the aim of determining the principles which underly distribution patterns in the sea, the species composition of the benthic fauna of the continental shelf of South Africa has been studied intensively. Hundreds of new species, belonging to several invertebrate phyla, have been described, and illustrated keys to the fauna and flora have been published (Day, 1969*a*). Particular attention has been devoted to the ecology of False

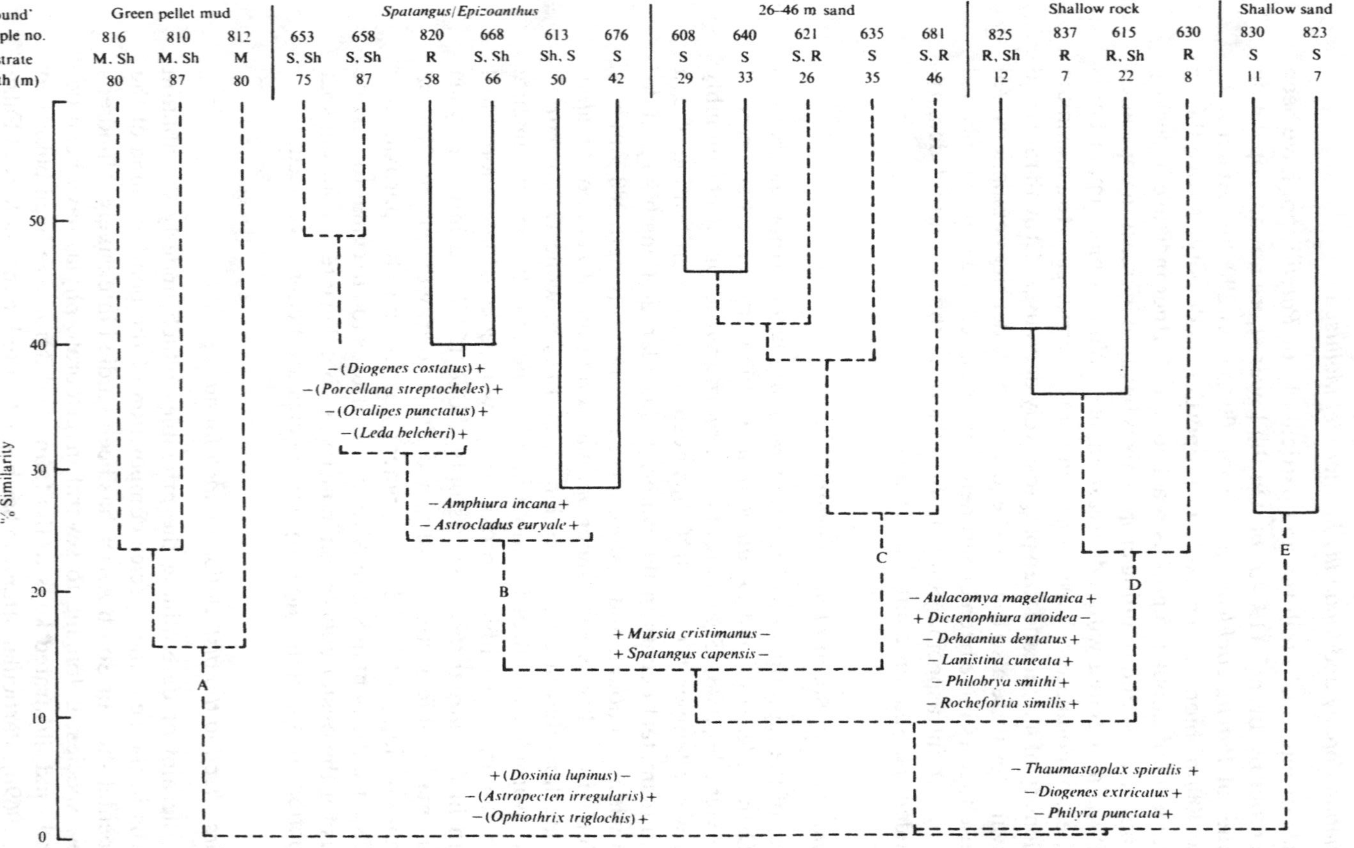

Fig. 5.5. Dendrogram superimposing the results of information statistic tests of significance on Czekanowski coefficient results. The scale shows percentage similarity. Unbroken lines indicate that no significant difference was found at the 1% level. Indicator species significant at the 1% level are given; the + and – on either side of each name indicate which group of samples the species is characteristic of or absent from, respectively. Some of the species significant at the 5% level are given in parentheses. Substrates given are mud (M), rock

Bay which receives contributions of both the Agulhas and Benguela Currents. The benthic fauna and fishes of False Bay have been studied and an analysis shows that over a third of the fauna is endemic. In addition there are many Indo-Pacific and Atlantic species but there is no distinct sub-Antarctic element. The warm-water species are mainly concentrated along the shallow northern shores and cold-water species are more common at the southern entrance near Cape Point or Cape Hangklip. Many of the fishes are seasonal migrants: warm-water species congregate along the northern shores in summer and cold-water migrants such as snoek visit the southern entrance in winter (Day, 1970; Day, Field & Penrith, 1970).

Numerical methods for the analysis of distribution records have been developed and tested, and it has been found that the Czekanowski or better the Bray–Curtis coefficient used in conjunction with the information statistic provides reliable patterns of distribution across the continental shelf (Field, 1969, 1970; Field & MacFarlane, 1968; Field & Robb, 1970) (Fig. 5.5).

Adult distribution is often determined by factors which affect the settlement of planktonic larvae. With this in mind, studies of larval development and metamorphosis have been undertaken. The breeding seasons of many invertebrates have been determined, artificial fertilisations have been made as well as experiments on the conditions necessary for successful settlement of some species (Cram, 1971*a*, *b*).

Studies of the stocks of the South African abalone, *Haliotis midae*, and the factors affecting its productivity have been carried out by Newman (1968*a*, *b*). The white mussel, *Donax serra*, has been studied by De Villiers (1973, 1974), while a study of the commercial potential of the alikreukel, *Turbo sarmaticus*, has been carried out.

An intensive study has been carried out on the biology of the langoustine *Nephrops andamanicus*, which is exploited commercially by means of trawling in the waters of Natal and Mozambique. A study of the distribution and affinities of the species within the genus *Nephrops* was included. Observations on the catch composition indicated that a non-migratory population was sampled. The biology of *N. andamanicus* was described in some detail, including the zoea larvae (Berry, 1969*a*).

Fish

Fish stocks

The assessment of fish stocks is basic to any determination of the productivity of a fishery or the understanding of factors affecting it. During the period of the IBP, fishery scientists devoted a large part of their effort towards the assessment of the stocks of several commercially important species of fish on the west coast of South Africa and South West Africa.

The most concentrated effort was aimed at assessing the stocks of the South West African pilchard, *Sardinops ocellata*. Data on catch composition were analysed routinely by the Sea Fisheries Branch, and age determination data, originally processed by scientists in Walvis Bay, were checked, and mortality estimates were made from these data. The work culminated in the publication of several reports on the state of stocks in this area (Newman, 1970*a*; Cram & Visser, 1972, 1973; Cram, 1974). The assessment of stocks in the western Cape was reviewed, updating the earlier analysis (Baird, 1973).

During 1971 an assessment of hake stocks in the south-east Atlantic was completed. This indicated that the hake were maximally exploited, and that the present yield of about 600000 tonnes could only be increased if a large mesh size was adopted (Botha & Mombeck, 1971). Factors for converting landed weight to whole weight for various species have been determined.

Various new methods have been used in the recent work on stock assessment. A calibrated high-frequency echo-sounder has been used to measure the density of pelagic fish shoals (Hampton, 1973). Various aircraft-borne remote sensors have also been employed to good effect (Cram, 1972). These have included airborne radiation thermometers (Cram, 1973) and low light level television cameras (Cram & Agenbag, 1974).

Commercial exploitation of fish stocks

Commercial exploitation is today perhaps the most important factor affecting the productivity of South African fisheries. During recent years the important stocks of pelagic fish off the west coast of southern Africa have declined to a disturbing degree (Fig. 5.6). In addition international fleets are exploiting the hake stocks off the Atlantic coast intensively. This led to the creation in 1972 of an International Commission for the regulation of fish stocks in the south-east Atlantic (ICSEAF).

The pelagic fishery was originally dependent primarily on the pilchard *Sardinops ocellata* and to a lesser degree on the maasbanker *Trachurus trachurus*. The contribution of the latter species has declined to an almost negligible level in recent years. Geldenhuys (1973) has shown that the good maasbanker catches in the 1950s were due to two exceptionally strong year classes which could stand the level of exploitation. The later decline in maasbanker catches can be attributed to poor recruitment in comparison with earlier years and continued pressure of exploitation.

The pilchard fishery in the 1950s was initially not too intensive, fluctuating between 66000 and 154000 tonnes. But during the next few years the yield was increased tremendously and a record catch of 410721 tonnes was landed from the west coast of South Africa in 1962. Thereafter

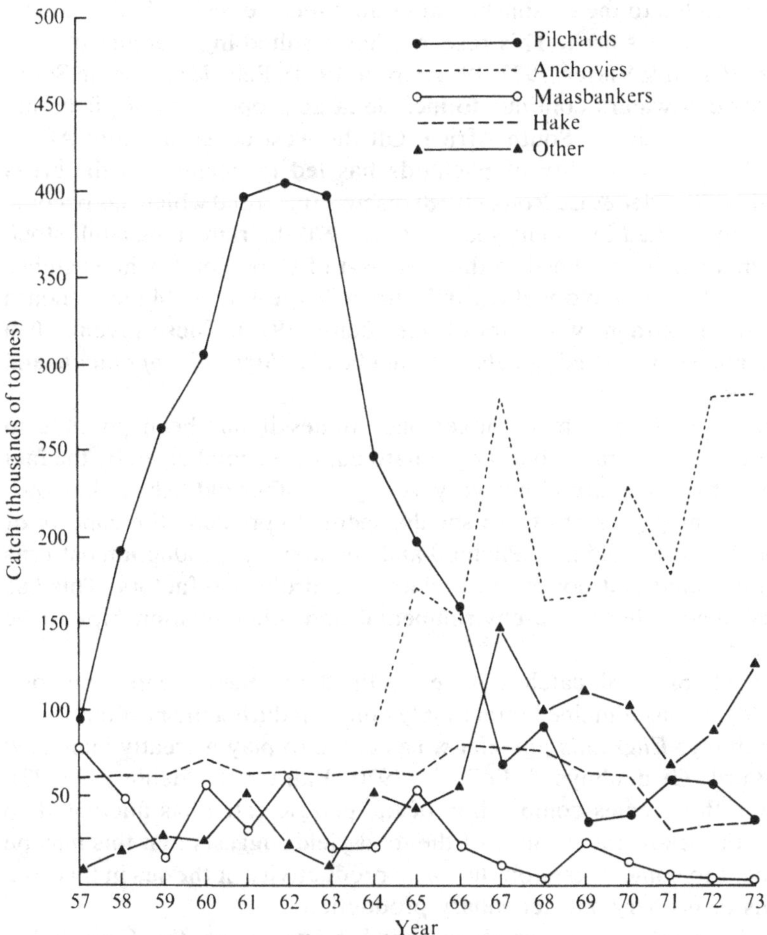

Fig. 5.6. Graph of catches of various species of fish for South Africa from 1957 to 1973. The relatively constant catches of Cape hake are contrasted here with the changing pilchard catches revealing the rise and fall of the pilchard fishery. The introduction of anchovy fishing in 1964 has served to keep up the total catches and they appear to have taken the place of the pilchards. (Revised from Grindley, 1969.)

catches declined rapidly. South African pilchard catches in more recent years have fluctuated between 94000 and 41000 tonnes. Changes in the length and age composition of the catches have been investigated and an increasing abundance of smaller individuals have appeared in catches in recent years. Since the smaller-meshed anchovy net was introduced into the fishery in 1966, length and age composition of the pilchard catches has shown a distinct bimodality (Baird, 1973).

Concern regarding the state of the pilchard fishery off the coast of South

West Africa led to the establishment of an intensive research programme in the Cape Cross area. This research has resulted in a number of publications (Cram & Visser, 1972, 1973; Cram, 1974). Pelagic catches in South West African waters continue to include large proportions of pilchards, unlike those made off South Africa. Off the west coast of South Africa the decline in availability of pilchards has led to changes in the areas for fishing. Stander & Le Roux (1968) discuss this trend which has become even more marked in recent years. Since 1970 the remaining adult stock has been virtually confined to the area east of Cape Point, while catches on the west coast have yielded only juveniles less than 14 cm standard length and less than two years of age (Baird, 1973). These juvenile fish usually appear in mixed shoals with anchovy, *Engraulis japonicus*, and other species.

From anchovy and maasbanker age studies it has been possible to calculate mortality rates, but the pilchard data are complicated by the fact that recent catches are either very young or rather old fish. Behaviour, either of fishing gear or of fish shoals, seems to preclude the capture of two- or three-year-old fish. Pilchard and anchovy egg production patterns are complicated and not easy to relate to hydrological factors. This has required many alternative environmental and other relationships to be examined.

Although pilchard catches have declined in recent years, the total pelagic catch has remained remarkably constant during the past ten years. The anchovy, *Engraulis japonicus*, has come to play a greatly increased role (Baird, Geldenhuys & Pollock, 1970; Baird & Geldenhuys, 1973). Although the species composition of the pelagic catch has fluctuated so greatly, the relative constancy of the total yield suggests that this may be dependent to some degree on the basic productivity of the sea in this area in terms of primary and secondary production.

The demersal fishery which is based primarily on the Cape hake, *Merluccius capensis*, has increased greatly in recent years as a result of a tremendously increased fishing effort by an international fleet. During the past few years there has been a steady decline in the catch per unit of effort (Botha & Mombeck, 1971). There has also been a trend towards a higher percentage of smaller and younger fish in catches. The trawling effort by foreign vessels off the coasts of South and South West Africa has expanded to the point where the local share of the catch has been falling while the total catch has increased. The local catch is estimated to be only 11% of the total. Not only are the foreign vessels larger, but the local vessels are restricted by law to the use of nets of a specified minimum mesh size. The disregarding of such conservation measures by foreign vessels is aggravating the over-exploitation of this resource and endangering its future.

Environmental factors affecting recruitment of pelagic fish

Hydrological conditions influence the recruitment, availability, and migration patterns of pelagic fish shoals to a marked extent. Low oxygen content in the coastal waters of South West Africa associated with the southward movement of warm oceanic water offshore tends to confine the fish to the area near Walvis Bay. Commercial catches are then highest in this area. When these anomalous hydrological conditions disappear, the northward and southward migration of fish shoals recovers.

While the unusually warm water is present, the pilchards, *Sardinops ocellata*, appear to be prevented from moving into their traditional spawning grounds. Both the pilchard and the anchovy, *Engraulis capensis*, prefer the cool, upwelled coastal water and appear to avoid water that is warmer than 16 °C or which has a low oxygen content. An influx of warm water will drive the fish shoals southwards and vice versa. The movement of the fish shoals is also influenced by the weak southward flowing counter-current which flows intermittently along the coast. The fish appear to utilise this current when migrating southwards to near Walvis Bay in the spawning season from August to October (Visser, Kruger & Coetzee, 1973).

Observations of movements and spawning of fish in conjunction with airborne radiation thermometer surveys provided reasonable correlations between breeding and migrations of the pilchards and sea temperature data. The aerial evaluation of quantities of pelagic fish including pilchards, *S. ocellata*, and maasbankers, *T. trachurus*, yielded results in good agreement with conventional methods and this revolutionary method of stock assessment appears valuable.

Length compositions of commercial catches of the pilchard, *S. ocellata*, indicate that recruitment occurs between January and March. In the period before 1963 when anomalous conditions appeared, the appearance of juvenile pilchards in catches was associated with strong northerly winds or calm periods, which lead to southerly counter-currents. Strong southerly winds and the upwelling of cool water lead to an exodus of juvenile pilchards. The temporary occurrence of juvenile pilchards during periods of temporary surface counter-currents suggests that the nursery grounds are found in the north and that these young fish are displaced by the southward-moving water. The northerly distribution of the main nursery stock is ascribed to a northerly drift of planktonic eggs and larvae (Schulein, 1973).

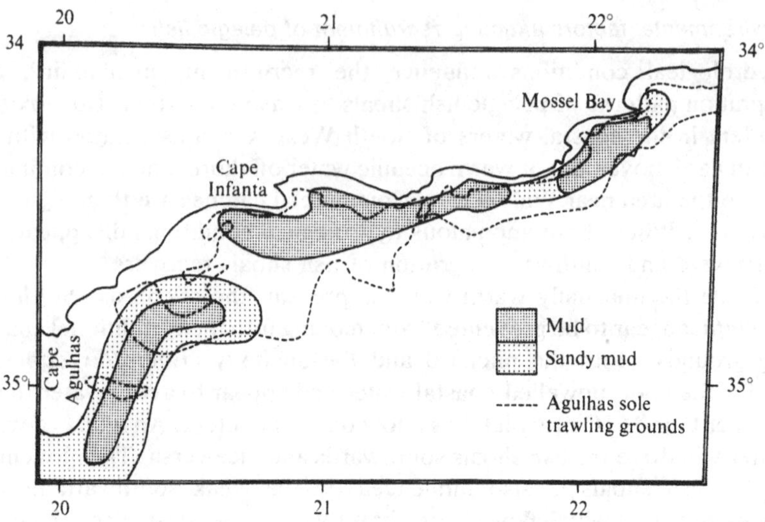

Fig. 5.7. The distribution of mud and sandy mud sediments over the Agulhas bank between Cape Agulhas and Mossel Bay and the Agulhas sole trawling grounds. (From Zoutendyk, 1973.)

Association of Agulhas sole with nepheloid layer

The Agulhas sole, *Austroglossus pectoralis*, which is endemic to the south coast of South Africa occurs associated with a bottom layer of turbid water (Fig. 5.7). This nepheloid layer was shown to be widely distributed over muddy areas of the Agulhas bank by SCUBA diving and the use of a transmissometer. The turbid bottom layer, 1–10 m thick, allowed near-zero light transmission while transmission in surface waters ranged from 50 to 70% (Zoutendyk, 1973). The areas of mud are fluvial in origin and probably represent relict estuarine sediments formed during a Pleistocene regression. *A. pectoralis* is apparently not found beyond these mud belts in significant numbers and is limited in distribution by them. The sole trawling grounds range from 36 to 120 m in depth. Studies by Zoutendyk (1973*a*, *b*) of zooplankton standing crops indicated that secondary productivity over the Agulhas bank may be greater than in Agulhas Current or Benguela Current waters.

There is evidence of migration from one ground to another and young soles predominate on certain grounds. The soles feed largely on polychaete worms; they breed in winter and reach maturity at an age of three to four years (Zoutendyk, 1974, 1975).

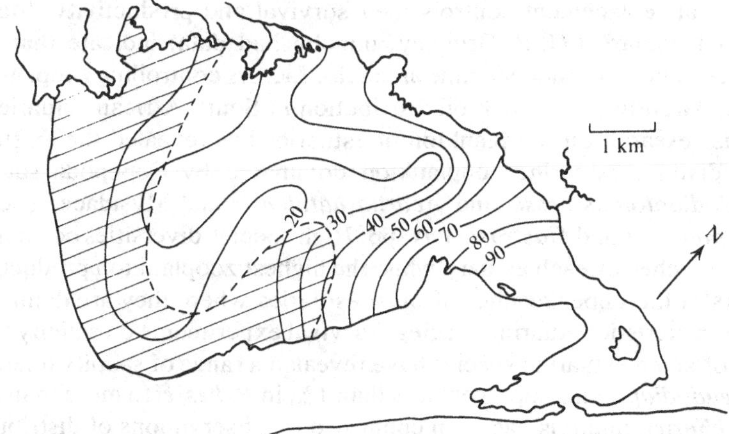

Fig. 5.8. Distribution of juveniles and ovigerous females of the copepod *Pseudodiaptomus charteri* in Richards Bay in January 1970. The percentage of juveniles in the population is indicated by contours which show the increasing percentage of juveniles as the limit of distribution of the species in the channel is approached. The area where ovigerous females make up more than 30% of the population is hatched. (From Grindley & Woolridge, 1974.)

Estuaries

Plankton

A study of the plankton of South African estuaries has been carried out (Grindley, 1970*a*). This study includes a survey of the distribution of plankton in 75 estuaries round the coast of southern Africa from Angola to Mozambique, and an investigation of the role of plankton in the productivity of estuaries.

Some interesting findings have been made regarding factors controlling the distribution of plankton in estuaries. In a study of the plankton of Richard's Bay (Grindley & Wooldrige, 1974) it was found that the patterns of distribution of zooplankton found in this study bore very little relation to the distribution of the physical and chemical variables conventionally investigated, such as salinity, temperature, dissolved oxygen and various nutrients. Tidal exchange appeared to be the single most important factor controlling the distribution of zooplankton in Richard's Bay. There is a clear correlation of plankton distribution with distance from the mouth. The estuarine zooplankton only survives beyond a point where the rate of tidal replacement is not too great. This boundary and areas of concentration appear to persist irrespective of the salinity regime prevailing (Fig. 5.8).

Earlier studies of vertical migration (Grindley, 1964, 1972) had already suggested that the problem of maintaining position within an estuary is important for the survival and productivity of estuarine plankton. In the case of *Pseudodiaptomus*, variations in the population structure indicate how tidal replacement controls their survival and productivity. Studies in other estuaries (J. R. Grindley, unpublished data) indicate that tidal replacement and residence time are major factors controlling zooplankton and apparently phytoplankton distribution in South African estuaries.

This research on the plankton of estuaries has revealed the existence of a distinct estuarine zooplankton dominated by Copepoda such as *Pseudodiaptomus hessei* and *Acartia natalensis* and Mysidacea such as *Mesopodopsis* and *Gastrosaccus* spp. High species diversities occur in the lower reaches of each estuary while the highest zooplankton productivity occurs in the upper reaches of most estuaries where they are dominated by characteristic estuarine species. Survival experiments on salinity tolerance of some estuarine species have revealed a range of salinity tolerance in *Pseudodiaptomus* spp. from less than 1‰ in *P. hessei* to more than 60‰ in *P. charteri* and this has been confirmed by observations of distribution in the field. Studies of vertical migration have revealed great differences in behaviour between species. Some plankton species including *P. hessei* have been found to go down to or even into the bottom (Grindley, 1972).

Regular monthly sampling to determine seasonal and other fluctuations in the plankton has been carried out in several South African estuaries. Marked seasonal fluctuations with a spring outburst of microplankton and an early summer peak of zooplankton have been found in several estuaries. The effects of floods and periods of hypersalinity have been observed in some estuaries. A series of samples from St Lucia was of particular interest in relation to the large-scale salinity fluctuations occurring there. A period of extreme hypersalinity led to a chain of ecological disturbances (Grindley & Heydorn, 1970).

General ecology

The Zoology Department of the University of Cape Town revisited the St Lucia estuary system to determine changes that had occurred since the original survey in 1948–51 (Millard & Broekhuysen, 1970). They found that the richness or poverty of this large estuary system varies greatly with variations in the inflow of fresh water. When the rivers flow strongly, the salinity in the upper lakes is low; aquatic vegetation flourishes and birds, fish and prawns are abundant throughout the system. During droughts the salinity increases in the upper lakes, the aquatic vegetation dies and the fish and penaeid prawns are driven down towards the mouth. The shortage of fresh water is aggravated by irrigation dams in the rivers flowing into the system. During droughts when the salinity rises to 70‰ mullet continue to exist in the northern lakes.

A study of the Morrumbene estuary in Mozambique (Day, 1974) re-

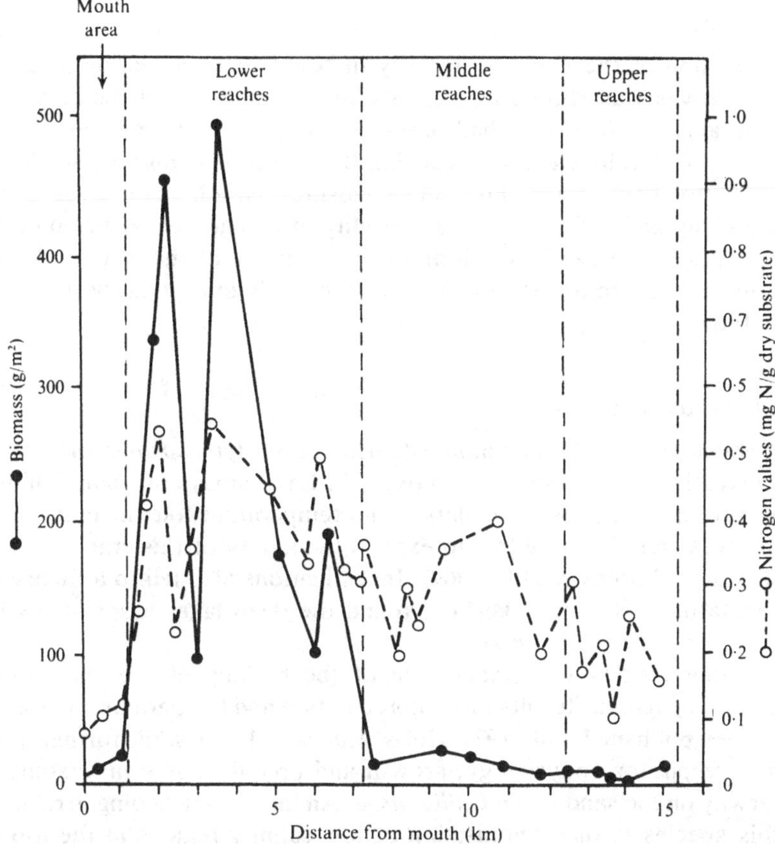

Fig. 5.9. Total biomass values recorded at stations along the Swartkops estuary. Included for comparison is a graph of nitrogen values recorded in the surface of the substrate along the estuary. Dashed vertical lines indicate boundaries between the different reaches of the estuary. (From McLachlan & Grindley, 1974.)

vealed that although the biomasses present were not very high there was a remarkably high diversity of fauna. In the estuary, 385 species of invertebrates and 120 species of fishes were recorded. In the dense mangrove swamps of the system, shade and high humidity allow species such as *Upogebia africana*, which are normally restricted to low tide levels, to extend high up the shore. The mangrove swamps were found to have 22.5% of the estuary's fauna. In contrast, the Swartkops estuary near Port Elizabeth has a much lower diversity, but biomasses of up to 500 g/m² of benthic fauna have been recorded (McLachlan & Grindley, 1974) (Fig. 5.9).

Mangroves

Investigation of the Umlalazi estuary in Natal showed that an invasion of mangroves was related to a change in the condition of the mouth of the estuary. The Umlalazi had previously been closed off from the sea for several months each year causing the death of mangrove seedlings through flooding. When this annual closure ceased, mangroves could successfully grow. When a mass mortality of mangroves occurred in the Kosi estuary, the cause was found to be flooding of the estuary due to an unusual closure of the mouth of this normally open estuary (Breen & Hill, 1969).

Benthic Crustacea

The physiology of the common estuarine crab *Hymenosoma orbiculare* was investigated following its discovery in Lake Sibaya in Natal. Salinity tolerance of larvae, osmoregulation and temperature tolerance of adults were investigated in an effort to explain the ability of this crab to live in fresh water (Forbes & Hill, 1969). Investigations also led to a theory of the estuarine origin of Lake Sibaya and the deep lakes associated with the Kosi estuary (Hill, 1969).

A detailed study was undertaken of the biology of the mud-prawn *Upogebia africana*. Results on osmoregulation and temperature tolerance have been published (Hill, 1971; Hill & Allanson, 1971), while further work covers respiratory physiology, growth and breeding. A similar study is underway on the sand-pawn *Callianassa*. An interesting finding in relation to this species is that colonisation of low-salinity regions at the top of estuaries is via adult migration. Breeding does not occur in these regions and larvae are intolerant of low salinity.

The crab *Scylla serrata* is found in many of the estuaries of the east coast. A research programme is currently investigating the growth, breeding and migration of *Scylla*. Initial findings confirm an annual seaward migration by mature females.

Estuarine fish

Among the most common estuarine fish along the east coast are species of *Rhabdosargus*. Research into one of these, *R. holubi*, is underway at present and includes studies of population size, growth, feeding and tolerance to temperature and salinity.

Emphasis has been placed by the Oceanographic Research Institute on species of teleost fish occurring in Natal which utilise the estuarine environment for part of their life cycle. *Johnius hololepidotus* (the kob or kabeljou), *Pomadasys commersoni* (the spotted grunter) *Rhabdosargus sarba* (the yellowfin bream) and a number of associated species have been studied (Wallace & Van Der Elst, 1975; Wallace, 1975*a*, *b*, *c*). All have been found to breed at sea, with adults often returning to feed in estuaries and juveniles recruiting into the estuarine 'nurseries' where some species grow to sexual maturity, before migrating back to sea. This stresses the need for estuarine conservation as a means of maintaining these stocks of fishes for angling and food. The marine phase of the lives of these fishes has also been investigated. Studies of recruitment have shown that six of the common angling fishes and four *Mugil* spp. migrate into estuaries as fry at lengths of 10–40 mm mainly between July and December (spring). Many of the smaller estuaries on the Natal coast are blocked by sand bars during the dry season and even during the wet months do not remain open for long. Migration into these estuaries therefore can only be sporadic. However, the middle of the recruitment period coincides with the onset of the spring rains and of the equinoctial tides which cause the river mouths to open. The fish that have matured during the winter are then released and the next generation of recruits migrates into the estuaries.

Concurrently with the biological research, a physiological study of the mechanisms of adaptation of teleost fish to waters of varying salinity has been carried out. Preliminary results were obtained using *Monodactylus argenteus* (the moonie), a euryhaline teleost fish from the estuaries and bays of the east coast, as test animals (Fearnhead & Fabian, 1971). The project has been expanded to include *Mugil cephalus*, *Rhabdosargus sarba* and *Pomadasys commersoni*.

Conservation

There has been an increasing interest and awareness of the need for conservation of South Africa's estuaries and lagoons. Estuaries are today an endangered habitat and the total area of estuaries around the coasts of South Africa is less than 600 km^2. Many factors are adversely affecting estuaries and lagoons and several papers (Grindley, 1970*b*, 1974; Heydorn, 1972*a*, *b*) have discussed the need for conservation of these ecosystems and for environmental planning procedures which will ensure wise management. An attempt has been made to establish some sort of economic valuation for the environmental importance of an estuary (Chmelik *et al.*, 1974). Without such a valuation decision makers have a purely abstract problem. The model established assesses how various economic parameters such as property values and business income for an area would be affected by particular environmental changes. The approach involves identifying the ecological and economic factors concerned and employing a discounted cash flow analysis technique to create a direct means for comparing in financial terms the effects of actions affecting the environment.

Bibliography

This bibliography is subdivided into five sections: hydrology, plankton, benthic invertebrates, fish and estuaries.

Hydrology

Anderson, F. P., Sharp, S. O. & Oliff, W. D. (1970). The reaction of coastal waters off Durban to changes in atmospheric pressure. In *Oceanography in South Africa. Durban, SANCOR, 1970.* Symposium paper H2, 1–22. Pretoria: South African National Committee for Oceanographic Research.

Andrews, W. R. H. (1973). The upwelling system off the Cape Peninsula. In *South African national oceanographical symposium abstracts, August 1973, Cape Town*, pp. 43–4 (abstract). Cape Town: Sea Fisheries Branch.

Andrews, W. R. H. (1974). The relevance of upwelling research to fisheries management off the west coast of Southern Africa. *South African Shipping News and Fishing Industry Review*, **29**(5), 60–1.

Andrews, W. R. H. & Cram, D. L. (1969). Combined aerial and shipboard upwelling study in the Benguela Current. *Nature*, **224**, 902–4.

Bang, N. D. (1971). The southern Benguela current region in February 1966. II. Bathythermography and air–sea interactions. *Deep-Sea Research*, **18**, 209–24.

Bang, N. D. & Andrews, W. R. H. (1974). Direct current measurements of a shelf-edge frontal jet in the southern Benguela system. *Journal of Marine Research*, **32**(3), 405–17.

Cram, D. L. (1970). A suggested origin for the cold surface water in central False Bay. *Transactions of the Royal Society of South Africa*, **39**, 129–37.

Cram, D. L. & Jansen Van Rensburg, C. H. (1973). Some surface features of the upwelled component of the Benguela Current system. In *South African national oceanographical symposium abstracts, August 1973, Cape Town*, pp. 1–26 (abstract). Cape Town: Sea Fisheries Branch.

De Decker, A. H. B. (1970). Notes on an oxygen-depleted subsurface current off the west coast of South Africa. *Investigational Report, Division of Sea Fisheries, South Africa*, **84**, 1–24.

Duncan, C. P. & Nell, J. H. (1969). Surface currents off the west cape coast. *Investigational Report, Division of Sea Fisheries, South Africa*, **76**, 1–19.

Henry, A. E. (1973). Hydrologie en voedingsoute van die Suidoos-Atlantiese en SuidwesIndiese Oseane in 1963. *Republiek van Suid-Afrika, Departement van Nywerheidswese, Afdeling Seevisserge Ondersoek, verslag*, **95**, 1–64.

Shannon, L. V. (1970). Oceanic circulation off South Africa. *Republic of South Africa, Division of Sea Fisheries, Fisheries Bulletin, Contributions to Oceanography and Fisheries Biology*, **6**, 27–33.

Shannon, L. V., Stander, G. H. & Campbell, J. A. (1973). Oceanic circulation deduced from plastic drift cards. *Republic of South Africa, Department of Industries, Sea Fisheries Branch, Investigational Report*, **108**, 1–31.

Shannon, L. V. & Van Ruswijck, M. (1969). Physical oceanography of the Walvis ridge region. *Investigational Report, Division of Sea Fisheries, South Africa*, **70**, 1.

Stander, G. H. & De Decker, A. H. B. (1970). Some physical and biological aspects of an oceanographic anomaly off south west Africa in 1963. *Investigational Report, Division of Sea Fisheries, South Africa*, **81**, 1–46.

Stander, G. H., Shannon, L. V. & Campbell, J. A. (1969). Average velocities of some ocean currents as deduced from the recovery of plastic drift cards. *Journal of Marine Research*, **27**, 293–300.

Visser, G. A. (1969*a*). Analysis of Atlantic waters off the west coast of southern Africa. *Investigational Report, Division of Sea Fisheries, South Africa*, **75**, 1–26.

Visser, G. A. (1969*b*). Hydrological observations in the south-east Atlantic Ocean. 1. The Schmidt-Ott Seamount area. *Investigational Report, Division of Sea Fisheries, South Africa*, **77**, 1–23.

Visser, G. A. (1970). The oxygen-minimum layer between the surface and 1000 m in the north-eastern South Atlantic. *Republic of South Africa, Division of Sea Fisheries, Fisheries Bulletin, Contributions to Oceanography and Fisheries Biology*, **6**, 10–22.

Visser, G. A., Kruger, I. & Coetzee, D. J. (1973). Environmental studies in south west African waters. *South African national oceanographical symposium abstracts, August 1973, Cape Town*, pp. 25–6 (abstract). Cape Town: Sea Fisheries Branch.

Welsh, J. G. & Visser, G. A. (1970). Hydrological observations in the south-east Atlantic Ocean. II. The Cape Basin. *Investigational Reports, Division of Sea Fisheries, South Africa*, **83**, 1–23.

Plankton

Andrews, W. R. H. & Cram, D. L. (1969). Combined aerial and shipboard upwelling study in the Benguela Current. *Nature*, **224**, 902–4.

Andrews, W. R. H., Cram, D. L. & Visser, G. A. (1970). An estimate of the potential production due to upwelling on the Cape Peninsula. In *Oceanography in South Africa. Durban, SANCOR, 1970.* Symposium paper B3: 1–14. Pretoria: South African National Committee for Oceanographic Research.

Berry, P. F. (1974). Palinurid and scyllarid lobster larvae of the Natal coast, South Africa. *South African Association for Marine Biological Research, Oceanographic Research Institute, Investigational Report*, **34**, 1–44.

Branch, M. L. (1974). Limiting factors for the gametophytes of three South African Laminariales. *Investigational Report, Division of Sea Fisheries, South Africa*, **104**, 1–38.

Burchall, J. (1968*a*). Primary production studies in the Agulhas Current region off Natal, June 1965. *South African Association for Marine Biological Research, Oceanographic Research Institute, Investigational Report*, **20**, 1–16.

Burchall, J. (1968*b*). An evaluation of primary productivity studies in the continental shelf region of the Agulhas current near Durban (1961–1966). *South African Association for Marine Biological Research, Oceanographic Research Institute, Investigational Report*, **21**, 1–44.

Cherry, R. D. & Shannon, L. V. (1974). The alpha radioactivity of marine organisms. *Atomic Energy Review*, **12**(1), 3–45.

De Decker, A. H. B. (1968). The Cape submarine canyon. II. Hydrology and plankton. *Republic of South Africa, Division of Sea Fisheries, Fisheries Bulletin, Contributions to Oceanography and Fisheries Biology*, **5**, 38–45.

De Decker, A. H. B. (1973*a*). Plankton distribution in the seas around South Africa. In *South African national oceanographical symposium abstracts,*

August 1973, Cape Town, pp. 42–3 (abstract). Cape Town: Sea Fisheries Branch.

De Decker, A. H. B. (1973*b*). Agulhas bank plankton. In *The biology of the Indian Ocean*, ed. B. Zeitzschel, pp. 189–219. Berlin: Springer–Verlag.

Grindley, J. R. (1969). The quaternary evolution of the Pseudodiaptomidae. *South African Archaeological Bulletin, Cape Town*, **24**, 149–50.

Grindley, J. R. (1970). Red water mussel poisoning. In *Oceanography in South Africa. Durban, SANCOR, 1970.* Symposium paper, pp. 1–15. Pretoria: South Africa National Committee for Oceanographic Research.

Grindley, J. R. (1971). Recent plankton studies in the seas around South Africa. In *Proceedings, Joint Oceanographic Assembly, IAPSO, IABO, CMG, SCOR, 1970 Tokyo Japan, 'The World Ocean'*, ed. Uda Michitaka, contribution in biological oceanography (D2-A-1), p. 429. Tokyo: Japan Society for the Promotion of Science.

Grindley, J. R. & Nel, E. (1968). Mussel poisoning and shelfish mortality on the west coast of South Africa. *South African Journal of Science*, **64**, 420–2.

Grindley, J. R. & Nel, E. H. (1970). Red water and mussel poisoning at Elands Bay, December 1966. *Republic of South Africa, Division of Sea Fisheries, Fisheries Bulletin, Contributions to Oceanography and Fisheries Biology*, **6**, 36–55.

Grindley, J. R. & Penrith, M. J. (1967). Notes on the bathypelagic fauna of the seas around South Africa. *Zoologica Africana*, **1**, 275–95.

Grindley, J. R. & Sapeika, N. (1969). The cause of mussel poisoning in South Africa. *South African Medical Journal*, **43**, 275–9.

Grindley, J. R. & Taylor, F. R. J. (1970). Factors affecting plankton blooms in False Bay. *Transactions of the Royal Society of South Africa*, **39**, 201–10.

Hoy, H. (1970). Chlorophyll in the sea off South Africa. *Republic of South Africa, Division of Sea Fisheries, Fisheries Bulletin, Contributions to Oceanography and Fisheries Biology*, **6**, 1–9.

Hutchings, L. (1973). Short-term variations in the zooplankton distribution off the Cape Peninsula. In *South African national oceanographical symposium abstracts, August 1973, Cape Town*, pp. 45–6 (abstract). Cape Town: Sea Fisheries Branch.

Hutchings, L., Robertson, A. A. & Allan, T. S. (1970). A plankton pump for fine scale sampling of zooplankton. In *Oceanography in South Africa. Durban, SANCOR, 1970.* Symposium paper, pp. 1–8. Pretoria: South African National Committee for Oceanographic Research.

Kensley, B. F. (1972). The family Sergestidae in waters around southern Africa. *Annals of the South African Museum*, **57**, 215–64.

Lazarus, B. O. (1967). The occurrence of phyllosomata off the Cape with particular reference to *Jasus lalandii. Investigational Reports, Division of Sea Fisheries, South Africa*, **63**, 1–38.

Masson, C. R. (1972). Biology of the Chaetognatha of the South African west and east coasts. MSc thesis, University of Stellenbosch.

Mitchell-Innes, B. A. (1967). Primary production studies in the south-west Indian Ocean, 1961–1963. *South African Association for Marine Biological Research Oceanographic Research Institute, Investigational Report*, **14**, 1–20.

Mostert, S. A. (1970). Analysis of particulate matter and nutrients in sea-water. *Republic of South Africa, Division of Sea Fisheries, Fisheries Bulletin, Contributions of Oceanography and Fisheries Biology*, **6**, 34–5.

Mostert, S. A. & Henry, A. E. (1973). Primary productivity in the north-western part of the Langebaan Lagoon, 1971–1972. In *South African national oceano-*

graphical symposium abstracts, August 1973. Cape Town, pp. 28–9 (abstract). Cape Town: Sea Fisheries Branch.

Nel, E. A. (1968*a*). The micro-plankton of the south-west Indian Ocean. *Investigational Reports, Division of Sea Fisheries, South Africa*, **62**, 1–178.

Nel, E. A. (1968*b*). Diatoms in the aphotic zone of the south-west Indian Ocean. *Republic of South Africa, Division of Sea Fisheries, Fisheries Bulletin, Contributions to Oceanography and Fisheries Biology*, **5**, 11–31.

Pieterse, F. & van Der Post, D. C. (1967). Oceanographical conditions associated with red tides and fish mortality in the Walvis Bay region. *Administration of South West Africa, Marine Research Laboratory, Investigational Report*, **14**, 1–125.

Rand, R. W. & Allan, T. S. (1973). A gravity-flow plankton sampler. *Antarctic Bulletin*, **2**(11/12), 100–2.

Reinecke, P. (1967). *Gonyaulax grindleyi* sp. nov.: A dinoflagellate causing a red tide at Elands Bay, Cape Province, in December 1966. *Journal of South African Botany*, **33**, 157–60.

Shannon, L. V. (1969). The alpha-activity of marine plankton. *Investigational Report, Division of Sea Fisheries, South Africa*, **68**, 1–38.

Shannon, L. V. & Cherry, R. D. (1967). Polonium-210 in marine plankton. *Nature*, **216**, 352–3.

Stander, G. H. & De Decker, A. H. B. (1969). Some physical and biological aspects of an oceanographic anomaly off South West Africa in 1963. *Investigational Report, Division of Sea Fisheries, South Africa*, **81**(1), 1–46.

Stone, J. H. (1966). The distribution and fecundity of *Sagitta enflata* Grossi in the Agulhas current. *Journal of Animal Ecology*, **35**, 533–41.

Stone, J. H. (1969). The Chaetognatha community of the Agulhas current: its structure and related properties. *Ecological Monographs*, **39**, 433–63.

Talbot, M. S. (1974). Distribution of euphausiid crustaceans from the Agulhas current. *Zoologica Africana*, **9**, 93–145.

Taylor, F. J. R. (1967). Phytoplankton of the South Western Indian Ocean. *Nova Hedwigia*, **12**, 433–76.

Thorrington-Smith, M. (1969). Phytoplankton studies in the Agulhas current region off the Natal coast. *South African Association for Marine Biological Research, Oceanographical Research Institute, Investigational Report*, **23**, 1–24.

Van Zyl, R. P. (1960). A preliminary study of the salps and Doliolids off the west and south coasts of South Africa. *Investigational Report, Division of Sea Fisheries, South Africa*, **40**, 1–31.

Williams, E. D. F. (1969). A submerged membrane filter apparatus for microbiological sampling. *Marine Biology*, **3**, 78–80.

Zoutendyk, P. (1968). Quantitative zooplankton sampling with the Indian Ocean standard net. *Transactions of the Royal Society of South Africa*, **38**, 23–7.

Zoutendyk, P. (1970). Zooplankton density in the south western Indian Ocean. In *Oceanography in South Africa. Durban, SANCOR, 1970*. Symposium paper E5, pp. 1–13. Pretoria: South African National Committee for Oceanographic Research.

Benthic invertebrates

Berrisford, C. D. (1969). Biology and zoogeography of the Vema seamount: a report on the first biological collection made on the summit. *Transactions of the Royal Society of South Africa*, **38**, 387–98.

Berry, P. F. (1969*a*). The biology of *Nephrops andamanicus* (Wood-Mason) (Decapoda, Reptantia). *South African Association for Marine Biological Research, Oceanographic Research Institute, Investigational Report*, **22**, 1–55.

Berry, P. F. (1969*b*). Occurrence of an external spermatophoric mass in the spiny lobster *Palinurus gilchristi* (Decapoda, Palinuridae). *Crustaceana*, **17**, 223–4.

Berry, P. F. (1970). Mating behaviour, oviposition and fertilization in the spiny lobster *Panulirus homarus* (Linnaeus). *South African Association for Marine Biological Research, Oceanographic Research Institute, Investigational Report*, **24**, 1–16.

Berry, P. F. (1971*a*). The biology of the spiny lobster *Panulirus homarus* (Linnaeus) off the east coast of southern Africa. *South African Association for Marine Biological Research, Oceanographic Research Institute, Investigational Report*, **28**, 1–75.

Berry, P. F. (1971*b*). The spiny lobsters (Palinuridae) of the east coast of southern Africa: distribution and ecological notes. *South African Association for Marine Biological Research, Oceanographical Research Institute, Investigational Report*, **27**, 1–23.

Berry, P. F. (1973). The biology of the spiny lobster *Palinurus delagoae*, Barnard, off the coast of Natal, South Africa. *South African Association for Marine Biological Research, Oceanographical Research Institute, Investigational Report*, **31**, 1–27.

Berry, P. F. & Hartnell, R. G. (1970). Mating in captivity of the spider crab *Pleistocantha moseleyi* (Miers) (Decapoda, Maiidae). *Crustaceana*, **19**, 214–15.

Berry, A. F. & Heydorn, A. E. F. (1970). A comparison of the spermatophoric masses and methods of fertilization in southern African spiny lobsters (Palinuridae). *South African Association for Marine Biological Research, Oceanographical Research Institute, Investigational Report*, **25**, 1–18.

Branch, G. M. (1971). The ecology of *Patella* Linnaeus from the Cape peninsula, South Africa. I. Zonation, feeding and movement. *Zoologica Africana*, **6**, 1–38.

Branch, G. M. (1973). The biology of *Patella cochlea* with reference to oil pollution. In *South African national oceanographical symposium abstracts, August 1973, Cape Town*, p. 18. Cape Town: Sea Fisheries Branch.

Branch, G. M. (1974*a*). The ecology of *Patella* Linnaeus from the Cape Peninsula, South Africa. II. Reproductive cycles. *Transactions of the Royal Society of South Africa*, **41**, 111–60.

Branch, G. M. (1974*b*). The ecology of *Patella* Linnaeus from the Cape Peninsula, South Africa. III. Growth rates. *Transactions of the Royal Society of South Africa*, **41**, 161–93.

Brown, A. C. (1971*a*). The ecology of the sandy beaches of the Cape Peninsula. Introduction. *Transactions of the Royal Society of South Africa*, **39**, 247–79.

Brown, A. C. (1971*b*). The ecology of sandy beaches of the Cape Peninsula. II. The mode of life of *Bullia* (Gastropoda: Prosobranchiata). *Transactions of the Royal Society of South Africa*, **39**, 281–330.

Brown, A. C. & Talbot, M. S. (1972). The ecology of the sandy beaches of the Cape Peninsula, South Africa. III. A study of *Gastrosaccus psammodytes* Tattersall (Crustacea: Mysidacea). *Transactions of the Royal Society of South Africa*, **40**, 309–33.

Brown, A. C. (1973). The ecology of the sandy beaches of the Cape Peninsula, South Africa. IV. Observations on two intertidal Isopoda *Eurydice longicornis* (Studer) and *Exosphaeroma truncatitelson* Barnard. *Transactions of the Royal Society of South Africa*, **40**, 381–404.

Champion, H. F. B. (1970). Aspects of the biology of *Penaeus indicus* (Milne

Edward) with notes on associated Penaeidae occurring off Natal on the east coast of South Africa. In *Oceanography in South Africa. Durban, SANCOR, 1970*. Symposium paper G1, pp. 1–17. Pretoria: South African National Committee for Oceanographic Research.

Christie, N. O. & Cutler, E. B. (1974). New distribution records for two species of *Siphonosoma* (Sipuncula) collected using a diver-operated suction sampler, *Transactions of the Royal Society of South Africa*, **41**, 109–10.

Christie, N. O. & Allen, J. C. (1972). A self-contained diver-operated quantitative sampler for investigating the macrofauna of soft substrates. *Transactions of the Royal Society of South Africa*, **40**, 299–308.

Cram, D. L. (1971*a*). Life history studies on South African echinoids. I. *Parechinus angulosus* (Leske). *Transactions of the Royal Society of South Africa*, **39**, 321–338.

Cram, D. L. (1971*b*). Life history studies on South African echinoids. II. *Echinolampas (Palaeolampus) crassa* (Bell). *Transactions of the Royal Society of South Africa*, **39**, 339–52.

Day, J. H. (1969*a*). *A guide to marine life on South African shores*. Cape Town: Balkema.

Day, J. H. (1969*b*). Feeding of the cymaliid gastropod *Argobuccinum argus* in relation to the structure of the proboscis and the secretions of the proboscis gland. *American Zoologist*, **9**, 909–14.

Day, J. H. (1970). The biology of False Bay, South Africa. *Transactions of the Royal Society of South Africa*, **39**, 211–21.

Day, J. H., Field, J. G. & Montgomery, M. P. (1971). The use of numerical methods to determine the distribution of the benthic fauna across the continental shelf of North Carolina. *Journal of Animal Ecology*, **40**, 93–125.

Day, J. H., Field, J. G. & Penrith, M. (1970). The benthic fauna and fishes of False Bay, South Africa. *Transactions of the Royal Society of South Africa*, **39**, 1–108.

Day, R. W. (1974). An investigation of *Pyura stolonifera* (Tunicata) from the Cape Peninsula. *Zoologica Africana*, **9**, 35–8.

De Villiers, G. (1970). Commercial potential of the alikreukel, *Turbo sarmaticus*, along the S.W. Cape coast, *South African Shipping News and Fishing Industry Review*, **25**, 87–9.

De Villiers, G. (1973). Voortplanting van die wit strandmossel *Donax serra* Röding. *Republick van Suid Afrika, Departement van Nyurerheidswese, Tak Seevisserye Ondersoekverslag*, **102**, 1–32.

De Villiers, G. (1974). Groei, bevolkingsdinamika, 'n grootskaalse sterfte en die rangskikking van die wit strandmossel, *Donax serra* Röding, op strande in die Suidwes-Kaap. *Republick van Suid Afrika, Departement van Nywerheidswese, Tak Seevisserye Ondersoekverslag*, **100**, 1–30.

Field, J. G. (1968). The 'turbulometer' – an apparatus for measuring relative exposure to wave action on shore. *Zoologica Africana*, **3**, 115–18.

Field, J. G. (1969). The use of the information statistic in the numerical classification of heterogeneous systems. *Journal of Ecology*, **57**, 565–9.

Field, J. G. (1970). The use of numerical methods to determine benthic distribution patterns from dredgings in False Bay. *Transactions of the Royal Society of South Africa*, **39**, 183–200.

Field, J. G. & McFarlane, G. (1968). Numerical methods in marine ecology. I. A quantitative similarity analysis of rocky shore samples in False Bay, South Africa. *Zoologica Africana*, **3**, 119–38.

Field, J. G. & Robb, F. T. (1970). Numerical methods in marine ecology. II.

Gradient analysis of rocky shore samples from False Bay. *Zoologica Africana*, **5**, 191–210.

Genade, A. B. (1973). A general account of certain aspects of oyster culture in the Knysna estuary. In *South African national oceanographical symposium abstracts. August 1973. Cape Town*, pp. 26–8 (abstract). Cape Town: Sea Fisheries Branch.

Grindley, J. R. (1967). Research on the Vema Seamount. *Commercial Fishing*, **2**, 14–19.

Heydorn, A. E. F. (1969*a*). Notes on the biology of *Panulirus homarus* and on length weight relationships of *Jasus lalandii*. *Investigational Reports, Division of Sea Fisheries, South Africa*, **69**, 1–26.

Heydorn, A. E. F. (1969*b*). The South African rock lobster, *Jasus tristani* at Vema Seamount, Gough Island and Tristan da Cunha. *Investigational Reports, Division of Sea Fisheries, South Africa*, **73**, 1–20.

Heydorn, A. E. F. (1969*c*). The rock lobster of the South African west coast, *Jasus lalandii* (H. Milne-Edwards). Population studies, behaviour, reproduction, moulting, growth and migration. *Investigational Reports, Division of Sea Fisheries, South Africa*, **71**, 1–52.

Heydorn, A. E. F. (1971). Distribution and ecology of Palinuridae in the southern African region. In *Proceedings of the joint oceanographic assembly, IAPSO, IABO, CMG, SCOR, 1970. Tokyo, Japan 'The World Ocean'*, ed. Uda Michitaka, contribution to biological oceanography. (D2B-6), pp. 458–460. Tokyo: Japan Society for the promotion of science.

Heydorn, A. E. F. & Newman, C. G. (1967). An oyster survey off the Cape coast. *Republic of South Africa, Division of Sea Fisheries, Fisheries Bulletin, Contributions to Oceanography and Fisheries Biology*, **4**, 25–7.

Hughes, D. A. (1970). The southern limits of distribution of commercially important penaeid prawns in South Africa. *Annals of the Cape Provincial Museums*, **8**, 79–83.

Kensely, B. F. (1972). *Shrimps and prawns of southern Africa*. Cape Town: Maskew Miller & South African Museum.

Kensley, B. F. (1973*a*). *Sea shells of southern Africa*. Cape Town: Maskew Miller & South African Museum.

Kensley, B. F. (1973*b*). Behavioural adaptations of the isopod *Tylos granulatus*. *Zoologica Africana*, **7**, 1–4.

Kensley, B. F. (1974). Preliminary observations on the biology and ecology of the genus *Tylos* in South Africa (Crustacea, Isopoda, Tylidae). *Annals of the South African Museum*, **65**, 401–71.

Lamoral, B. H. (1968). On the ecology and habitat adaptations of two intertidal spiders, *Desis formidabilis* (O.P. Cambridge) and *Amaurdriodes africanus* Hewitt, at 'The Island' (Kommetjie, Cape Peninsula), with notes on the occurrence of two other spiders. *Annals of the Natal Museum*, **20**, 151–93.

Mostert, S. A. (1972). Preliminary report on black mussel culture in the Langebaan Lagoon. *South African Shipping News and Fishing Industry Review*, **27**, 59, 61, 63.

Newman, G. G. (1968*a*). Growth of the South African abalone, *Haliotis midae*. *Investigational Reports, Division of Sea Fisheries, South Africa*, **67**, 1–21.

Newman, G. G. (1968*b*). Distribution of the abalone (*Haliotis midae*) and the effect of temperature on productivity. *Investigational Reports, Division of Sea Fisheries, South Africa*, **74**, 1–8.

Newman, G. G. & Pollock, D. E. (1971). A mass stranding of rock lobsters, *Jasus lalandii*, at Elands Bay, South Africa. *Crustaceana*, **26**(1), 1–4+Pl. (1974).

Newman, G. G. & Pollock, D. E. (1969). The efficiency of rock lobster fishing gear. *South African Shipping News and Fishing Industry Review*, **24**, 79–81.

Newman, G. G. & Pollock, D. E. (1970). Migration and availability of the rock lobster, *Jasus lalandii* at Elands Bay, South Africa. In *Oceanography in South Africa Durban, SANCOR, 1970*. Symposium paper 1–26. Pretoria: South African National Committee for Oceanographic Research.

Newman, G. G. & Pollock, D. E. (1971). Biology and migration of rock lobster *Jasus lalandii* and their effect on availability at Elands Bay, South Africa. *Investigational Reports, Division of Sea Fisheries, South Africa*, **94**, 1–24.

Newman, G. G. & Pollock, D. E. (1973). Growth of the rock lobster *Jasus lalandii* and its relationship to benthos. In *South African national oceanographical symposium abstracts, August 1973, Cape Town*, pp. 11–12 (abstract). Cape Town: Sea Fisheries Branch.

Newman, G. G. & Pollock, D. E. (1974). Biological cycles, maturity and availability of rock lobster *Jasus lalandii* on two South African fishing grounds. *Investigational Reports, Division of Sea Fisheries, South Africa*, **107**, 1–16.

Paterson, N. F. (1968). The anatomy of the Cape rock lobster, *Jasus lalandii* (H. Milne-Edwards). *Annals of the South African Museum*, **51**, 1–232.

Penrith, M. L. & Kensley, B. F. (1971*a*). The constitution of the intertidal fauna of rocky shores of S.W.A. I. Luderitzbucht. *Cimbebasia*, (A) **1**, 191–238.

Penrith, M. L. & Kensley, B. F. (1971*b*). The constitution of the intertidal fauna of rocky shores of S.W.A. II. Rocky Point. *Cimbebasia*, (A) **1**, 243–68.

Pollock, D. E. (1973). Growth of juvenile rock lobster *Jasus lalandii*. *Investigational Reports, Division of Sea Fisheries, South Africa*, **106**, 1–16.

Silberbauer, B. I. (1971*a*). The biology of the South African rock lobster *Jasus lalandii* (H. Milne-Edwards). I. Development. *Investigational Report Division of Sea Fisheries, South Africa*, **92**, 1–10.

Silberbauer, B. I. (1971*b*). The biology of the South African rock lobster *Jasus lalandii* (H. Milne-Edwards). II. The reproductive organs, mating and fertilization. *Investigational Report, Division of Sea Fisheries, South Africa*, **93**, 1–46.

Fish

Baird, D. (1970*a*). Age and growth of the South African pilchard, *Sardinops ocellata*. *Investigational Report, Division of Sea Fisheries, South Africa*, **91**, 1–16.

Baird, D. (1970*b*). The Natal 'sardine run'. *South African Shipping News and Fishing Industry Review*, **25**, 70–1.

Baird, D. (1970*c*). A preliminary report on the Natal 'sardine run'. In *Oceanography in South Africa. Durban, SANCOR, 1970*. Symposium paper E2, pp. 1–13. Pretoria: South African National Committee for Oceanographic Research.

Baird, D. (1971). Seasonal occurrence of the pilchard *Sardinops ocellata* on the east coast of South Africa. *Investigational Report Division of Sea Fisheries, South Africa*, **96**, 1–19.

Baird, D. (1972). Notes on the South African mackerel. *South African Shipping News and Fishing Industry Review*, **27**(10), 63, 65.

Baird, D. (1973). The pilchard fishery in South Africa, 1930–72. In *South African national oceanographic symposium abstract. August 1973. Cape Town*, pp. 39–40 (abstract). Cape Town: Sea Fisheries Branch.

Baird, D. & Geldenhuys, N. D. (1973). Biology and fishery of the anchovy in South

Africa. *South African Shipping News and Fishing Industry Review*, **28**(10), 43, 45, 47, 49.

Baird, D., Geldenhuys, N. D. & Pollock, D. E. (1970). Growth of three pelagic fish species in South African west coast waters. In *Oceanography in South Africa. Durban, SANCOR, 1970.* Symposium paper 1–8. Pretoria: South African National Committee for Oceanographic Research.

Bohi, H., Botha, L. & Van Eck, T. H. (1971). Selection of Cape hake (*Merluccius merluccius capensis* Castenau and *Merluccius merluccius paradoxus* Franca) by bottom trawl cod-ends. *Journal du Conseil Permanent International pour l'Exploration de la Mer*, **33**, 438–70.

Botha, L. (1969). The growth of the Cape hake *Merluccius capensis. Investigational Report, Division of Sea Fisheries, South Africa*, **82**, 1–9.

Botha, L. (1970). S.A. Trawlfish landings from 1955 to 1968 with special reference to hake. *South African Shipping News and Fishing Industry Review*, **25**, 70–1, 73, 75.

Botha, L, (1971). Growth and otolith morphology of the Cape hakes *Merluccius capensis* Cast. and *M. paradoxus* Franca. *Investigational Report, Division of Sea Fisheries, South Africa*, **97**, 1–32.

Botha, L. (1973*a*). Migration and spawning behaviour of the Cape hakes. *South African shipping News and Fishing Industry Review*, **28**(4), 62–3, 65, 67.

Botha, L. (1973*b*). Depth distribution, migration, migration and spawning of the Cape hakes as related to the fishery. In *South African national oceanographical symposium abstracts. August 1973. Cape Town*, pp. 24–5 (abstract). Cape Town: Sea Fisheries Branch.

Botha, L., Lucks, D. K. & Chalmers, D. S. (1971). Mesh selectivity experiments of the east coast sole. *South African Shipping News and Fishing Industry Review*, **26**, 50–7.

Botha, L. & Mombeck, F. (1971). Research and conservation on the Cape hake. *South African Shipping News and Fishing Industry Review*, **26**, 59–63.

Centurier-Harris, O. M. (1974). The appearance of lantern-fish in commercial catches. *South African Shipping News and Fishing Industry Review*, **29**(1), 45.

Centurier-Harris, O. M. & Crawford, R. J. (1974*a*). Distribution of major species in South African pelagic fishery. *South African Shipping News and Fishing Industry Review*, **29**(3), 67, 69, 71.

Centurier-Harris, O. M. & Crawford, R. J. (1974*b*). Yield from the Western Cape pelagic resource. *South African Shipping News and Fishing Industry Review*, **29**(12), 54–5.

Cram, D. L. (1972). The role of aircraft-borne remote sensors in South African fisheries research. In *South African Council for Scientific and Industrial Research, Remote Sensing Symposium 3–5 May 1972, Pretoria*, p. 562.

Cram, D. L. (1973). How A.R.T. would assist the search for Cape mackerel. *South African Shipping News and Fishing Industry Review*, **28**(1), 82–3.

Cram, D. L. (1974). South West African pilchard stock continues to recover: summary of results of Phase IV of the Cape Cross Programme. *South African Shipping News and Fishing Industry Review*, **29**(9), 74–5.

Cram, D. L. & Agenbag, J. J. (1974). Low light level television – an aid to pilchard research. *South African Shipping News and Fishing Industries Review*, **29**(7), 52–3.

Cram, D. L. & Schülein, F. H. (1974). Observations on surface shoaling Cape hake off South West Africa. *Journal du Conseil Permanent International pour l'Exploration de la Mer*, **35**, 272–5.

Cram, D. L. & Visser, G. A. (1972). Cape Cross Programme Phase II. *South African Shipping News and Fishing Industry Review*, **27**(2), 40–1, 43.

Cram, D. L. & Visser, G. A. (1973). S.W.A. pilchard stocks show first signs of recovery. Summary of results of Phase III of the Cape Cross pelagic research programme. *South African Shipping News and Fishing Industry Review*, **28**(3), 56–7, 59, 61, 63.

Geldenhuys, N. D. (1972). The maasbanker fishery. *South African Shipping News and Fishing Industry Review*, **27**(4), 54–5.

Geldenhuys, N. D. (1973). Growth of the South African maasbanker *Trachurus trachurus* Linnaeus and age composition of the catches, 1950–1971. *Investigational Report, Division of Sea Fisheries, South Africa*, **101**, 1–24.

Grindley, J. R. (1969). *Riches of the sea.* Cape Town: Caltex.

Haigh, E. H. (1972*a*). Larval development of three species of economically important South African fishes. *Annals of the South African Museum*, **59**, 47–70.

Haigh, E. H. (1972*b*). Development of *Trachurus trachurus* (Carangidae) the South African maasbanker. *Annals of the South African Museum*, **59**, 139–50.

Hampton, I. (1973). Fish-counting with an echo-sounder. In *South African national oceanographical symposium abstracts, August 1973, Cape Town*, pp. 1–35 (abstract). Cape Town: Sea Fisheries Branch.

Henning, H. F. K. P. (1974). The effect of a larval *Anisakis* (Nematoda: Ascaroidea) on the South West African anchovy, *Engraulis capensis*. *Journal du Conseil Permanent International pour l'Exploration de la Mer*, **35**, 185–8.

Hulley, P. A. (1972*a*). The origin, interrelationships and distribution of southern African Rajidae (Chondrichthyes, Batoidea). *Annals of the South African Museum*, **60**, 1–103.

Hulley, P. A. (1972*b*). A report on the mesopelagic fishes collected during the deep-sea cruises of R.S. *Africana II*, 1961–1966. *Annals of the South African Museum*, **60**, 197–236.

Jones, B. W. & Van Eck, T. H. (1967). The Cape hake: its biology and the fishery. *South African Shipping News and Fishing Industry Review*, **22**, 80–97.

King, D. P. F. & Robertson, A. A. (1973*a*). Variability in Bongo net catches of pilchard eggs. *South African national oceanographic symposium abstracts, August 1973, Cape Town*, pp. 1–35 (abstract). Cape Town: Sea Fisheries Branch.

King, D. P. F. & Robertson, A. A. (1973*b*). Methods of pelagic fish egg and larval research in South West Africa. *South African Shipping News and Fishing Industry Review*, **28**(5), 57, 59, 61.

Lucks, D. K. (1972). Mesh selectivity studies on sole off South West Africa. *South African Shipping News and Fishing Industry Review*, **27**(9), 54–5, 57.

Lucks, D. K., Payne, A. I. L. & Maree, S. (1973). The trawl fishery of South West Africa. *South African Shipping News and Fishing Industry Review*, **28**(9), 65, 67, 69, 71.

Lucks, D. & Payne, A. (1973). Trends in the demersal fishery of South West Africa. In *South African national oceanographical symposium, abstracts. August 1973. Cape Town*, pp. 1–28 (abstract). Cape Town: Sea Fisheries Branch.

Nepgen, C. S. de V. (1970*a*). Exploratory fishing for tuna off the South African west coast. *Investigational Report, Division of Sea Fisheries, South Africa*, **87**, 1–26.

Nepgen, C. S. de V. (1970*b*). The Japanese longline fishery off the South African coast 1964–1967. *Investigational Report, Division of Sea Fisheries, South Africa*, **90**, 1–13.

Newman, G. G. (*1970a*). A stock assessment of the pilchard (*Sardinops ocellata*) at Walvis Bay in South West Africa. *Investigational Report, Division of Deep Sea Fisheries, South Africa*, **85**, 1–13.

Newman, G. G. (1970*b*). Migration of the pilchard (*Sardinops ocellata*). *Investigational Report, Division of Sea Fisheries, South Africa*, **86**, 1–16.

O'Toole, M. J. (1973). Fish larvae of the Benguela Current system. *South African national oceanographical symposium abstracts, August 1973, Cape Town*, pp. 1–19 (abstract). Cape Town: Sea Fisheries Branch.

O'Toole, M. J. (1974). Fish larval investigations off South West Africa: Summary of results. *South African Shipping News and Fishing Industry Review*, **29**(11), 53, 55, 57, 59.

Penrith, M. J. (1972). The behaviour of reef-dwelling sparid fishes. *Zoologica Africana*, **7**, 43–8.

Penrith, M. J. & Cram, D. L. (1972). The Cape of Good Hope – a hidden barrier to billfish? In *Proceedings of the international billfish symposium Hawaii, August 1972*, part 2, reviews and contributed papers, NOAA Technical Report, NMFS, SSRF-675, pp. 175–87.

Ribbink, A. J. (1971). The jaw mechanism and feeding of the holocephan, *Callyorhynchus capensis* Dumeril. *Zoologica Africana*, **6**, 45–74.

Schulein, F. (1973). Recruitment studies on the South West African pilchard, 1958–1972. In *South African national oceanographical symposium abstracts, August 1973, Cape Town*, pp. 1–29 (abstract). Cape Town: Sea Fisheries Branch.

Stander, G. H. (1967). Trends in the pilchard fishery, 1950–1965. *Republic of South Africa, Division of Sea Fisheries, Fisheries Bulletin, Contributions to Oceanography and Fisheries Biology*, **4**, 35–9.

Stander, G. H. & Le Roux, P. J. (1968). Notes on fluctuations of the commercial catch of the South African pilchard (*Sardinops ocellata*) 1950–1965. *Investigational Report, Division of Sea Fisheries, South Africa*, **65**, 1–14.

Stander, G. H. & Nepgen, C. S. de V. (1968). Some facts about linefish – with special reference to False Bay. *South African Shipping News and Fishing Industry Reviews*, **23**, 108–11.

Talbot, F. H. & Penrith, M. J. (1968). The tunas of the genus *Thunnus* in South African waters. I. Introduction, systematics, distribution and migrations. *Annals of the South African Museum*, **52**, 1–41.

Thompson, D. & Mostert, G. (1974). Muscle esterase genotypes in the pilchard, *Sardinops ocellata*. *Journal du Conseil Permanent International pour l'Exploration de la Mer*, **36**, 50–3.

Van Eck, T. H. (1970). Fish population estimates by acoustic means. In *Oceanography in South Africa. Durban, SANCOR, 1970.* Symposium paper, pp. 1–16. Pretoria: South African National Committee for Oceanographic Research.

Van Eck, T. H., Botha, L., Von Brandt, A. & Bohl, H. (1968). The selectivity of synthetic fibre codends for the capture of South African hake. *South African Shipping News and Fishing Industry Review*, **23**, 124D–35.

Wallace, L. (1972). Reactions of the sharks *Carcharhinus leucas* (Müller and Henie) and *Odontaspis taurus* (Rafinesque) to gill net barriers under experimental conditions. *Investigational Report, Oceanographical Research Institute*, **30**, 1–24.

Welsh, J. G. (1968). A new approach to research on tuna distribution in South African waters. *Republic of South Africa, Division of Sea Fisheries, Fisheries Bulletin, Contributions to Oceanography and Fisheries Biology*, **5**, 32–4.

Zoutendyk, P. (1973). The biology of the Agulhas sole, *Austroglossus pectoralis*. I. Environment and trawling grounds. *Transactions of the Royal Society of South Africa*, **40**, 349–66.

Zoutendyk, P. (1974*a*). The biology of the Agulhas sole, *Austroglossus pectoralis*. II. Age and growth. *Transactions of the Royal Society of South Africa*, **41**, 33–41.

Zoutendyk, P. (1974*b*). The biology of the Agulhas sole, *Austroglossus pectoralis*. III. Length–weight relationships. *Transactions of the Royal Society of South Africa*, **41**, 99–110.

Estuaries

Breen, C. M. & Hill, B. J. (1969). A mass mortality of mangroves in the Kosi Estuary. *Transactions of the Royal Society of South Africa*, **38**, 285–303.

Day, J. H. (1967). The biology of Knysna Estuary. In *Estuaries*, ed. G. Lauff, pp. 398–417. Washington: American Association for the Advancement of Science.

Day, J. H. (1974). The ecology of Morrumbene estuary, Mozambique. *Transactions of the Royal Society of South Africa*, **41**, 43–97.

Chmelik, F. B., Van Loggerenberg, B., Darracott, A. & Grindley, J. R. (1974). Ecomonic model for estuarine valuation. In *Proceedings of the 10th annual conference of the Marine Technology Society, National needs and ocean solutions. September 23–25, 1974, Washington D.C.*, pp. 233–75. Washington: Marine Technology Society.

Fearnhead, E. A. & Fabian, B. C. (1971). The ultrastructure of the gill of *Monodactylus argenteus* (an euryhaline teleost fish) with particular reference to morphological changes associated with changes in salinity. *Investigational Reports, Oceanographical Research Institute*, **26**, 1–39.

Forbes, A. T. & Hill, D. J. (1969). The physiological ability of the marine crab *Hymenosoma orbiculare* Desm. to live in a subtropical freshwater lake. *Transactions of the Royal Society of South Africa*, **38**, 271–84.

Grindley, J. R. (1964). Effect of low salinity water on the vertical migration of estuarine plankton. *Nature*, **203**, 781–2.

Grindley, J. R. (1970*a*). The plankton of South African estuaries. In *Oceanography in South Africa. Durban, SANCOR, 1970.* Symposium paper H1, pp. 1–16. Pretoria: South African National Committee for Oceanographic Research.

Grindley, J. R. (1970*b*). The role of freshwater in the conservation of South African estuaries. In *Republic of South Africa, water year 1970, convention: water for the future*. Pretoria: Department of water affairs.

Grindley, J. R. (1972). The vertical migration behaviour of estuarine plankton. *Zoologica Africana*, **7**, 13–20.

Grindley, J. R. (1974). Estuaries, an endangered habitat. *African Wildlife*, **28**(1), 23–6.

Grindley, J. R. & Heydorn, A. E. F. (1970). Red water and associated phenomena in St Lucia. *South African Journal of Science*, **66**, 210–13.

Grindley, J. R. & Wooldridge, T. (1974). The plankton of Richards Bay. *Hydrobiological Bulletin. (Amsterdam)*, **8**, 201–12.

Hemens, J. & Warwick, R. J. (1972). The effects of fluoride on estuarine organisms. *Water Research*, **6**, 1301–8.

Hemens, J., Warwick, R. J. & Oliff, W. D. (1976). The effect of extended exposure to low fluoride concentration on estuarine fish and Crustacea. *Progress in Water Technology*, **7**, 579–85.

Heydorn, A. E. F. (1972*a*). The interdependence of marine and estuarine ecosystems in relation to the utilization and conservation of the living resources of the sea. *South African Journal of Science*, **69**(1), 18–23.

Heydorn, A. E. F. (1972*b*). South African estuaries – an economic asset or a natural resource being squandered? *Scientiae*, **13**(5), 2–6.

Hill, B. J. (1969). The bathymetry and possible origin of Lakes Sibaya, Nhlange and Sifungwe in Zululand (Natal). *Transactions of the Royal Society of South Africa*, **38**, 205–16.

Hill, B. J. (1971). Osmoregulation by an estuarine and marine species of *Upogebia*. *Zoologica Africana*, **6**, 229–36.

Hill, B. J. & Allanson, B. R. (1971). Temperature tolerance of the estuarine prawn *Upogebia africana*. *Marine Biology*, **11**, 337–43.

McLachlan, A. & Grindley, J. R. (1974). Distribution of macrobenthic fauna of soft substrata in Swartkops estuary. *Zoologica Africana*, **9**, 211–33.

Millard, N. A. H. & Broekhuysen, G. J. (1970). The ecology of South African estuaries. X. St Lucia: a second report. *Zoologica Africana*, **5**, 227–308.

Penrith, M. & Penrith, M. L. (1967). A new genus of mullet from St Lucia Estuary, Zululand. *Durban Museum Novitates*, **8**, 69–75.

Wallace, J. H. (1975*a*). The estuarine fishes of the east coast of South Africa. I. species composition and length distribution in the estuarine and marine environments. II. Seasonal abundance and migrations. *Investigational Reports, Oceanographical Research Institute*, **40**, 1–72.

Wallace, J. H. (1975*b*). The estuarine fishes of the east Coast of South Africa. III. Reproduction. *Investigational Reports, Oceanographical Research Institute*, **41**, 1–51.

Wallace, J. H. (1975*c*). Aspects of the biology of *Mugil cephalus* in a hypersaline estuarine lake on the east coast of South Africa. (International symposium on the aquaculture of grey mullet, Haifa, 2–8 July, 1974.) *Aquaculture*, **5**(1), 111.

Wallace, J. H. & Van Der Elst, R. P. (1975). The estuarine fishes of the East coast of South Africa. IV. Occurrence of juveniles in estuaries. V. Ecology, estuarine dependence and status. *Investigational Reports, Oceanographical Research Institute*, **42**, 1–63.

6. The Strait of Georgia Programme

T. R. PARSONS

Introduction

The purpose of the IBP Strait of Georgia Programme was to obtain a general description of the level of plankton production in the strait and to determine where possible how the production of the strait was governed by various oceanographic factors, particularly those affecting the early-life stages of fish. Since the programme did not have an integrated physical oceanographic component, the examination of factors governing plankton production became largely a study of biological oceanographic processes with particular emphasis on trophic relationships leading to the survival of larval and juvenile fish. This gave rise to a large experimental component within the programme; in particular, experiments were performed with zooplankton feeding on phytoplankton, and juvenile fish (salmonids) feeding on zooplankton. Two features of these feeding experiments were that they were conducted with natural prey items and secondly, particular attention was paid to the relationship between predator and prey size. Other experiments related to the descriptive oceanographic programme included studies on marine bacteria and detritus, and the use of mathematical models to better diagnose the overall effect of simulated changes in trophic relationships.

General geographic and hydrographic features of the Strait of Georgia

The Strait of Georgia is located on the west coast of British Columbia between the mainland of Canada and Vancouver Island (Fig. 6.1). It is approximately 200 km long, has an average width of 30 km and a mean depth of 156 m. The area is studded with numerous small islands, two larger islands and is open to the sea in the north through Seymour Narrows, and in the south through the San Juan Archipelago to the Strait of Juan de Fuca and the ocean. The major source of fresh water in the area comes from the Fraser River which flows into the southern half of the strait. Freshwater is also contributed from the surrounding coastline which is characterized by numerous inlets extending up to 80 km into the coast.

A general description of hydrographic conditions in the Strait of Georgia has been given by Tully & Dodimead (1957) and Waldichuck (1957). The major factors influencing the hydrographic conditions in the Strait of Georgia are wind, tide, insolation and freshwater runoff. The influence of wind, which reaches its maximum effect in January and December, and a minimum from June to September, is primarily to cause an anticlockwise

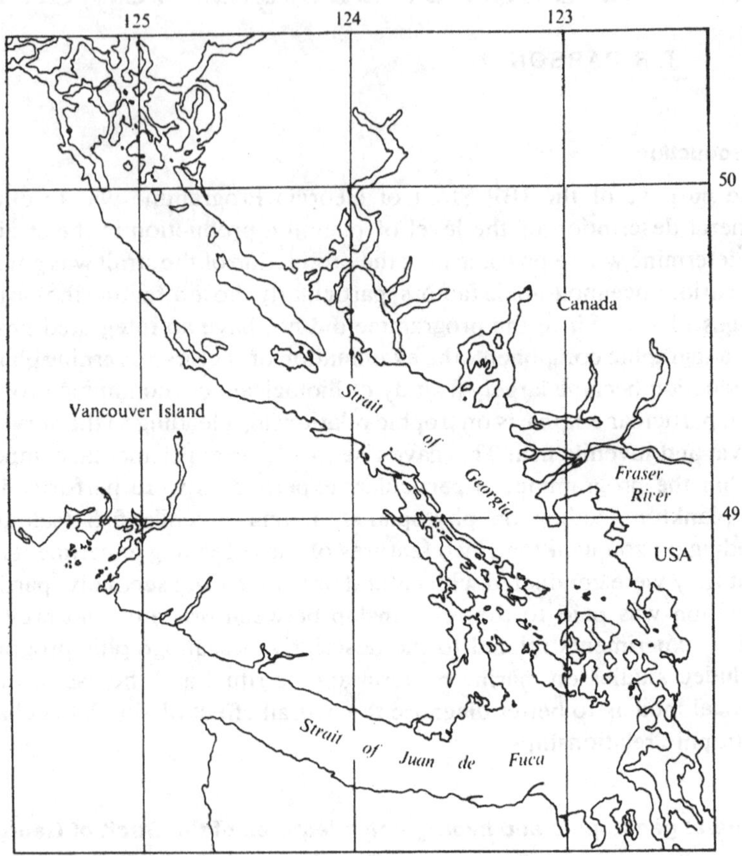

Fig. 6.1. The Strait of Georgia.

circulation of water around the strait which is strongest along the mainland shore. This circulation is complicated by the tidal cycle. The effect of wind on mixing is apparent during the winter months but is strongly suppressed by stratification during the summer.

Tully & Dodimead (1957) have described three tidal regions for the strait. These are the northern and southern regions and a large central region of the strait where tidal currents are generally less than 1 knot except in the immediate vicinity of the Fraser River. The northern and southern regions are identified as regions of turbulent mixing where rapid tidal currents extend from 10 to 15 km into the strait during flooding and retreat on the ebb tide into the passages.

The effects of heating and freshwater runoff have been interpreted by Waldichuck (1957) in terms of the stability expression $\partial\sigma_t/\partial z \times 10^3$

(Sverdrup, Johnson & Fleming, 1946) in the upper 100 m. Seasonal variations in this value have shown that in the immediate vicinity of the Fraser River high stability is maintained throughout the year by the direct effect of low-salinity surface layers. In the southern part of the strait and over a large area north of the Fraser River, stability is reduced to nearly zero during the winter months but is established during the summer by insolation and by freshwater from the Fraser River which reaches a maximum discharge during the period from May to July.

Salinities in the region of the Fraser River vary from 10 to 20 ‰ in the surface layers and from 28 to 31‰ in the deeper water. Over the rest of the area, surface salinities vary from 24 to 28‰ except in the areas of tidal mixing which are virtually homogeneous with salinities of 31 to 32‰ from the top to bottom. Seasonal temperatures in the area vary from about 7 to 22 °C at the surface and from 8 to 9 °C in deeper water.

Descriptive biological oceanographic features of the Strait of Georgia

A general description of the biological oceanographic features of the Strait of Georgia is given by Parsons, LeBrasseur & Barraclough (1970). Seasonal data contained in this reference cover a four year period, 1965–8, and are reproduced in Figs. 6.2, 6.3 and 6.4. From these data (Fig. 6.2) it is apparent that primary productivity starts to increase in February, reaches a maximum of 1.2 g C/m^2/day in May and then declines to 0.2 g C/m^2/day in late summer. Changes in the average standing stock of phytoplankton in the first 20 m as measured by chlorophyll *a* analysis show a smaller range of values from 1 μg/l during the winter to a maximum of 4 μg/l during June. There is a strong correlation between chlorophyll *a* and particulate nitrogen (Fig. 6.3) which indicates that the phytoplankton are responsible for the primary production of protein in the food chain. However the correlation between chlorophyll *a* and particulate carbon is less direct and an intercept at zero chlorophyll and 140 μg C/l indicates that there is an allochthonous source of organic carbon. This material is probably derived from the land as indicated in separate studies (Stephens, Sheldon & Parsons, 1967; Seki, Stephens & Parsons, 1969) which showed that even a relatively small river transported appreciable quantities of organic materials into the Strait.

The biomass of herbivorous zooplankton is closely related to the primary productivity and both reach a maximum before the annual freshet of the Fraser River in June. Although the river is a predominant feature in the dynamics of phytoplankton production in the strait (see following section on primary productivity in the strait) it does not contribute any appreciable nitrate or phosphate to the strait; on the other hand silicate is higher in Fraser River water than in seawater and a substantial increase

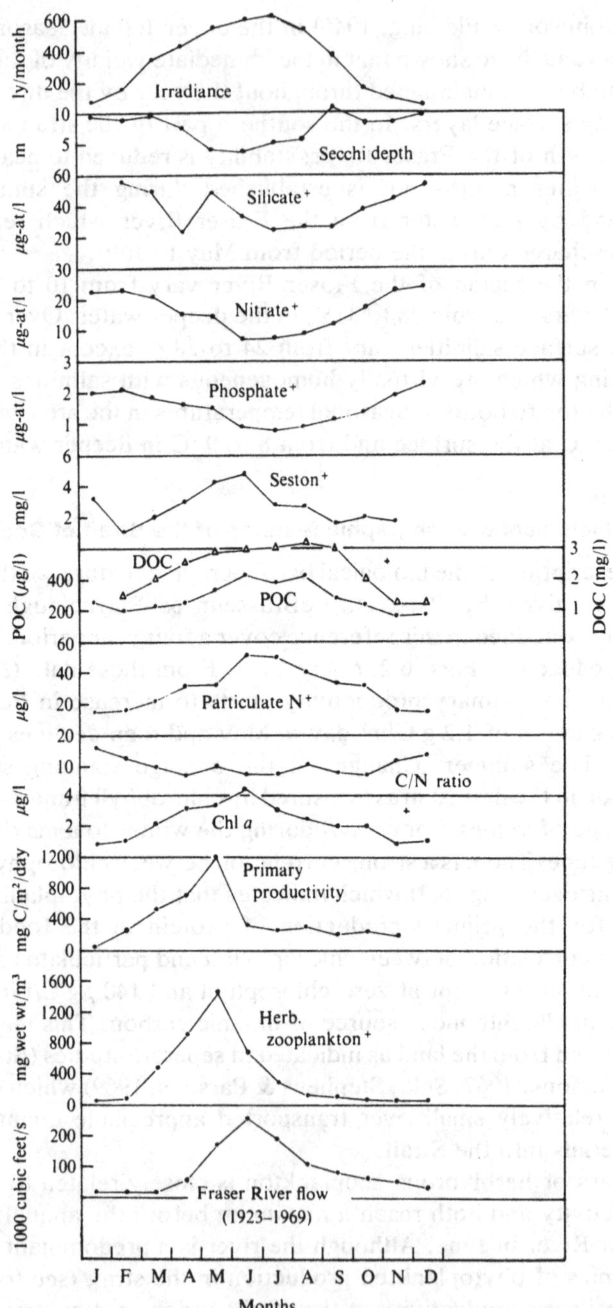

Fig. 6.2. Summary of production data for the Strait of Georgia expressed as monthly averages for the period, 1965–8 (+ denotes average value for the water column 0–20 m). Dissolved and particulate organic carbon, DOC and POC respectively.

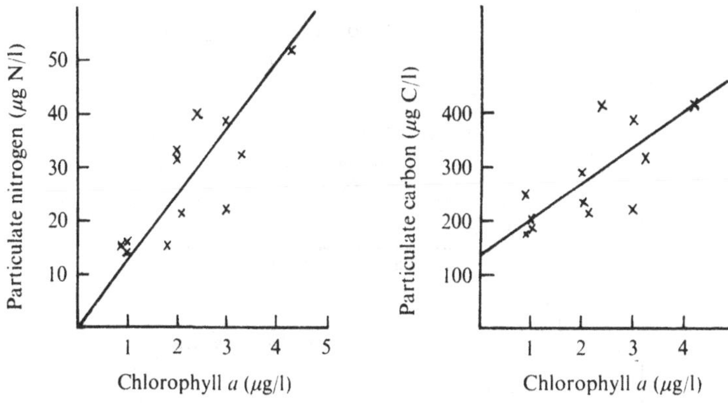

Fig. 6.3. Monthly average values of particulate carbon and nitrogen versus chlorophyll *a* in the Strait of Georgia.

in silicate content of the strait can be seen from April to May during the start of the annual Fraser River freshet. Another nutrient that appears to be higher in Fraser River water than seawater is vitamin B_{12} (Cattell, 1973).

Carbon to nitrogen ratios of less than ten, together with data on the concentration of chlorophyll *a*, indicate that the particulate material was dominated by phytoplankton from March to October, and by detritus from November to February. Soluble organic carbon showed a general increase throughout the summer reaching a maximum in August, three months after the maximum in primary production. This indicates that the increase in soluble organic carbon is more closely associated with the decomposition of phytoplankton by zooplankton grazing and bacteria, than with exudation of soluble organic carbon from actively growing phytoplankton.

The seasonal cycle of zooplankton was heavily predominated by a large bloom of *Calanus plumchrus* which reached a maximum in April (Fig. 6.4) and declined in the surface layers as the stage V copepodites migrated to greater depths (see Fulton, 1973 for a life history of *Calanus plumchrus*). Two other species of herbivorous copepods were relatively abundant throughout the period from March to October. Unlike *Calanus plumchrus*, these species (*C. pacificus* and *Pseudocalanus minutus*) have several generations per year. The major species and genera of carnivorous zooplankton were *Sagitta elegans*, *Pleurobrachia* and *Philidium*; data on these animals were sporadic but they appear in Fig. 6.4 to reach maximum numbers during July and August. During the period from April to July large numbers of juvenile salmon enter the strait and feed off the zooplankton community. Most of the juvenile salmon remain in the strait for two or three months before migrating to the open waters of the Pacific. Larval fish of other commercially important species, including herring, cod and

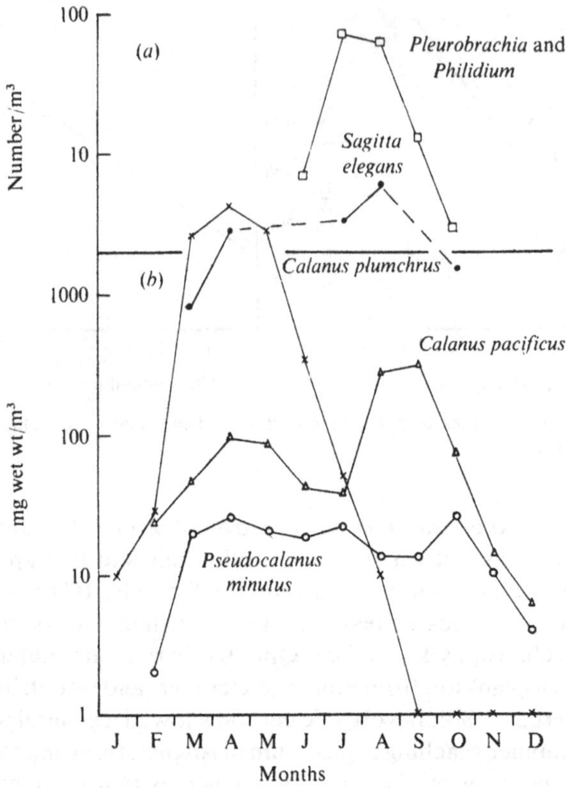

Fig. 6.4. Changes in the principal species of zooplankton in the Strait of Georgia (1965–8 average values for 20 m vertical haul using a 350 μm Hensen net). (*a*) Carnivores; (*b*) herbivores.

various flat fish also show maximum abundance during the early spring, from February to June (Parsons *et al.*, 1970).

Processes governing the production of plankton in the Strait of Georgia

Primary productivity

The principal feature of the primary productivity of the Strait of Georgia is the influence of the Fraser River on the spring-bloom (Hutchinson & Lucas, 1931; Parsons, Stephens & LeBrasseur, 1969). The effect of a large zone of brackish water, known as the Fraser River plume, is to impart a strong stability which prevents mixing of the water column below the compensation depth during most of the year. This effect can be examined with the use of classical models for the prediction and measurement of primary productivity. Using Sverdrup's (1953) critical depth model it is

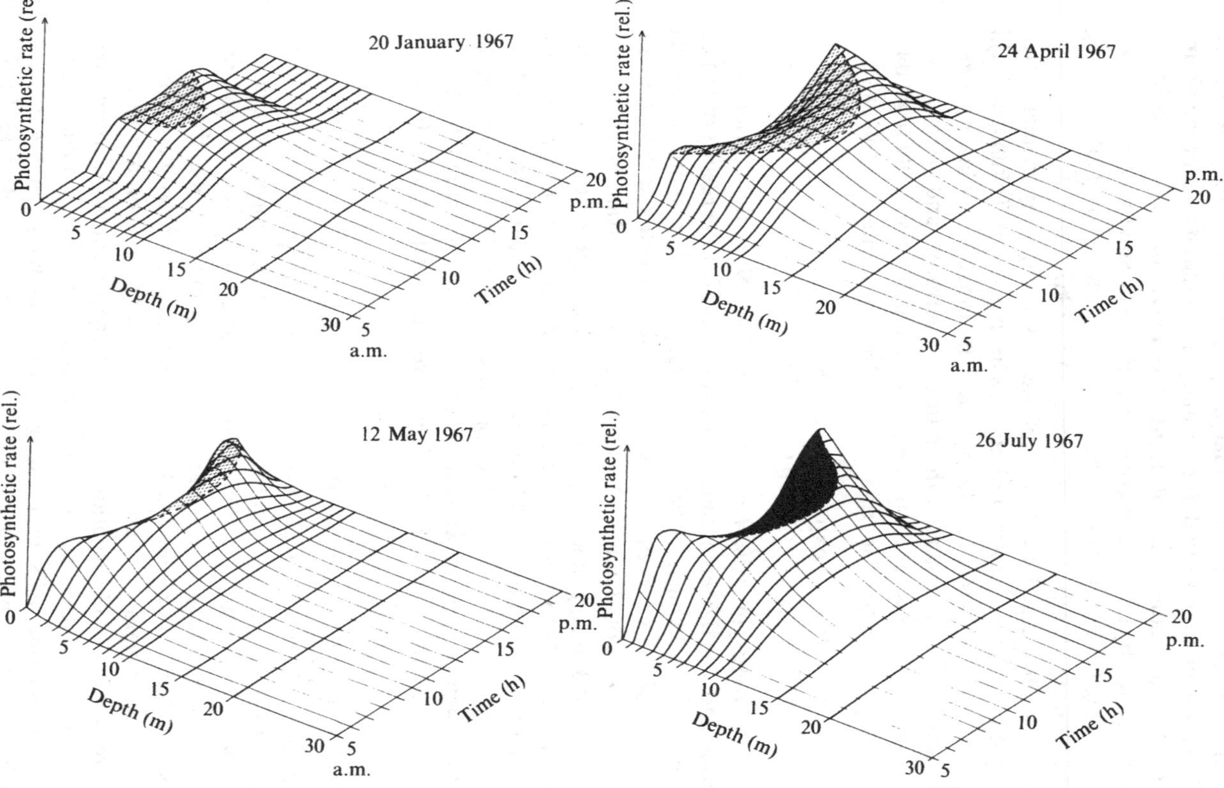

Fig. 6.5. Diurnal changes in vertical photosynthetic rate. The relative photosynthetic rate is determined by the light intensity, including photo-inhibition at high light intensities (shaded areas). Two kinds of photo-inhibition are apparent; one associated with low temperatures (light shading) and the other with low nutrients (dark shading). Areas not shaded indicate photosynthetic rate limitation due to light only. From Takahashi, Fujii & Parsons (1973).

possible to show that some production could occur below the Fraser River plume throughout the year but that only from March to October are conditions sufficiently settled for primary productivity to proceed unrestricted by the amount of light in the mixed layer of the water column. This is illustrated by a tenfold increase in primary productivity values from 124 to 140 mg C/m²/day during February, to values of 1135–2132 mg C/m²/day during March/April, as measured in 1967.

A more sophisticated treatment of factors controlling primary productivity under the Fraser River plume has been given by Takahashi, Fujii & Parsons (1973). In this study the productivity model described by Steele (1962) was used together with nutrient (nitrate) and temperature restrictions. The two latter factors were formulated empirically from data collected in the Strait of Georgia and can be represented as follows:

$$P = a \cdot I \exp\,[1 - a \cdot I\,(1 + h \cdot [N]/g\,[N])]$$

and,

$$P = a \cdot I \exp\,(1 - a \cdot I/c + d \cdot T)$$

where P is photosynthesis, I is light intensity, $[N]$ is the nutrient concentration and T is the temperature. An illustration of the results obtained are given in Fig. 6.5 which shows diurnal changes in photosynthesis during four months of the year. From this study it was concluded that the lack of solar radiation could inhibit photosynthesis in the surface layers during December and January, and at greater depths throughout the year. Temperature caused photosynthetic inhibition from February through to April and nutrients (nitrate) restricted photosynthesis from July to August. In addition to these factors, the effect of Fraser River silt in increasing the extinction coefficient of the water and decreasing primary productivity was apparent during the annual Fraser River freshet, from May to June.

Secondary and tertiary production

During the study period of this programme the most abundant secondary producer to occur in the Strait of Georgia was the copepod, *Calanus plumchrus*. The biomass of this organism started to increase from February to March in response to the increase in primary productivity. The greatest mortality took place in the naupliar stages which were present in the surface layers over an extended period from February to April. During the period of observation *in situ*, the growth rate of *C. plumchrus* ranged from 3.5 to 14% per day with food intake values ranging from 6 to 60% of the animal's body weight/day depending on the type of phytoplankton present in the water at any one time. It was further apparent that except for the high naupliar mortality, estimates of zooplankton productivity matched actual zooplankton production *in situ* to an order of magnitude. From these observations it was concluded that the

Table 6.1. *Occurrence of larval (L) and anadromous juvenile (J) fish in the Strait of Georgia*

Species caught	Jan.	Feb.	Mar.	Apr.	May	June	July	Aug.	Sept.	Oct.	Nov.	Dec.
Lampetra ayresi	—	—	—	J	J	J	—	—	—	—	—	—
Clupea pallasi	—	—	L	L	L	L	L	—	—	—	—	—
Engraulis mordax	—	—	—	—	—	J	J	—	—	—	—	—
Oncorhynchus sp[a]	—	—	—	J	J	J	J	—	—	—	—	—
Thaleichthys pacificus	—	—	L	L	L	L	L	—	—	—	—	—
Hypomesus pretiosus	—	—	—	—	—	—	L	L	L	—	—	—
Mallotus villosus	L	L	—	—	—	—	—	—	—	L	L	L
Leuroglossus stilbius	L	L	L	L	L	—	—	—	—	—	—	—
Merluccius productus	—	—	L	L	L	L	L	—	—	—	—	—
Theragra chalcogrammus	—	L	L	L	L	L	—	—	—	—	—	—
Pleuronectidae[b]	—	—	L	L	L	L	L	L	L	—	—	—
Hexagrammos decagrammus	L	L	L	L	L	—	—	—	—	—	—	—
Hexagrammos stelleri	L	L	L	L	L	—	—	—	—	—	—	—
Ophiodon elongatus	L	L	L	L	L	—	—	—	—	—	—	—
Sebastodes sp.	—	—	—	L	L	L	L	—	—	—	—	—
Scorpaenichthys marmoratus	—	L	L	L	L	L	—	—	—	—	—	—
Hemilepidotus hemilepidotus	—	L	L	L	L	L	—	—	—	—	—	—
Gilbertidia sigalutes	—	L	L	L	L	L	—	—	—	—	—	—
Agonopsis emmelane	—	—	—	L	L	L	—	—	—	—	—	—
Gasterosteus aculeatus	—	—	J	J	J	J	J	—	—	—	—	—
Syngnathus griseolineatus	—	—	—	—	—	—	L	L	L	—	—	—
Coryphopterus nicholsii	—	—	—	—	—	L	L	—	—	—	—	—
Bathymaster signatus	—	—	L	L	—	—	—	—	—	—	—	—
Stichaeidae	L	L	L	L	L	L	L	—	—	—	—	—
Lumpenus sagitta	—	—	L	L	L	—	—	—	—	—	—	—
Lyconectes aleutensis	—	L	L	L	L	L	—	—	—	—	—	—
Ammodytes hexapterus (larval)	—	L	L	L	L	L	—	—	—	—	—	—

[a] Including all five species of juvenile salmon.
[b] Including *Platichthys stellatus*, *Hippoglossoides elassodon*, *Isopsetta isolepis*, and *Lyopsetta exilis*.

zooplankton exerted considerable pressure on the growth of the phytoplankton population. But the zooplankters themselves were not heavily predated by the next trophic level, which included many of the larval and juvenile stages of commercially important fish (Parsons, Le Brasseur, Fulton & Kennedy, 1969).

An interesting sequel to these events has occurred since the programme was concluded in 1968. From 1970 to 1974 there has been a gradual decline in the numbers of *Calanus plumchrus* occurring in the Strait of Georgia to a point where they now appear to be no longer the predominant species during the spring-bloom (Gardner, 1977). Their place appears to have been taken by another species of copepod, *Calanus marshallae*. This species has a different life cycle to *C. plumchrus* in that it can produce

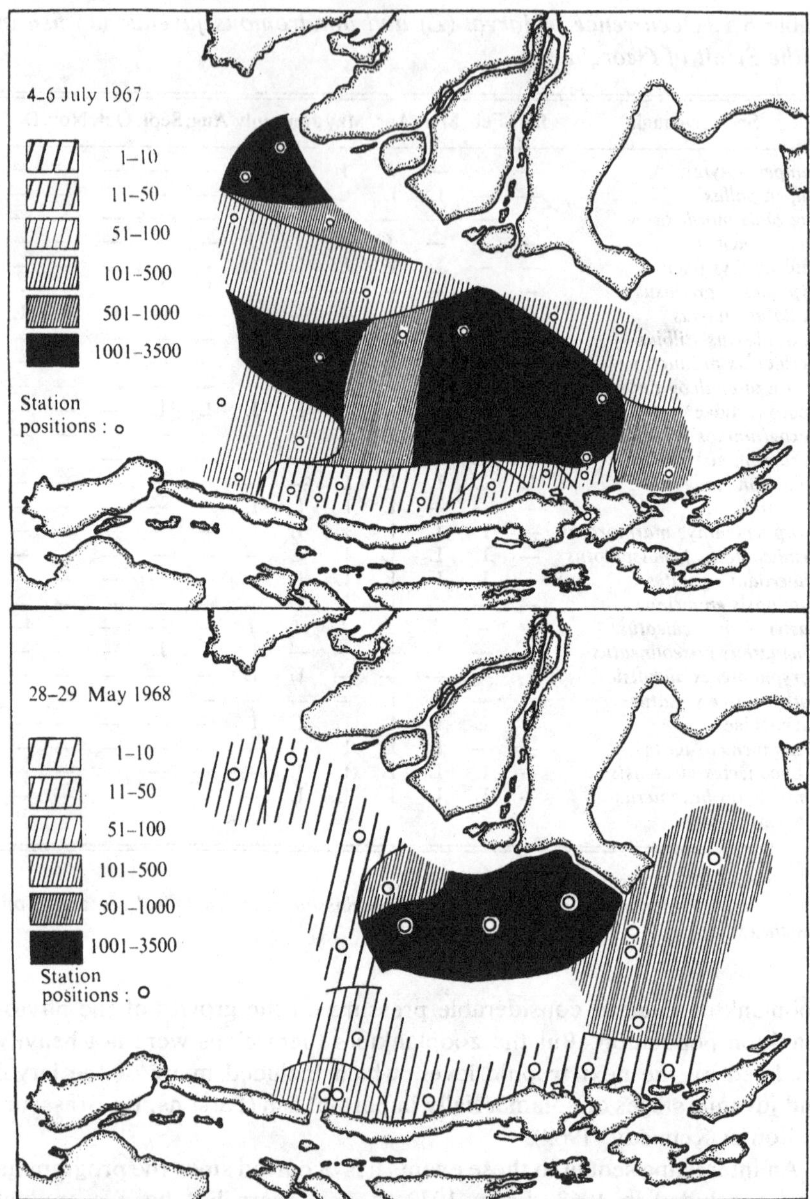

Fig. 6.6. Numbers of herring larvae, *Clupea pallasii* (10–35 mm), encountered off the Fraser River during July 1967 and May 1968 (Numbers per 10 min tow using a surface trawl). From Barraclough.

several generations per year and is not confined to a single period in which eggs are laid, as is the case with *C. plumchrus*. In this sense *C. marshallae* might be described as a more opportunistic species, taking advantage of any increases in plankton productivity regardless of whether it occurs in the early spring or later in the year.

The occurrence of larval and juvenile fish in the Strait of Georgia is shown in Table 6.1 from Parsons, Le Brasseur & Barraclough (1970). The data illustrate that the maximum numbers of fish occur during April and May which coincides with the maximum biomass of zooplankton. The importance of the Fraser River plume in the distribution of larval and juvenile fish has been illustrated by Barraclough. Fig. 6.6 has been taken from these studies as the one example which shows the maximum concentration of larval herring as occurring in the plume during July 1967 and May 1968. Juvenile fish (salmonid) feeding experiments conducted in the Fraser River plume during the spring of 1967 (LeBrasseur *et al.*, 1969) further indicated that the high standing stock of zooplankton was an integral part of the food requirements of young fish during their first few weeks of sea life.

The relationships between the young stages of fish and their food was examined further by Parsons & LeBrasseur (1970) and LeBrasseur (1969). In the latter reference it was possible to show that feeding of young salmonids was concentration-dependent up to a maximum ration which allowed for a growth rate of 5.4% per day. Prey size selectivity for these juvenile salmon was shown to favour *Calanus plumchrus* as opposed to larger organisms (*Euphausia pacifica*), or smaller organisms (*Pseudocalanus minutus*). Further it was apparent that growth efficiency of young salmonids declined with increasing ration as suggested by Palaheimo & Dickie (1966). From later work by Parker (1971) it is also apparent that the need for high growth rates based on adequate plankton is important in the case of pink salmon survival so that the animals grow fast enough to outstrip the growth rate of their predators. Thus becoming large, as soon as possible, appears to be an important mechanism for survival to juvenile salmon.

Bacterial interactions and detritus

Since it is generally recognized that a large portion of the organic material in the sea is in the form of detritus, some effort was made to determine the role of detrital material and associated bacterial populations in the trophic ecology of the strait. The input of organic carbon to a benthic community was studied over a period of one year (Stephens, Sheldon & Parsons, 1967) and the material collected in sediment jars was used in feeding experiments on *Artemia* (Seki, Skelding & Parson, 1968). From these studies it was apparent that two periods of maximum detrital input to the benthic community occurred during the late-spring and in the autumn. The first rain of detrital material was largely phytogenous and was associated with the spring-bloom of phytoplankton. The second was associated with terrigenous material washed off the land with the first autumn rains. In laboratory feeding tests it was found that both these sources of organic carbon could be utilized by *Artemia*, but only if they were first degraded by bacteria.

In other studies on bacteria the difference in salt tolerance between halophiles and non-halophiles in an estuarine environment was determined to be at a salinity of 19‰, which agreed closely with Larsen's (1962) definition. The role of bacteria as food items for zooplankton (Seki & Kennedy, 1969; Seki, 1971), or fish (Seki, 1969), in the pelagic environment was not generally considered to be as important as their role in the benthic community in association with the rain of detritus as described above.

An important discovery associated with yeasts in the Strait of Georgia was the presence of an infectious species (*Metschnikowia* sp.) which is believed to cause high mortality among the population of *Calanus plumchrus* during their period of hibernation in deep water (Seki & Fulton, 1969).

Trophic models

During the period of the IBP Strait of Georgia Programme, McAllister (1969, 1970) developed a number of equations which have helped to better define the relationship between phytoplankton and zooplankton production. In particular the question of continuous versus nocturnal grazing was examined with respect to the total effect on estimates of marine production. For a given set of data and coefficients it could be shown that effective phytoplankton production was least under conditions of continuous zooplankton feeding compared with nocturnal feeding. Furthermore, a relatively small difference in primary production under these two grazing regimes could be shown to have large effects on estimates of secondary production.

In another assessment of an important problem in plankton studies, McAllister (1970) determined the effect of averaging patchiness in plankton distributions on estimates of production. From these simulations it was apparent that any attempt to average concentrations in order to project the eventual production of a body of water could theoretically lead to serious errors within a time period of *c.* 10 days. However in actual studies on patchiness in the sea, (Platt, Dickie & Trites, 1970) showed that in practice major seasonal trends could be studied from a single point since patchiness variations are in a sense self-correcting, and seasonal abundance in fact predominates in data on biological changes in temperate latitudes.

Discussion of IBP Strait of Georgia Programme in relation to other aquatic systems

Twenty years ago the general pattern of plankton production studies was to determine geographical and seasonal differences in the rate and standing stock of primary and secondary producers. The most popular methodology available at the time involved using $^{14}CO_2$ uptake to measure primary productivity and chlorophyll *a* to measure the standing stock of primary producers. Zooplankton measurements were generally confined to wet weight estimates of standing stock and a general description of life cycles. The net result of these studies was the development of a number of models (e.g. Riley, 1946; Steele, 1962) which were correct to an order of magnitude in their assessment of pelagic productivity. It became apparent, however, that both the data and the models were insufficient in their explanation when it came to serving as a basis for the assessment of higher trophic levels (Steele, 1964).

With the advent of better techniques in nutrient analysis, such as the development of the Autoanalyser and ^{15}N studies on nitrogen metabolism, differences in the nutrient uptake kinetics of different species of phytoplankton became apparent (e.g. Dugdale & Goering, 1967; Eppley, Rogers & McCarthy, 1969). At the same time feeding experiments with herbivorous zooplankton (e.g. Mullin & Brooks, 1967; Conover, 1966) as well as carnivorous plankton (e.g. Reeve, 1964) gave some insight into species preferences in predator–prey relationships. Ivlev's (1961) study on the feeding of fishes, although not conducted with species of particular interest to marine scientists, gave some indication of the trophic relationships that may exist between food availability and the feeding requirements of planktonivorous fish.

Within this framework of data there was, however, an apparent lack of ways in which to express community structure. While the most satisfying means of expressing the organization of a community would be to account for all the interactions at a species level (Hairston, 1959) it is quite apparent that the amount of data required for this exercise would be impossible to conceive in terms of current and future research efforts. The solution to this problem may lie in being able to express the principal components of a community in terms of size groups. An original idea in this direction was given by Isaacs (1966) in considering the size spectrum of food items available to sardines. The further development of data on the size spectrum of organisms in the plankton community has been an integral part of the Strait of Georgia Programme (e.g. Parsons, 1969;

Parsons & LeBrasseur, 1970). Other recent reports have also shown the feasibility of expressing the size of pelagic marine organisms in terms of various environmental, physiological and feeding requirements (e.g. Sheldon, Prakash & Sutcliffe, 1972; Poulet, 1973; Parsons & Takahashi, 1973). From metabolic and growth studies it can be shown, for example, that functional size differences exist throughout the plankton community. Thus the total metabolic needs of planktonic organisms are higher per unit body weight for the smallest animals (Ikeda, 1970) which leads to a higher growth efficiency during exponential growth of zooplankton (Mullin & Brooks, 1970; Makarova & Zaika, 1971; see Parsons, 1976, for discussion). From a combination of these observations it may be possible to develop more respresentative simulation models of actual events without having to resort to the complexity of individual species-interactions.

Further progress on the role of detritus and feedback mechanisms in the marine food web have been discussed in recent reviews by Nishizawa (1969) and Riley (1963). In this respect it is apparent from studies initiated in the Strait of Georgia that there is a considerable input of organic detritus from the land to the sea in some nearshore environments. It is further apparent from studies conducted on the east coast of Canada (e.g. Mann, 1972) that seaweed communities may produce quantities of organic material similar in amount to the primary productivity of phytoplankton in coastal waters and that a large part of the macrophyte production may enter a detrital food chain. The kinetics of detrital decay has been discussed recently by Nakajima & Nishizawa (1972). From environmental data on the distribution of detritus with depth they postulated an exponential decay in which the rate of decay (k) was proportional to the initial concentration of detritus. The general form of their expression appears to be equally true for the disappearance of detritus in time following a single, large input to a system, such as following a phytoplankton bloom or during periods of high runoff from the land. This could then be represented as

$$C = C_0 \exp^{-k(t-t_0)}$$

where C_0 is the maximum concentration of detritus at time t_0 and C is the concentration at time t.

In summary it appears that studies on the productivity of the lower trophic levels in the marine environment are becoming more detailed and moving from the concept of a food chain (in which one considers plants, zooplankton and fish as single trophic levels) to concepts of food webs (in which the major components within a trophic level are considered). Greater attention is now being paid not only to the productivity but also to the stability of aquatic ecosystems. In this respect it is apparent that mathematical models of trophodynamic relationships in the plankton community still fall far short in their accountability to actual events. Too often the model that 'works' is the one that has had just the 'right' coefficients added and small changes cause the computerized system to collapse. Yet nature, under constant attack from various environmental and man made perturbations, does not readily show the same tendency towards total extinction. The ability of a plankton community to maintain itself can be measured in terms of a stability index (e.g. Patten, 1962). However the representation of stability in finite terms, equivalent to those used to define productivity, appears to be a new area of research for aquatic scientists.

I would like to thank all the scientists and technicians who took part in the Strait of Georgia Programme for showing such good cooperation with their associates. Also a great deal of credit for the field programme must go to the captains and crews of vessels which were used during this study.

References

Barraclough, W. E. (1974). Distribution maps of larval and juvenile fish in the Strait of Georgia, British Columbia during the spring and summer months, 1966–1969. *Fish. Res. Bd. Tech. Rept.* (in preparation).

Cattell, S. A. (1973). The seasonal cycle of vitamin B_{12} in the Strait of Georgia, British Columbia. *Journal of the Fisheries Research Board of Canada*, **30**, 217–22.

Conover, R. J. (1966). Assimilation of organic matter by zooplankton. *Limnology and Oceanography*, **11**, 338–45.

Dugdale, R. C. & Goering, J. J. (1967). Uptake of new and regenerated forms of nitrogen in primary productivity. *Limnology and Oceanography*, **12**, 196–206.

Eppley, R. W., Rogers, J. M. & McCarthy, J. J. (1969). Half saturation constants for uptake of nitrate and ammonium by marine phytoplankton. *Limnology and Oceanography*, **14**, 912–20.

Fulton, J. (1973). Some aspects of the life history of *Calanus plumchrus* in the Strait of Georgia. *Journal of the Fisheries Research Board of Canada*, **30**, 811–15.

Gardner, G. (1977). Analysis of zooplankton population fluctuations in the Strait of Georgia, British Columbia. *Journal of the Fisheries Research Board of Canada*, **34**, 1196–1206.

Hairston, N. G. (1959). Species abundance and community organization. *Ecology*, **40**, 403–16.

Hutchinson, A. H. & Lucas, C. C. (1931). The epithalassa of Georgia Strait. *Canadian Journal of Research*, **5**, 231–84.

Ikeda, T. (1970). Relationship between respiration rate and body size in marine plankton animals as a function of temperature of the habitat. *Bulletin of the Faculty of Fisheries, Hokkaido University*, **21**, 91–112.

Isaacs, J. D. (1966). Larval sardine and anchovy interrelationships. *Reports of the Californian Cooperative Oceanic Fisheries Investigations*, **10**, 102–13.

Ivlev, V. S. (1961). *Experimental ecology of the feeding of fishes*, transl. D. Scott. New Haven: Yale University Press.

Larsen, H. (1962). Halophilism. In The bacteria: A treatise on structure and function. Vol. IV, *the physiology of growth*, ed. I. C. Gunsalus & R. Y. Stanier, pp. 297–342. New York: Academic Press.

LeBrasseur, R. J. (1969). Growth of juvenile chum salmon (*Oncorhynchus keta*) under different feeding regimes. *Fisheries Research Board of Canada*, **26**, 1631–45.

LeBrasseur, R. J., Barraclough, W. E., Kennedy, O. D. & Parsons, T. R. (1969). Production studies in the Strait of Georgia. III. Observations on the food of larval and juvenile fish in the Fraser River plume, February to May, 1967. *Journal of Experimental Marine Biology and Ecology*, **3**, 51–61.

McAllister, C. D. (1969). Aspects of estimating zooplankton production from phytoplankton production. *Journal of the Fisheries Research Board of Canada*, **26**, 199–220.

McAllister, C. D. (1970). Zooplankton rations, phytoplankton mortality and the estimation of marine production. In *Marine food chains*, ed. J. H. Steele, pp. 419–57. Edinburgh: Oliver & Boyd.

Makarova, N. P. & Ye. Zaika, V. (1971). Relationship between animal growth and quantity of assimilated food. *Hydrobiological Journal*, **7**, 1–8.

Mann, K. H. (1972). Ecological energetics of the seaweed zone in a marine bay on the Atlantic coast of Canada. *Marine Biology*, **12**, 1–10.

Mullin, M. M. & Brooks, E. R. (1967). Laboratory culture, growth rate and feeding behaviour of a planktonic copepod. *Limnology and Oceanography*, **12**, 657–66.

Nakajima, K. & Nishizawa, S. (1972). Exponential decrease in particulate carbon concentration in a limited depth interval in the surface layer of the Bering Sea. In *Biological oceanography of the northern North Pacific Ocean*, ed. A. Y. Takensuti, pp. 495–505. Tokyo: Idemitsu Shoten.

Nishizawa, S. (1969). Suspended material in the sea. II. Re-evaluation of the hypotheses. *Bulletin of the Plankton Society of Japan*, **16**, 1–42.

Palaheimo, J. E. & Dickie, L. M. (1966). Food and growth of fishes. III. Relation among food, body size and growth efficiency. *Journal of the Fisheries Research Board of Canada*, **23**, 1209–48.

Parker, R. R. (1971). Size selective predation among juvenile salmonid fishes in a British Columbia inlet. *Journal of the Fisheries Research Board of Canada*, **28**, 1503–10.

Parsons, T. R. (1969). The use of particle size spectra in determining the structure of a plankton community. *Journal of the Oceanographical Society of Japan*, **25**, 172–81.

Parsons, T. R. (1976). The structure of life in the sea. In *The ecology of the sea*, ed. D. H. Cushing & J. Walsh, pp. 81–97. Toronto: W. B. Saunders Co.

Parsons, T. R., LeBrasseur, R. J., Fulton, J. D. & Kennedy, O. D. (1969). Production studies in the Strait of Georgia. II. Secondary production under the Fraser River plume, February to May, 1967, *Journal of Experimental Marine Biology and Ecology*, **3**, 39–50.

Parsons, T. R. & LeBrasseur, R. J. (1970). The availability of food to different trophic levels in the marine food chain. In *Marine food chains*, ed. J. H. Steele, pp. 325–43. Edinburgh: Oliver & Boyd.

Parsons, T. R., LeBrasseur, R. J. & Barraclough, W. E. (1970). Levels of production in the pelagic environment of the Strait of Georgia, British Columbia. A review. *Journal of the Fisheries Research Board of Canada*, **27**, 1251–64.

Parsons, T. R., Stephens, K. & LeBrasseur, R. J. (1969). Production studies in the Strait of Georgia. I. Primary production under the Fraser River plume, February to May, 1967. *Journal of Experimental Marine Biology and Ecology*, **3**, 27–38.

Parsons, T. R. & Takahashi, M. A. (1973). Environmental control of phytoplankton cell size. *Limnology and Oceanography*, **18**, 511–15.

Patten, B. C. (1962). Improved method for estimating stability in plankton. *Limnology and Oceanography*, **7**, 266–8.

Platt, T., Dickie, L. M. & Trites, R. W. (1970). Spatial heterogeneity of phytoplankton in a near shore environment. *Journal of the Fisheries Research Board of Canada*, **27**, 1453–1473.

Poulet, S. A. (1973). Grazing of *Pseudocalanus minutus* on naturally occurring particulate matter. *Limnology and Oceanography*, **18**, 564–73.

Reeve, M. R. (1964). Feeding of zooplankton with special reference to some experiments with *Sagitta. Nature*, **201**, 211–13.

Riley, G. A. (1946). Factors controlling phytoplankton populations on Georges Bank. *Journal of Marine Research*, **6**, 54–73.

Riley, G. A. (1963). Organic aggregates in seawater and the dynamics of their formation and utilization. *Limnology and Oceanography*, **8**, 372–81.

Seki, H. (1969). Marine microorganisms associated with the food of young salmon. *Applied Microbiology*, **17**, 252–5.

Seki, H. (1971). Microbial clumps in seawater in the euphotic zone of Saanich Inlet. (British Columbia). *Marine Biology*, **9**, 4–8.

Seki, H. & Fulton, J. (1969). Infections of marine copepods by *Metschnikowia* sp. *Mycopathologia et Mycologia Applicata*, **38**, 61–70.

Seki, H. & Kennedy, O. D. (1969). Marine bacteria and other heterotrophs as food for zooplankton in the Strait of Georgia during the winter. *Journal of the Fisheries Research Board of Canada*, **26**, 3165–73.

Seki, H., Skelding, J. & Parsons, T. K. (1968). Observations on the decomposition of a marine sediment. *Limnology and Oceanography*, **13**, 440–7.

Seki, H., Stephens, K. V. & Parsons, T. R. (1969). The contribution of allochthonous bacteria and organic materials from a small river into a semi-enclosed sea. *Archiv für Hydrobiologie*, **66**, 37–47.

Sheldon, R. W., Prakash, A. & Sutcliffe, W. H. Jr (1972). The size distributon of particles in the ocean. *Limnology and Oceanography*, **17**, 327–40.

Steele, J. H. (1962). Environmental control of photosynthesis in the sea. *Limnology and Oceanography*, **7**, 137–50.

Steele, J. H. (1964). Some problems in the study of marine resources. ICNAF Environmental Symposium Rome, 1964. Contrib. No. *C-4*, pp. 11.

Stephens, K., Sheldon, R. W. & Parsons, T. R. (1967). Seasonal variations in the availability of food for benthos in a coastal environment. *Ecology*, **48**, 852–5.

Sverdrup, H. V., Johnson, M. W. & Fleming, R. H. (1946). The oceans – their physics, chemistry and general biology. New York: Prentice Hall Inc.

Sverdrup, H. V. (1953). On conditions for the vernal blooming of phytoplankton. *Journal du Conseil Permanent International pour l'Exploration de la Mer*, **18**, 287–95.

Takahashi, M., Fujii, K. & Parsons, T. R. (1973). Simulation study of phytoplankton photosynthesis and growth in the Fraser River estuary. *Marine Biology*, **19**, 102–16.

Tully, J. P. & Dodimead, A. J. (1957). Properties of the water in the Strait of Georgia, British Columbia, and influencing factors. *Journal of the Fisheries Research Board of Canada*, **14**, 241–319.

Waldichuck, M. (1957). Physical oceanography of the Strait of Georgia, British Columbia. *Journal of the Fisheries Research Board of Canada*, **14**, 321–486.

7. Biological production in the Gulf of St Lawrence

M. J. DUNBAR

Introduction

One of the marine projects sponsored by the Canadian Committee for IBP was the study of the biological production of the Gulf of St Lawrence. This project, planned and supervised by D. M. Steven and M. J. Dunbar and carried out under the field direction of Steven, involved four field seasons, 1969–1972, many cruises in the 96-foot vessel *Ambrose Foote* from May to September of each year, and the operation of five shore stations at Rimouski, Grande Rivière, Cap Rouge in the Magdalen Islands (Quebec), Ellerslie (Prince Edward Island), and Bonne Bay (Newfoundland) (Fig. 7.1). A sixth shore station, operated only in 1970, was at Matamek, near Seven Islands on the north shore.

Five manuscript reports contain the raw data produced by this work, and a general comprehensive summary of results (Bulleid & Steven, 1972; Steven *et al.*, 1973; El-Sabh, Glombitza & Johannessen, 1971). Published papers arising from the material, in whole or in part, include Hoffer (1972) on the hyperiid amphipods, Berkes (1975) on feeding mechanisms in euphausiids, and Spence & Steven (1974) on phytoplankton pigments. Other papers are in press, and a number of graduate student theses have come out of this study. A summary of results has recently appeared (Steven, 1975) as a contribution to the Canadian IBP volume published by the Royal Society of Canada. Dr Steven died in July 1974, while he was planning to write another paper with the present writer for this volume, a task which now has fallen to me. In this paper I shall summarize the results and dwell upon work done within the program since Steven's paper went to press.

First a word on the economic significance of the Gulf of St Lawrence. The Gulf is by far the most productive, in biological terms, of all the seas of the Canadian coasts. Until quite recently it produced about 40% of all Canadian sea fish landings, by weight, and some 27% by dollar value. Approximately 65% of Canadian shipping sails through it. It provides the sea access to the largest Canadian cities, and to a very large proportion of the population and industrial centres of the United States. It has great recreational value and potential. It displays environmental conditions and fauna ranging from the subtropical to the sub-Arctic. There are good prospects in its floor for exploration for minerals, including oil and gas.

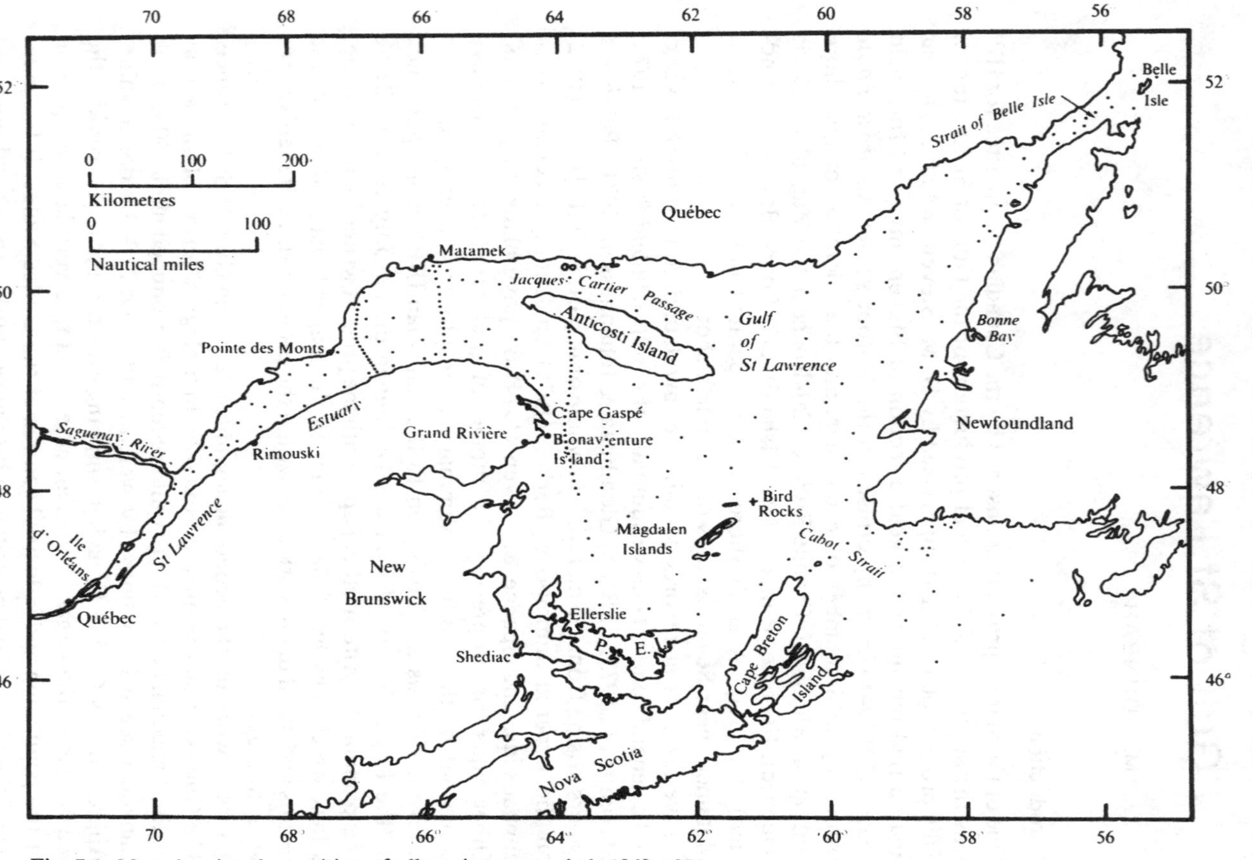

Fig. 7.1. Map showing the position of all stations occupied, 1969–1972.

It serves also as an open sewer for very large upstream populations. It has been declared an exclusive Canadian fishing zone. In short, this marginal sea, with an area equal to that of the Maritime Provinces (about 214000 km^2) has a far-reaching impact on the economic and social life of the eastern and central regions of the country, and vice versa.

Water masses and circulation

The 1915 Canadian Fisheries Expedition (Hjort, 1919), the first significant study of the Gulf, established the existence of the celebrated 'cold layer', which in summer separates the warm upper layer from the deep water below. This cold layer extends from about 50 m down to some 150 m, varying somewhat in time and place. The salinities in it are of the order of 31–33‰, and the temperatures lie at or below 0 °C. It extends outside the Gulf through Cabot Strait to the Scotian Shelf. It was long thought that the cold layer consisted of Arctic or sub-Arctic water entering through the Strait of Belle Isle, or more probably through Cabot Strait, but more recent study has shown that most or all of it must be formed *in situ* by a process of winter cooling (Forrester, 1964). It is interesting, however, that this cold layer contains planktonic species of distinctly Arctic affinities.

Although the water mass pattern shows a three-layered structure in summer, the basic pattern is two-layered, the pattern shown in winter, when there is a cold layer varying in thickness from the surface to 100 or 150 m with temperatures as low as −1.7 °C and salinities from about 25 to 33‰. This is underlain by the deep water extending to the bottom of the Laurentian Channel, with temperatures mainly between 4 and 5 °C (and up to 6 °C) and salinities close to 34.6‰. The temperature of the deep water varies, not seasonally, but rather in response to changes in the same water mass outside the Gulf (Trites, 1971; Lauzier & Bailey, 1957).

The St Lawrence estuary, as would be expected, shows the typical estuarine circulation and entrainment effect. Fresh water from the river outflow entrains salt water from below, and this process of entrainment and mixing increases seaward, causing a return counter-current at depth and upwelling of deeper water. Most of the upwelling appears to occur opposite the mouth of the Saguenay fjord, where the bottom shelves rapidly, and upwelling in that area has been amply demonstrated. It is not quite clear yet, however, whether the deep layer, below the cold layer, is involved in this upwelling or not. Tidal forces are also of first rank importance in engendering and maintaining the upwelling in the sill region. Upwelling of the cold layer water is clearly shown by the temperature regime alone, and the deep, warmer layer is very probably also involved at certain times.

Thus there exists a nutrient pump in the estuary, a feature normal to estuaries; nevertheless the question of the sources of nutrients to the Gulf system as a whole, and even to the estuary itself, is by no means closed. Following the completion of the IBP Gulf Project, there was an almost general acceptance of the view that the upwelling mechanism described above could account for all the nutrient supply to the estuary (or 'lower estuary', to be more precise); the region between the mouth of the Saguenay and Pointe des Monts (Fig. 7.1). But nutrient measurements in the middle and upper estuaries (up to Quebec City) have not been made to our complete satisfaction. More recent work has shown that nutrients coming down the river are by no means insignificant, and possibly, sometimes, dominant. To quote from the abstract of Greisman & Ingram (1978):

> The relationship of nutrient concentration to other water mass properties has been examined over the entire St Lawrence estuary. Average nutrient flux values have been computed over a semi-diurnal tidal cycle at a number of sites. Evidence of nutrient upwelling associated with strong vertical mixing at the head of the Laurentian Channel was found. However, in contrast to some earlier findings, the major source of nutrients to the estuarine system during the period of our study appeared to be from the river input and not from the upwelling of deep, nutrient rich water at the head of the Laurentian Channel. The relative importance of the two nutrient sources is thought to vary with the fortnightly tides. The upwelling mechanism is probably dominant during periods of spring tides (a few days each month), while the land derived nutrients are probably the major source at other times and, on the average, the more important.

It may well be surmised that this importance of the river in the supply of nutrients to the Gulf is not a natural phenomenon unaided by man. The St Lawrence system is highly polluted. There is a great deal of agricultural activity on the banks of the Great Lakes and the river, and also of industrial development and domestic settlement, all of which supplies the system with much chemical effluent, not least nitrates and phosphates.

The work described by Greisman & Ingram is part of the results of a cruise in the CS *Dawson*, in June/July 1975, a joint Bedford Institute–McGill University enterprise with Dr K. Kranck as chief scientist. Nitrate concentrations, both at the surface and at 20 m were higher upstream of the Saguenay (8–12 mg-at/m^3) than below the Saguenay (6–10 mg-at/m^3), and the highest values were found at Ile d'Orléans, just below Quebec City. The phosphate and silicate concentrations, measured by the present writer, have not yet been published. In agreement with the nitrate pattern, the values in the river and upper estuary are high, approximately equal to the concentrations downstream (between 5 and 14 μg-at/l upstream as compared with 4 to 13 μg-at/l downstream, for silicate; and between 0.5 and 1 μg-at/l upstream compared with 0.6 to 1 μg-at/l downstream, for phosphate – all surface values).

The horizonal surface circulation of water in the Gulf is well known, at least during the ice-free period. The winter circulation during the period of ice cover (approximately January–April) is less well understood. For the summer pattern, I quote the summary given by Steven (1975):

> A strong surface current flows eastward from the St Lawrence Estuary along the north shore of the Gaspé Peninsula through the Gaspé Passage and continues as a slower drift in the anticlockwise direction across the Magdalen Shallows and eventually to Cabot Strait. Most of the water leaving the Gulf passes through the southern part of Cabot Strait, where strong surface currents are found. The circulation in the eastern basin of the Gulf is also generally anticlockwise, proceeding north along the Newfoundland coast and west along the north shore of Quebec. A smaller anticlockwise gyre is situated to the west of Anticosti. Current speeds are highly variable, being greatest in the Gaspé Current (10–20 miles/day) and through Cabot Strait and least on the Magdalen Shallows and the northeastern Gulf (0.24–3.5 miles/day). A consequence of this system of gyres is that a large though unknown proportion of the surface water is recirculated repeatedly within the Gulf, thus contributing to its distinctive biological characteristics.

The subsurface circulation is dominated by the two lower summer layers of water described above, both of which move slowly inward from Cabot Strait. The deep layer fills the Laurentian Channel, which at its deepest toward Cabot Strait is some 500 m deep. The estimated seaward transport of water through Cabot Strait is 50–60 times the total of fresh water discharged into the Gulf by all the rivers, which gives a measure of the importance of the entrainment and mixing processes already mentioned. This outflow must be balanced by an equivalent amount of water entering the Gulf from the Atlantic, the greatest proportion of which enters by Cabot Strait, not by way of the Strait of Belle Isle, through which the flow is intermittent and changeable; the Strait of Belle Isle is a heat sink in the system. 'Estimates of the mean flushing time of the Gulf vary from about 200 to more than 500 days, the lower value being thought to be more probable, whereas the mean time for a particle to be carried from the mouth of the Saguenay to Cabot Strait has been estimated at 80–90 days. (Steven, 1975.)

What is known of the winter circulation has been brought together by El-Sabh & Johannessen (1971) and El-Sabh (1976) from the results of six cruises between 1956 and 1970. Using the Defant method for determining a variable reference layer, these authors found three depressions in the reference layer, west and south-east of Anticosti Island and in the middle of Cabot Strait. The main features of the circulation were the existence of two gyres, anticlockwise west and clockwise south-east of Anticosti, and outflow on both sides of Cabot Strait and inflow in the middle. This inflow is diverted to both sides of the Strait and joins the outflowing water. The anticlockwise gyre west of Anticosti is a feature also of the summer circulation, and the Gaspé Current is maintained throughout the winter, but the clockwise gyre southeast of Anticosti has not been observed so far in summer.

Internal subsurface waves of considerable amplitude are a feature of the Gulf of St Lawrence (El-Sabh, Glombitza & Johannessen, 1971), and have considerable biological effects which have not yet been given much attention in the study of biological productivity. There is also good reason to suppose that there is constant, or intermittent, upwelling along the north shore of the Gulf, judging from the pattern of prevailing winds, and it is to be assumed, from the primary production pattern described below, that vertical exchange of water in autumn and winter is important in the fertilization of the euphotic zone, in addition to the 'nutrient pump' at the head of the lower estuary.

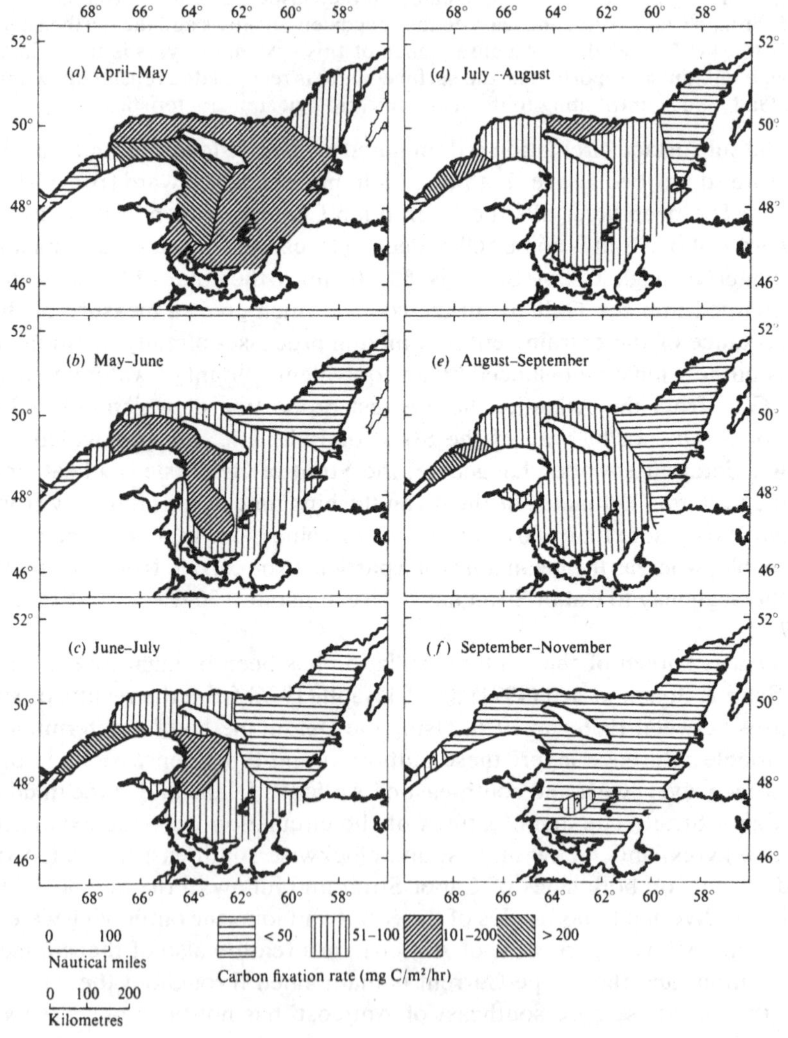

Fig. 7.2. Average carbon fixation rates for 30-day periods from mid-April to mid-September, and for mid-September to November, 1969–72. (From Steven 1976.)

Table 7.1. *Phytoplankton production in various regions*

Location	Area (km²)	Phytoplankton production (tons wet wt/km²/yr)
Gulf of St Lawrence	214000	4670
North Sea	544000	4300
Black Sea	430000	3060
Caspian Sea	461000	5540
Sea of Azov	37600	9040

Primary production

The pattern of primary production demonstrated by the IBP Gulf Project is shown in summary in Fig. 7.2, based on some 600 individual measurements using the ^{14}C technique. No significant early phytoplankton bloom was recorded in the lower estuary, where the highest nutrient concentrations were found, in April/May. High carbon fixation, over 100 mg $C/m^2/h$, was not found there until June/July. What is termed the 'Gaspé Current System' dominates the western part of the Gulf, from the estuary round the Gaspé Peninsula and into the Magdalen Shallows in the southern part of the Gulf. These areas are the direct legatees of the high nutrient levels engendered as described above. In the central and eastern parts of the Gulf, beyond the reach of the Gaspé Current, conditions are oceanic, or maritime, rather than estuarine; there is high production in the early spring (April/May) using nutrients made available by winter vertical exchange, followed by lower values for the rest of the season. The lowest production levels are found in the north-east corner, toward the Strait of Belle Isle, which effectively disposes of the former widely-held belief that the Labrador Current was an important source of nutrient supply to the Gulf.

Total annual primary production in the Gulf, based on the four years' work of the IBP Project, was estimated at 53.37 million tonnes carbon, divided between the three main regions as follows: Estuary (lower estuary), 6.62 million tonnes; Gaspé Current, 9.05 million tonnes; Main Gulf, 37.7 million tonnes. These estimates put the Gulf of St Lawrence high among comparable regions of the world oceans, shown in the table taken from Steven (1975) (Table 7.1). Figures for regions other than the Gulf of St Lawrence are from Moiseev (1971).

Sinclair, El-Sabh & Brindle (1976) estimate the production in the lower estuary as an order of numbers lower than Steven's estimate, but their figure is based on measurement of nutrient transport across a section, not on direct production measurement. These authors also point out that the 'nutrient pump' at the head of the estuary cannot produce the supply of nutrients necessary for the measured production of the whole Gulf, a point which Steven also makes, though with less assurance.

Secondary production

Using the material of the IBP Gulf study and from earlier cruises, Hoffer (1972) published an account of the life cycle of *Parathemisto abyssorum*, the most abundant pelagic amphipod in the Gulf, finding a one-year life cycle and an extended spawning period in spring and summer, with the peak of activity in late April and May. This is in keeping with work on others of the larger zooplankton, in which the life span is normally two years in the Arctic, one year in the boreal waters farther south (see for example, Dunbar, 1941, 1946, 1957, 1962; Weinstein, 1973, for chaetognaths, euphausiids, amphipods). *Parathemisto gaudichaudi*, in both its forms (*compressa* and *bispinosa*) is also found in the Gulf, but less abundantly than *P. abyssorum*. Both species are carnivorous, feeding mostly on copepods. Euphausiid remains were occasionally found in the gut of the *compressa* form of *P. guadichaudi* but in general the feeding habits of *compressa* and *bispinosa* were found to be very similar, leaving the exaggerated length of the fifth leg unexplained. Morphological evidence was obtained questioning the present classification of the two forms as belonging to the same species.

Berkes (1976) has analysed the euphausiid populations of the IBP material in the Gulf and has found that spawning occurs at one year of age, but that individuals may live to spawn a second time, at two years of age. The three species of *Thysanoessa*, *T. inermis*, *T. raschii* and *T. longicaudata*, are spring spawners, starting in April; *Meganyctiphanes norvegica* spawns in summer, mainly from July to September.

Adults of *M. norvegica*, *T. inermis* and *T. raschii* occurred most abundantly in the western part of the Gulf and their larvae in the Magdalen Shallows, showing passive transport of the eggs and larvae by the Gaspé Current. *T. longicaudata*, on the other hand, which is an oceanic species, occurred mainly in the eastern Gulf all year round. All four species were found to be omnivorous, but *Meganyctiphanes* and *T. longicaudata* appeared to eat more animal matter than the other two species (Berkes, 1976).

Berkes (1974 and in preparation) calculates the mean annual biomass of *M. norvegica*, *T. raschii*, *T. inermis* and *T. longicaudata* as 1.28, 0.47, 0.88 and 0.077 mg dry wt/m^3 respectively. The production of *T. raschii*, which constituted about 17% of the euphausiid and 1% of the total zooplankton biomass in the Gulf, was estimated from the data on growth and population as 1.962 mg dry wt/m^3/yr, giving a production/biomass (P/B) ratio of 4.2. In terms of carbon the production rate was estimated at 0.36 mg C/m²/day. Moulting, which was not included in these estimates, may require the increase of these estimates by about two-thirds. Berkes considers that the energy from the moulted exoskeletons probably returns to the planktonic community while, judging from their sinking and disintegration rates, most of the faecal pellets probably reach the benthic community. He adds the following:

> Since all four species of euphausiids are omnivorous and since their predators occupy different levels in the food chain, none is confined to a single trophic level. It is believed that they are most heavily fed upon by redfish, herring and cod, which together account for 84% of the commercial fish catch in the Gulf, and also by capelin, but several important links in the food web remain to be elucidated. Direct exploitation of euphausiids may become technically feasible in the next few years but seems unlikely to be economically viable in the Gulf of St Lawrence. Should this take place, it could result in a serious reduction of the food of the commercially important species of fish.

Wright (1972) used some of the IBP material to make the first detailed study of Mysidacea of the Gulf of St Lawrence. Mysids are not well sampled with ordinary plankton nets, most of them being bathyplanktonic or hyperbenthic in habit. Wright recorded 18 species of the group, including seven which are new for the Gulf: *Amblyops abbreviata*, *A. kempii*, *Erythrops abyssorum*, *E. microps*, *Mysidetes farrani*, *Parerythrops obesa* and *Pseudomma affine*.

The relatively warm (4 °C–6 °C), saline (> 34‰), deep water of the Laurentian Channel appears to support sympatric populations of *Boreomysis arctica*, *Erythrops abyssorum*, *Mysidetes farrani*, *Pseudomma roseum*, *Amblyops kempii* and possibly *Amblyops abbreviata*, the last two species being not so much sympatric as overlapping in the regions sampled. There is also a slope fauna found along the margins of the deep channels where the boundary layer (1 °C–4 °C and 33‰–34‰) meets the bottom. The mysid fauna characteristic of this zone consisted of *Parerythrops obesa*, *Erythrops microps* and *Pseudomma affine*.

The Magdalen Shallows is the major area in the Gulf of St Lawrence where the cold intermediate layer (< 0 °C–1.5 °C and 30‰–32‰) meets the bottom. The Cape Breton, Shediac and Chaleur troughs at the extremities of the Shallows are characterized by silt-clay bottoms and by *Erythrops erythropthalma*, *Pseudomma truncatum* and *Meterythrops robusta*. Temperature and salinity appear to define the broad limits of their distributions while sediment type, particularly for *P. truncatum*, defines the areas of highest population densities. None of these three species showed strict temperature tolerances, and they were occasionally found in the upper portion of the boundary layer. *Stilomysis grandis*, although encountered rarely, appeared to occur in the cold intermediate layer.

Mysis mixta was the only species present over the major part of the sublittoral portion of the Magdalen Shallows, and showed a mode of abundance between 45 and 85 m. It was found in many areas to occur in concentrations between 10 and 25 individuals/m^3. *Boreomysis nobilis*, an Arctic relict species, was found in the cold, sub-euphotic waters of Bonne Bay east arm, West Newfoundland. Distributional evidence suggests that it may be common in the deep, cold fjord environments found in parts of Eastern Canadian waters. *Boreomysis arctica*, *B. tridens* and *B. nobilis* were compared morphologically and morphometrically; the first two species are more similar to one another than either is to *B. nobilis*. *B. arctica* and *B. tridens*, apparently sibling species, have been found together but in this study it was found that, although they inhabited the same geographical locations, they were separated vertically in the water column, *B. arctica* being more pelagic than hyperbenthic.

Mysis mixta on the Magdalen Shallows exhibited two distinct year classes which were separated spatially from one another. Breeding of *M. mixta* was deduced to take place during the winter or spring, lending support to this finding for the species in other areas. *Boreomysis nobilis* exhibited a two-year life cycle with continuous breeding throughout spring, summer and autumn but with fewer offspring produced during the winter. Data on *Pseudomma roseum* imply that the population consists of two overlapping cohorts representing spring/summer and summer/autumn broods respectively. *Boreomysis arctica* appeared to breed in the autumn and exhibited a two-year life cycle (Wright, 1972). *Boreomysis nobilis* and *Mysis litoralis* were recorded for the first time for the Saguenay Fjord (Judkins & Wright, 1974).

Not much work has yet been done on the very large material of Copepoda. The calanoid copepods of the eastern Canadian Seaboard are being studied for publication by D. C. Maclellan, C-T. Shih & G. Harding (personal communication), and the IBP collections will form a very important part of that study. Maclellan & Shih (1974) have published a study of the copepodite stages of *Chiridius gracilis*.

Birds

The distribution and abundance of the sea-bird population was studied by Pilote (1974), who reported significantly higher concentrations of birds in four regions: the Bonaventure area off the Cape Gaspé, the region round and to the northeast of the Magdalen Islands, Cabot Strait and the Strait of Belle Isle. This pattern of abundance is not quite in accordance with the patterns of either primary or secondary production that came out of the study as a whole; very probably factors of depth, upwelling over sills, and of the geographic distribution of the species themselves must be taken into account. Twenty-six species were recorded.

Commercial fishery

Steven (1975) published a table comparing fish catch/primary production ratios for various parts of the world, derived from Hempel (1973) and Moiseev (1971). The North Sea, coastal Peru, Gulf of Thailand and Sea of Japan, according to this information, show ratios greater than 1/800 (1/100 in the cases of the North Sea and the Peru upwelling area); the Sea of Okhotsk and the Sea of Azov, 1/1000 and 1/1100 respectively, and the Gulf of St Lawrence (IBP study) ratios varying between 1/2000 and 1/4000. Steven concluded that 'it seems likely. . . that the present catch is not much more than 25% of fish production, which is low compared with the Baltic, North or Black Sea, in all of which it is 50% or more', and went on to suggest that the Gulf of St Lawrence is not being fished to its full potential, or that 'some special factor or set of factors, such as the proportion of relatively deep to shallow water, or the migration patterns of some species of fish or the quantity consumed by seals, whales and sea birds, prevents a higher level of exploitation. Regional differences may be important; for instance it seems possible that the Magdalen Shallows are being fished at or above the safe level but other regions are considerably under-exploited.' It is indeed quite likely that over-fishing is the main problem; the total output from the Gulf has decreased considerably in recent years, in spite of increased fishing effort, a matter bitterly complained of by the fishermen themselves. But it is also possible that part at least of the discrepancy may be the result of too high an estimate of the total primary production (see above).

Oxygen

Oxygen concentrations were measured by the Winkler method at all stations and standard depths during the *Ambrose Foote* IBP cruises of 1969 and at some stations in 1970. We have also a considerable material of oxygen determinations done by the present writer and colleagues from pre-IBP cruises in 1966, 1967 and 1968, some of which have been reported in preliminary form (d'Anglejan & Dunbar, 1968; Dunbar, 1971). Very little work on oxygen had been done in the Gulf before 1966.

The pattern of distribution of oxygen is interesting. A summer oxygen maximum is found toward the bottom of the thermocline just above the summer intermediate cold layer. The narrow range of density encountered there has a mean value which is close to the density of the winter surface layer ($\sigma_t = 25.76$). Supersaturation in the summer mixed layer and in the thermocline is frequent and it is suggested that heating of waters once in equilibrium with the atmosphere has taken place. These conditions in the summer months support the now orthodox view, mentioned above, that the summer intermediate cold waters are a residue of the winter cooled water. An oxygen minimum observed above the bottom in the deep part of the Gulf (Laurentian Trough) decreases from values typical of Slope Water, outside the Gulf, approximately 4.5 ml/l in Cabot Strait, to concentrations lower than 2 ml/l in the western and northern Gulf.

It is these minimum values that are particularly interesting. From the measurements made on three cruises in 1966 (*Theta* cruise, 18–28 July), 1967 (CSS *Hudson* cruise No. 2467, 10–20 August), and 1968 (CHS *Baffin* cruise No. 1768, 24 April–10 May), it became apparent that the lowest oxygen values occurred within the deep layer, and that the oxygen minima at all stations occurred in water of density between 27.03 and 27.56. There was a marked decrease in the minimum oxygen values from Cabot Strait toward the western and northern ends of the Gulf where the minimum values were approximately 2 ml/l, sometimes less. Lauzier & Trites (1958) have shown that the origin of the deep water is the inflow through Cabot Strait of water of mixed Slope and Labrador Current source. Published sections across the Grand Banks and the continental slope of Nova Scotia (Mann, Grant & Foote, 1965) show values of dissolved oxygen at depths between 200 and 500 m which are in the same range as the lowest values recorded in Cabot Strait. The reduction in the oxygen content of the deep water in the direction of transport indicates that the water entering through Cabot Strait penetrates into the Gulf and remains there long enough to be most significantly affected by local oxidative processes, including the breakdown of organic detritus and the respiration of animals. Oxygen saturation values in the deep layer vary from about 60% in Cabot Strait to 40% and less in the Gaspé region and in the Esquiman Trough.

The IBP cruises of the *Ambrose Foote*, 1969 and 1970, confirm and elaborate this pattern. The lowest concentrations are in the Esquiman Channel, the Jacques Cartier Passage (north of Anticosti Island), and west of Anticosti; that is, at the extremities of the arms of the deep water. It follows that the study of the flushing time of the deep layer would be valuable, and would help to establish the dynamics of the oxygen balance, from which again an estimation of the total oxygen demand and hence the animal production, might be forthcoming.

The vertical profiles of dissolved oxygen concentration show that out of 111 stations (all stations in all cruises which covered deep layer water) 57 stations showed the oxygen minimum at or close to the bottom in the deepest sample that could be taken; and that 54 stations showed the minimum well above the bottom, with water below the minimum significantly higher in oxygen than at the minimum. The oxygen minimum is normally in the 200–300 m depth region, and if the water is of that order of depth or shallower (but still within the deep layer) the oxygen minimum is usually at the bottom. If the water is deeper, then the water below the

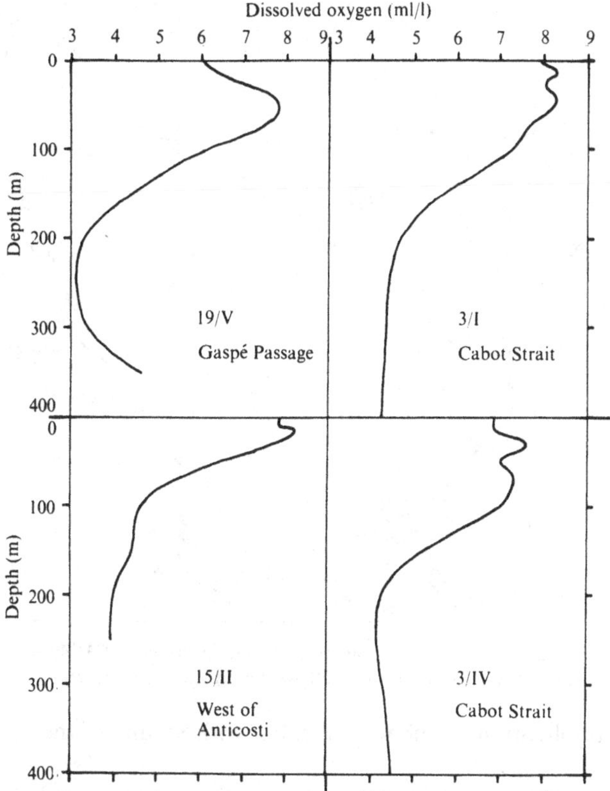

Fig. 7.3. Typical profiles of oxygen concentration, *Ambrose Foote* cruises, 1969. (Station number/cruise number in arabic and roman respectively.)

200–300 m depth is higher in oxygen content than the water above it. In other words, oxygen is not a simple function of depth. Examples of typical vertical profiles are shown in Fig. 7.3. A definitive paper on the oxygen distribution in the Gulf is at present in preparation.

Ice biota

It had been intended to include winter work in the IBP Gulf Project, but funds and circumstances made it impossible to obtain an appropriate vessel. As a follow-up to the Gulf Project, however, studies of the ice biota were made by the present writer in 1973, and again in 1974, using space made available on the CSS *Hudson* by the Bedford Institute of Oceanography, Dartmouth, Nova Scotia. The 1973 material, by far the largest body of results, is summarized here; part of it has been given

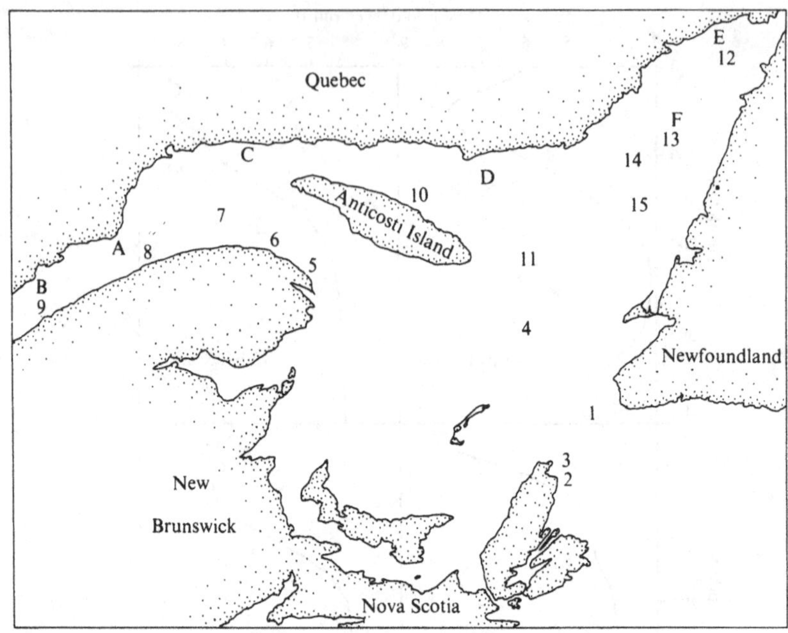

Fig. 7.4. Stations occupied for ice sampling, *Hudson* winter cruise, 1973. Stations 1–15, ice coring stations. Stations A–F, stations at which sampling was done by dipnet.

preliminary publication (Dunbar, 1973), but without any of the systematic results.

Sea-ice coloured (or discoloured) by algae has been recorded for over a century, and was no doubt observed long before that by ice navigators. The scientific investigation of it, however, is of very recent vintage. Apollonio (1961, 1965) measured the concentration of chlorophyll in the sea ice of Jones Sound, north of Devon Island, in 1961, 1962, and 1963; Bunt (1963) and Bunt & Wood (1963) followed with similar and more elaborate studies in the Antarctic. From the Beaufort Sea, Meguro, Ito & Fukushima (1966, 1967) contributed important papers, and more recently Horner & Alexander (1972) and Clasby, Horner & Alexander (1973) have examined the importance of heterotrophic activity in the ice biota, and have described a method of measuring the rate of carbon assimilation *in situ*. Grainger (1977) measured nutrient and chlorophyll concentrations in the ice of Frobisher Bay, Baffin Island, throughout one whole winter. Buinitzky (1977) and Hoshiai (1977) have contributed papers on recent Antarctic studies to the polar oceans conference (SCOR/SCAR) in Montreal, May 1974. The Gulf of St Lawrence is the most southerly region in the northern hemisphere in which sea ice occurs

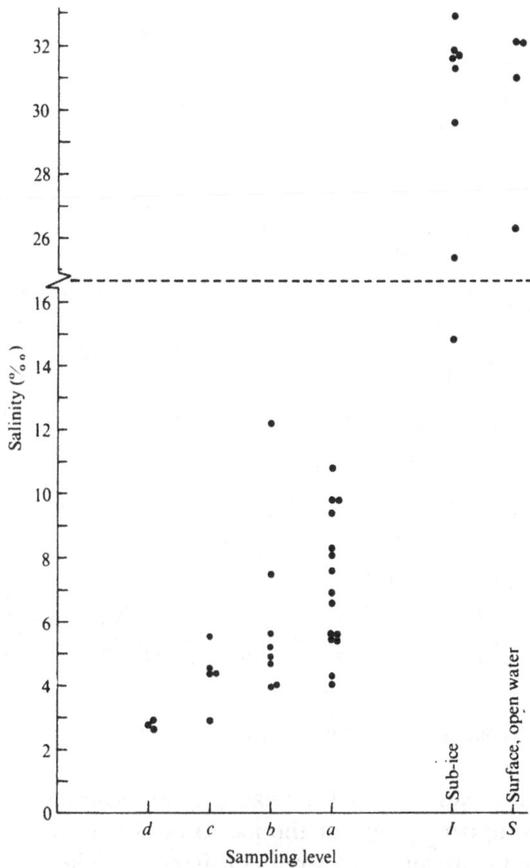

Fig. 7.5. Salinities, *Hudson* winter cruise, 1973. Sampling levels: *a*, bottom of ice core (8-in section); *b*, *c*, and *d*, next levels upwards. 'Sub-ice' water (*I*) sampled by pump immediately under the ice through the corer hole. Surface (*S*); samples from open water not covered by ice. Symbols *a*, *b*, *c*, *d*, *I* and *S* are used in these same senses in Fig. 7.6 and 7.7.

to any extent, and the present study has proved interesting partly for that reason.

The 1973 work extended from 14 February to 7 March. The ice was sampled by means of a standard SIPRE ice corer, diameter 7.5 cm. Ice thickness varied from a few centimetres to 100 cm, being usually between 20 and 70 cm. Sampling stations are shown in Fig. 7.4, salinities in Fig. 7.5, chlorophyll concentrations in Fig. 7.6, and nitrates in Fig. 7.7. Phosphates, silicates, and phaeophytin were also measured.

Chlorophyll *a* concentrations were highest in Cabot Strait and the eastern part of the Gulf, lowest in the region west of Anticosti Island; the

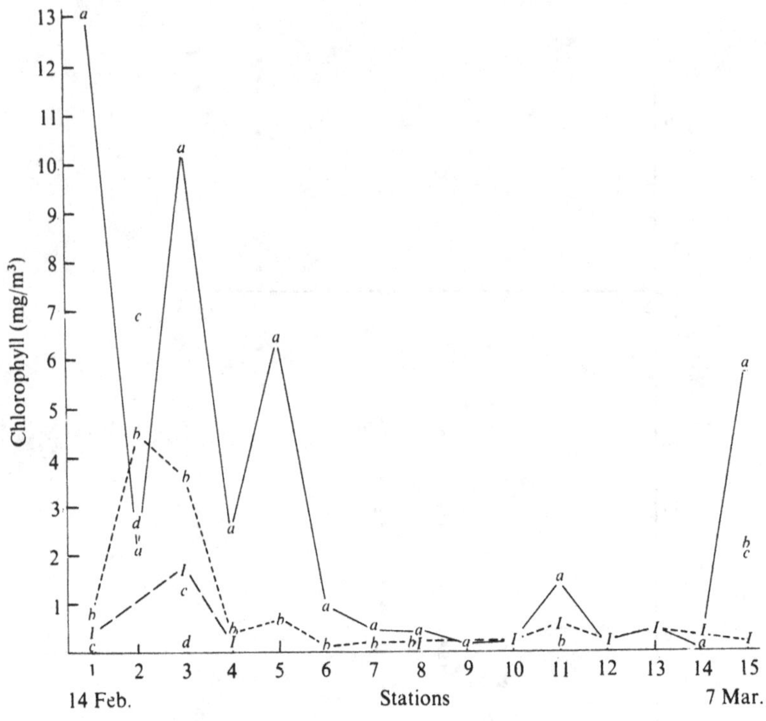

Fig. 7.6. Chlorophyll concentration in mg/m³. Symbols as in Fig. 7.5.

reverse distribution was found for phaeophytin. Salinities varied from about 3‰ in the upper section of the ice to between 4 and 10‰ in the lowest ice section. The nutrients were consistently higher in the western Gulf than in the eastern part (the reverse of the chlorophyll pattern); this may be due to differences in time of freezing and in ice thickness. Chlorophyll concentrations per unit volume were of the same order as is found in spring and summer in the Gulf water. The standing crop of chlorophyll (per unit area) in the ice was between one-tenth and one-hundredth of the levels measured in the Arctic and Antarctic, probably in relation to the much shorter period of ice coverage in the Gulf. A very high level of phosphate concentration was found at one station in the north-eastern Gulf (over 6 μg-at/l); this high value was probably caused by a concentration process involving melting and re-freezing of the ice, but it was not paralleled by the nitrate and silicate measurements at the same station.

Some 45 species of diatoms have been identified in the Gulf ice samples of the 1973 material, some of them tentatively only. The most abundant genera are *Chaetoceros*, *Achnanthes*, *Fragilaria*, *Navicula*, *Nitzschia*,

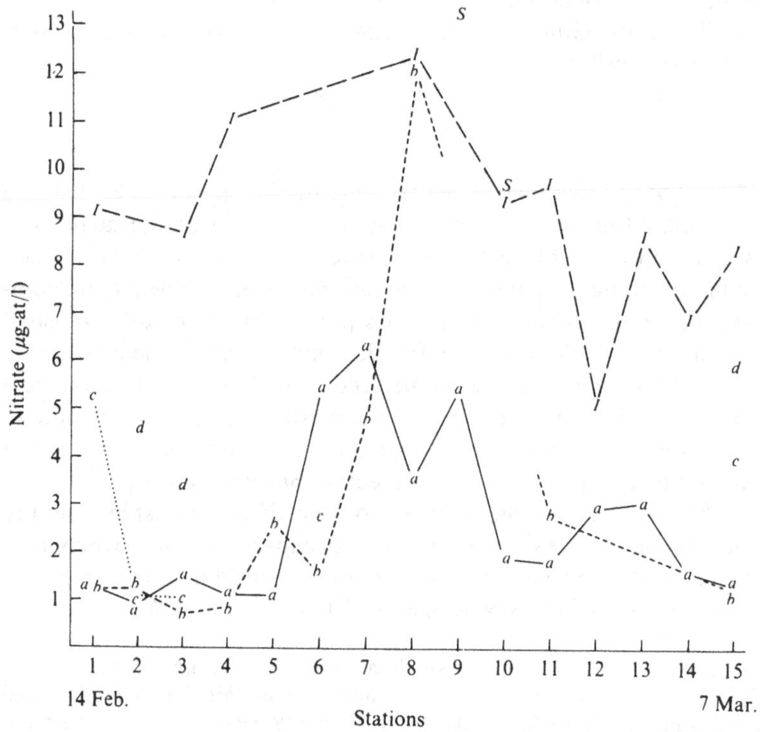

Fig. 7.7. Nitrate concentration in µg-at/1. Symbols as in Fig. 7.5.

Porosira and *Thalassiosira*. Three of these genera, which made up a very large proportion of the total, are centric diatoms, namely *Chaetoceros*, *Porosira* and *Thalassiosira*. Species of these three genera dominate the biota, which is in striking contrast to the results obtained by workers in the Arctic and Antarctic, in which pennate diatoms, and mainly benthic forms, are dominant and centric species are not common at all in the ice. The explanation for the difference is most probably to be found in the much shorter seasonal life of the ice, and therefore of the ice diatoms, in the Gulf of St Lawrence as compared with polar waters. In the Arctic and Antarctic the ice forms a more or less permanent upper solid substrate, ecologically comparable with the sea floor, in fact a 'ceiling', which has allowed the evolution of a substrate community, analogous to the benthos. It is remarkable, for instance, that the dominant crustacean species of the higher trophic level in the Arctic waters is not the pelagic hyperiid amphipod *Parathemisto libellula*, which is dominant in the north where there is open water during the spring and summer, but the gammarid amphipod *Gammarus wilkitzkii*, belonging to a benthic genus. The benthic diatoms have taken over the Arctic ice biotope in a manner which has not been possible in the Gulf of St Lawrence, where the ice lasts only from January to late April.

Conclusions

The IBP Gulf of St Lawrence Project has greatly increased our knowledge of the Gulf planktonic ecosystem. Much of the resulting material has still to be worked up for publication, as is evident from this summary. When the results, including the physical oceanographic measurements, are combined and put in the context of past and present cruise results, we shall be in a position to achieve a fairly full detailed picture of the production system and its metabolism as a whole, and to understand the ecological signals to which the constituent species are attuned; hence to be able to manage the resources of the Gulf in a manner that has not been possible hitherto. The piece-meal, species-by-species approach of the past has not given us this power, and the emphasis on the theory of fishing and the population dynamics of exploited species has caused a failure to recognize the importance of environmental influences on the production mechanisms and on the survival of the several parts of the living system.

So many people were involved in the shipboard work that it is impossible to thank them all individually. The programme was directed from McGill University, and help is gratefully acknowledged from the Fisheries Research Board of Canada, the Bedford Institute of Oceanography, le Ministère de l'Industrie et du Commerce de la Province du Québec, and the Woods Hole Oceanographic Institution. Certain of the identification work was done or confirmed by the Canadian Oceanographic Identification Centre, Ottawa. Miss Carol Spence, Miss Judy Acreman and Mr Gary Borstad did most of the laboratory analytical work.

References

Apollonio, S. (1961). The chlorophyll content of Arctic sea-ice. *Arctic*, **14**, 197–200.

Apollonio, S. (1965). Chlorophyll in Arctic sea-ice. *Arctic*, **18**, 118–22.

Berkes, F. (1974). Production and comparative ecology of euphausiids. In *Primary and secondary production in the Gulf of St Lawrence*, ed. D. M. Steven, manuscript report 26, pp. 85–7. Montreal: Marine Sciences Centre, McGill University.

Berkes, F. (1975). Some aspects of feeding mechanisms of euphausiid crustaceans. *Crustaceana*, **29**(3), 266–70.

Berkes, F. (1976). Ecology of euphausiids in the Gulf of St Lawrence. *Journal of the Fisheries Research Board of Canada*, **33**(9), 1894–905.

Buinitsky, V. K. (1977). Organic life in sea ice. In *Polar Oceans: Proceedings of the polar oceans conference (SCOR/SCAR), Montreal, May 1974*, pp. 301–6, Montreal: Arctic Institute of North America.

Bulleid, E. R. & Steven, D. M. (1972). *Measurements of primary and secondary production in the Gulf of St Lawrence*, manuscript report 21. Montreal: Marine Sciences Centre, McGill University.

Bunt, J. S. (1963). Diatoms of Antarctic sea-ice as agents of primary production. *Nature*, **199**, 1255–7.

Bunt, J. S. & Wood, E. J. F. (1963). Microalgae and Antarctic sea-ice. *Nature*, **199**, 1254–5.

Clasby, R. C., Horner, R. & Alexander, V. (1973). An *in-situ* method for measuring primary productivity of Arctic sea ice algae. *Journal of the Fisheries Research Board of Canada*, **30**, 835–8.

d'Anglejan, B. F. & Dunbar, M. J. (1968). *Some observations of oxygen, pH and total alkalinity in the Gulf of St Lawrence, 1966, 1967, 1968*, manuscript report 7. Montreal: Marine Sciences Centre, McGill University.

Dunbar, M. J. (1940). On the size distribution and breeding cycles of four marine planktonic animals from the Arctic. *Journal of Animal Ecology*, **9**(2), 215–26.

Dunbar, M. J. (1941). The breeding cycle in *Sagitta elegans arctica* Aurivillius. *Canadian Journal of Research, Series D*, **19**(9), 258–66.

Dunbar, M. J. (1946). On *Themisto libellula* Mandt in Baffin Island coastal waters. *Journal of the Fisheries Research Board of Canada*, **6**(6), 419–34.

Dunbar, M. J. (1957). The determinants of production in northern waters. *Canadian Journal of Zoology*, **35**(6), 697–719.

Dunbar, M. J. (1962). The life cycle of *Sagitta elegans* in Arctic and sub-Arctic seas, and the modifying effects of hydrographic differences in the environment. *Journal of Marine Research*, **20**(1), 76–91.

Dunbar, M. J. (1971). *The Gulf of St Lawrence; past and future. Second Gulf of St Lawrence Workshop*, Bedford Institute, Dartmouth, Nova Scotia, Nov. 30–Dec. 3, 1970, pp. 1–31. MS Report. Bedford Institute.

Dunbar, M. J. (1973). Chlorophyll and nutrient measurements in sea ice, Gulf of St Lawrence. In *Proceedings of the Workshop on the Gulf and Estuary of the St Lawrence, Rimouski, Université du Québec, Oct. 11–12, 1973*, pp. 106–127.

El-Sabh, M. I. (1976). Surface circulation pattern in the Gulf of St Lawrence. *Journal of the Fisheries Research Board of Canada*, **33**(1), 124–38.

El-Sabh, M. I., Glombitza, R. & Johannessen, O. M. (1971). *On the vertical fluctuations of hydrochemical parameters in the Gulf of St Lawrence, 1969*, manuscript report 19. Montreal: Marine Sciences Centre, McGill University.

El-Sabh, M. I. & Johannessen, O. M. (1971). *Winter geostrophic circulation in the Gulf of St Lawrence*, manuscript report 18. Montreal: Marine Sciences Centre, McGill University.

Forrester, W. D. (1964). A quantitative temperature–salinity study of the Gulf of St Lawrence. Canada, Department of the Environment; *Bedford Institute of Oceanography, Report*, **64–11**, 16 pp.

Grainger, E. H. (1977). The annual nutrient cycle in sea ice. In *Polar Oceans: Proceedings of the polar oceans conference (SCOR/SCAR), Montreal, May 1974*, pp. 285–300. Montreal: Arctic Institute of North America.

Greisman, P. & Ingram, G. (1978). Nutrient distribution in the St Lawrence estuary. *Journal of the Fisheries Research Board of Canada*, **34**(11), 2117–23.

Hempel, G. (1973). Productivity of the oceans. *Journal of the Fisheries Research Board of Canada*, **30**, 2184–9.

Hjort, J. (ed.) (1919). *Canadian fisheries expedition, 1914–1915*. Ottawa: Canadian Department of the Naval Service.

Hoffer, S. A. (1972). Some aspects of the life cycle of *Parathemisto abyssorum* (Amphipoda: Hyperiidea) in the Gulf of St Lawrence. *Canadian Journal of Zoology*, **50**(9), 1175–8.

Horner, R. & Alexander, V. (1972). Algal populations in Arctic sea ice: an investigation of heterotrophy. *Limnology and Oceanography*, **17**(3), 454–8.

Hoshiai, T. (1977). Seasonal change of ice communities in the sea ice near Syowa Station, Antarctica. In *Polar Oceans: Proceedings of the polar oceans conference (SCOR/SCAR), Montreal, May 1974*, pp. 307–18. Montreal: Arctic Institute of North America.

Judkins, D. C. & Wright, R. (1974). New records of the mysids *Boreomysis nobilis* G.O. Sars and *Mysis litoralis* (Banner) in the Saguenay Fjord (St Lawrence estuary). *Canadian Journal of Zoology*, **52**(8), 1087–90.

Lauzier, L. M. & Bailey, W. B. (1957). Features of the deeper waters of the Gulf of St Lawrence. *Bulletin of the Fisheries Research Board of Canada*, **111**, 213–50.

Lauzier, L. M. & Trites, R. W. (1958). The deep water in the Laurentian Channel. *Journal of the Fisheries Research Board of Canada*, **15**(6), 1247–57.

Maclellan, D. C. & Shih, C-T. (1974). Descriptions of copepodite stages of *Chiridius gracilis* Farran 1908 (Crustacea: Copepoda). *Journal of the Fisheries Research Board of Canada*, **31**, 1337–49.

Mann, C. R., Grant, A. B. & Foote, T. R. (1965). Atlas of oceanographic sections, northwest Atlantic Ocean, Feb. 1962–July 1964. *Bedford Institute of Oceanography Report*, No. 65–16, 51pp.

Meguro, H., Ito, K. & Fukushima, H. (1966). Diatoms and the ecological conditions of their growth in sea ice in the Arctic Ocean. *Science*, **152**, 1089–90.

Meguro, H., Ito, K. & Fukushima, H. (1967). Ice flora (bottom type): a mechanism of primary production in polar seas and the growth of diatoms in sea ice. *Arctic*, **20**, 114–33.

Moiseev, P. A. (1971). *The living resources of the world ocean.* English translation by N. Kamer & W. E. Ricker. Jerusalem: Israel Programme of Scientific Translations.

Pilote, S. (1974). Densité et distribution des populations d'oiseaux marins du Golfe de St.-Laurent. In *Primary and secondary production in the Gulf of St Lawrence*, ed. D. M. Steven, manuscript report 26, pp. 80–5. Montreal: Marine Sciences Centre, McGill University.

Sinclair, M., El-Sabh, M. & Brindle, J. R. (1976). Seaward nutrient transport in the lower St Lawrence Estuary. *Journal of the Fisheries Research Board of Canada*, **33**(6), 1271–7.

Spence, C. & Steven, D. M. (1974). Seasonal variation of the chlorophyll *a*: phaeopigment ratio in the Gulf of St Lawrence. *Journal of the Fisheries Research Board of Canada*, **31**, 1263–8.

Steven, D. M. (1975). In *Energy flow – its biological dimensions. A summary of the IBP in Canada 1964–74*, ed. T. W. M. Cameron, pp. 229–48. Ottawa: Royal Society of Canada.

Steven, D. M., Acreman, J., Axelsen, F., Brennan, M. & Spence, C. (1973). *Measurements of primary and secondary production in the Gulf of St Lawrence*, manuscript reports 23, 24 & 25. Montreal: Marine Sciences Centre, McGill University.

Trites, R. W. (1971). The Gulf as a physical oceanographic system. Manuscript report of the Second Gulf of St Lawrence Workshop, Bedford Institute of Oceanography, Dartmouth, Nova Scotia, 30 Nov.–3 Dec. 1970, pp. 32–63.

Weinstein, M. (1973). Studies on the relationship between *Sagitta elegans* Verrill and its endoparasites in the southwestern Gulf of St Lawrence. PhD thesis, McGill University (Marine Sciences Centre).

Wright, R. A. (1972). Occurrence and distribution of the Mysidacea of the Gulf of St Lawrence. MSc thesis, McGill University (Marine Sciences Centre).

8. Patterns of the vertical distribution of phytoplankton in typical biotopes of the open ocean

H. J. SEMINA

Introduction

We know from investigations in which cells of phytoplankton were counted down to great depths (Lohmann, 1920; Hentschel, 1936; Ohwada, 1960, 1972; Kozlova, 1964) that in the ocean the layer rich in phytoplankton extends to a certain depth, below which the phytoplankton is never as abundant as in the upper layer, especially if only living cells are taken into account (Ohwada, 1960, 1972).

The vertical distribution of phytoplankton within the rich layer has been investigated by many authors and it is known that phytoplankton tend to concentrate in the discontinuity layers. Theoretical explanations for the lack of uniformity in the vertical distribution of phytoplankton are presented in the works of Riley, Stommel & Bumpus (1949) and Steele & Yentsch (1960).

We now have at our disposal a very large amount of material on the vertical distribution of cell numbers of phytoplankton in different regions of the Pacific and one region in the tropical Atlantic Ocean. This allows us to present the patterns of the quantitative distribution of phytoplankton with depth, and to relate these patterns to the peculiarities of the biotopes.

Material and methods

The materials used consisted of collections made during the cruises of the R/SS *Vityaz* and *Akademik Kurchatov* of the Shirshov Institute of Oceanology, Academy of Sciences of the USSR. Phytoplankton was collected at 143 stations in the Pacific and 84 stations in the Atlantic; one litre of water was taken at each station. Samples were taken at depths of 0, 10, 25, 50, 75, 100, 150 and 200 m. On the cruises of the *Vityaz* samples were also taken from depths of 5 m (cruise 26) and 35 m (cruise 20). During cruise 4 of *Akademik Kurchatov*, the depths of 20 m and 30 m were sampled instead of the standard 25 m, and additional samples were made in the main pycnocline whose depth was estimated before sampling by the use of a bathythermograph.

The samples were preserved by adding 10 ml of 40% formalin to one-litre samples. The formalin was neutralized with borax. The samples were treated by the settling method with subsequent centrifugation. The volume of the sample was reduced to 1 or 3 or 5 cm^3 depending on the amount of phytoplankton. From this volume 0.05 cm^3 was taken and examined in a counting cell. In very poor samples the phytoplankton cells were counted in the entire concentrate of a one-litre sample. Magnifications of 10×10 and 10×20 were used for counting. The material from the Pacific Ocean was worked on by the author and I. A. Tarkhova, and the material from the Atlantic Ocean by Truong Ngoc An.

The main pycnocline was determined from density data obtained from temperature and salinity. The density gradients for layers between standard depths were computed by dividing the difference between the conventional densities in the underlying and overlying layers by the distance between the layers (in metres). The values were computed for the upper 500 m layer. The differences between the layers were evaluated by their density gradients, a difference exceeding 0.001 $g/cm^3/m \cdot 10^3$ or sigma-t units/m being considered as significant. At most stations a layer could be recognized showing a marked density gradient. Small density gradients were infrequent through the 0–500 m layer.

We assumed that at each station there were as many pycnoclines as there were layers with a density gradient higher than in the adjacent layers. The main pycnocline is the layer of density discontinuity nearest to the surface, existing in the same place during the whole year. Since no all-year-round observations were conducted on our vessels we had to use data from the literature on seasonal variations with depth, temperature, salinity and density in the water structures we were interested in (see references in the text).

Seasonal variations of temperature and salinity are well-known for the northern part of the Pacific, but are less well-known for the subtropical and equatorial regions. Particular difficulties were encountered in the western equatorial part of the ocean where at some stations several pycnoclines occurred simultaneously in shallow depths, close to each other. The depths at which they occur in different seasons are unknown. At these stations no main pycnocline could be recognized.

Geographical variations of the properties of the biotope operative in the vertical distribution of plant cells

A phytoplankton biotope in the ocean may be described as a water layer extending from the surface to the bottom in shallow depths, and to the depth of the main pycnocline in greater depths beyond the shelf (Semina, 1966). Horizontally, the biotopes are large-scale gyrals formed by systems of currents. The term 'main pycnocline' is applied to a permanent (not seasonal) layer of density discontinuity which exists in the ocean during the whole year. In waters of different structure the main pycnocline may be of different origin. Thus in the northern Pacific it is a halocline, and in the tropical zone it is most often a thermocline (Dobrovolsky & Arseniev, 1961; Seckel, 1962; Uda, 1963; Tully, 1964; Tully & Giovando, 1964; Wyrtki, 1964; Arseniev, 1967; Fleming, 1958; etc.).

The importance of this permanent pycnocline to phytoplankton is a result of it being a layer of density discontinuity which limits the maximum depth of the upper mixed layer. The upper mixed layer in the open ocean cannot lie at a depth greater than that of the main pycnocline, but may occur at lesser depths in the presence of seasonal (or other temporary) pycnoclines or, if in the water layers overlaying the main pycnocline, there are layers stable enough to hamper the vertical mixing (see Fig. 8.6).

The effect of discontinuity layers on phytoplankton has been studied for a long time, but usually the authors writing on this problem did not attach much importance to exactly what type of discontinuity layer they were studying, and, as a rule, confined their investigations to the uppermost discontinuity.

Of importance for the vertical distribution of phytoplankton are such factors as the depth of the main pycnocline, the relationship between this depth and the depth of the euphotic zone, the presence of seasonal (or other temporary) discontinuity layers, the general stability of the water in the biotope, as well as the abruptness of the lower boundary of the biotope, or, in other words, the density gradient in the main pycnocline. The vertical distribution of phytoplankton is influenced also by the general concentration of nutrients in the biotope and at the depth of the main pycnocline and at the depths of seasonal or other temporary pycnoclines.

The vertical distribution of phytoplankton in different regions of the Pacific Ocean

Let us now consider the vertical distribution of phytoplankton in different regions of the Pacific Ocean (the boundaries of these regions are shown in Fig. 8.1).

Sub-Arctic Region

This region is rich in nutrients (more than 1 μg-at/l of phosphate at a depth of 100 m in winter; Sapozhnikov & Mokievskaya, 1966). The upper boundary of the main pycnocline lies at a depth of 75–200 m. The density

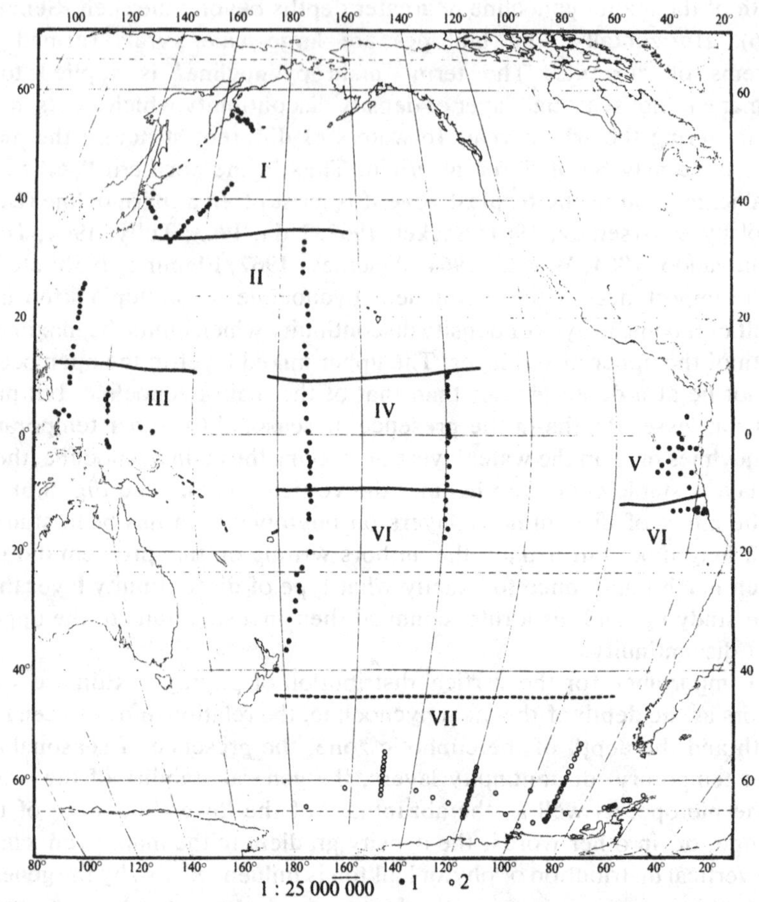

Fig. 8.1. Map of stations in the Pacific Ocean. The stations are designated by points. 1 (solid points) *Vityaz* and *Ak. Kurchatov;* 2 (open circles) *Brategg*. Regions, I–VII. I, sub-Arctic; II, northern subtropical; III, southern subtropical; IV, eastern equatorial; V, central equatorial; VI, western equatorial; VII, Antarctic.

gradient in the main pycnocline approximates 0.015 sigma-t units/m in the eastern part of the region and 0.02–0.006 sigma-t units/m in the western part. In winter the water in the biotope is mixed by convection from the surface down to the depth of the main pycnocline. Between spring and autumn a seasonal density discontinuity layer develops at a depth of 10–25 m. In spring phytoplankton occurs in quantities averaging tens of thousands of cells per litre in the 0–100 m layer, while maximum concentrations near the shore may exceed 10^9 cells/l.

The vertical distribution of phytoplankton varies considerably depending on the season. It is rather uniform in winter due to intensive

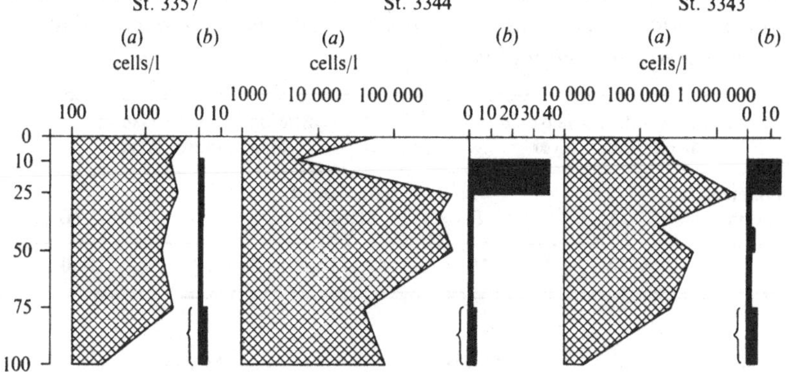

Fig. 8.2. Vertical distribution of phytoplankton in the sub-Arctic region in June 1955, at a station with no seasonal discontinuity layer (st. 3357, 52° 26′ N, 170° 54′ E), and with seasonal discontinuity layers (st. 3343, 55° 06′ N, 163° 29′ E; and st. 3344, 55° 34′ N, 162° 01′ E). (*a*) number of cells; (*b*) stability of water layers in arbitrary units. Main pycnocline is marked by a bracket.

convective mixing of water in the biotope, and in early spring, before stratification sets in. A fairly uniform distribution was observed in spring in the north-western Pacific, far offshore, at stations with no seasonal discontinuity layer. Phytoplankton was found to decrease markedly below the upper boundary of the main pycnocline (Fig. 8.2, st. 3357). This evenness of distribution of phytoplankton in spring in the northern part of the Pacific Ocean has been reported by several other authors (Kuzmina, 1959; Taniguchi & Kawamura, 1972). Uniform vertical distribution in association with intensive mixing of water was observed in the Bering Sea (Semina, 1955; Taniguchi, 1969), the Japan Sea (Ohwada, 1972) and in the sub-Antarctic and Antarctic (Hasle, 1969).

This relatively uniform distribution is, however, of short duration. Maxima of cell numbers soon begin to appear at the surface despite the still-existing isothermy and isohalinity. Thus in the spring of 1955, in a section running parallel to the Kurile Islands, and in the region of Kamchatka, when no seasonal discontinuity layers had as yet appeared at some of the stations, half of these stations already had appreciable maxima of cell numbers, usually located near the surface (Table 8.1). The occurrence of surface maxima under conditions of isothermy and isohalinity was observed by Kuzmina (1959), Taniguchi & Kawamura (1972) in the Pacific Ocean and by Ryther & Hulburt (1960) in the Atlantic.

It is noteworthy that in early spring some of these stations had maxima confined to the main pycnocline simultaneously with surface maxima

Table 8.1. *Frequency of occurrence of maxima of cell numbers (%) in the main pycnocline (A), and in the layers above the main pycnocline (B), in the sub-Arctic region (June 1955)*

Stations with no seasonal discontinuity layer			Stations with a seasonal discontinuity layer		
Number of stations	*A*	*B*	Number of stations	*A*	*B*
17	88	50	13	68	100

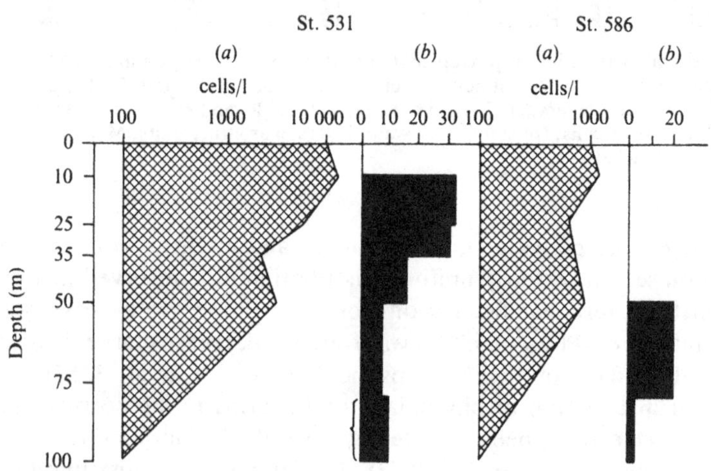

Fig. 8.3. Vertical distribution of phytoplankton in September 1950, in the Bering Sea at stations with a seasonal discontinuity layer (st. 531, 54° 50′ N, 170° 17′ E; st. 586, 61° 17′ N, 178° 58′ W). Symbols as in Fig. 8.2.

(Table 8.1). The maxima in the main pycnocline were either equal to those situated nearer to the surface, or slightly smaller; very rarely were they greater than the latter. Despite the existence of the maxima, phytoplankton is generally more evenly distributed during the periods with no seasonal discontinuity layer, than it is during the period of seasonal discontinuity layer.

Later in the season, with increasing stability of water and the development of a seasonal discontinuity layer, the unevenness of vertical distribution becomes still more pronounced. Even when the seasonal pycnocline is but slightly expressed, phytoplankton may tend to concentrate in the layer above it (Marumo , Kitou & Ohwada, 1958). In spring when the seasonal discontinuity layer has barely appeared, other maxima, along with those in the seasonal layer and above it, may appear at the depth of the main pycnocline (Table 8.1; Fig. 8.2, st. 3343, 3344), usually by an order smaller than in the seasonal discontinuity layer.

During the summer and autumn, aggregations of phytoplankton are frequent in the seasonal discontinuity layer and above it (Semina, 1955; Taniguchi, 1969). Below it phytoplankton usually is scarce (Fig. 8.3). In the Antarctic aggregations of phytoplankton were discovered by Hasle (1969) in the seasonal pycnocline and above it. No maxima were observed at this time at the depth of the main pycnocline.

In late autumn, with the destruction and sinking of the seasonal discontinuity layer, the vertical distribution of phytoplankton within the biotope becomes more even, approaching the uniform distribution of early spring. This is well illustrated by the figures given by Ohwada (1972), showing the distribution of the phytoplankton in the Japan Sea during October.

Subtropical regions

These regions are characterized by low concentrations of nutrients (less than 0.25 μg-at/l of phosphate at a depth of 100 m*, Sapozhnikov & Mokievskaya, 1966). The main pycnocline lies in shallow depths in the eastern part of the northern subtropical region (75–100 m) but deepens to 150 m and even more in the central and western part of the northern subtropical region. The density gradient in the main pycnocline is small: 0.010–0.020 sigma-t units/m in the southern subtropical region and on the periphery of the northern, and less than 0.010 sigma-t units/m in the central part of the latter. In the absence of seasonal discontinuity layers, the water layers in the biotope are rather unstable. The stability of water under the main pycnocline is also low. Winter mixing is not intensive in these regions. As in the sub-Arctic region a seasonal discontinuity layer develops here in summer at a depth of approximately 50 m. These regions are poor in phytoplankton (tens, hundreds, rarely thousands of cells per litre).

In the subtropical regions two or three maxima of cell number are usually observed in the 0–200 m layer, confined to different depths. During the period when no seasonal discontinuity layers develop in the subtropical regions, the vertical distribution of phytoplankton is characterized by the presence of two or three maxima in the biotope, one of them occurring at the depth of the main pycnocline and the others above its upper boundary (Table 8.2, Fig. 8.4, st. 230). Most often the numbers of cells in the main pycnocline and above it differ but little.

The vertical distribution of phytoplankton in the subtropical regions in

* Phosphate concentration increases markedly in the east (over 0.5 μg-at/l).

Table 8.2. *Frequency of occurrence of maxima of cell numbers (%) in the main pycnocline (A), and of depths above the main pycnocline (B), in sub-tropical regions*

Stations with no seasonal discontinuity layer			Stations with a seasonal discontinuity layer		
Number of stations	*A*	*B*	Number of stations	*A*	*B*
25	72	68	22	30	60

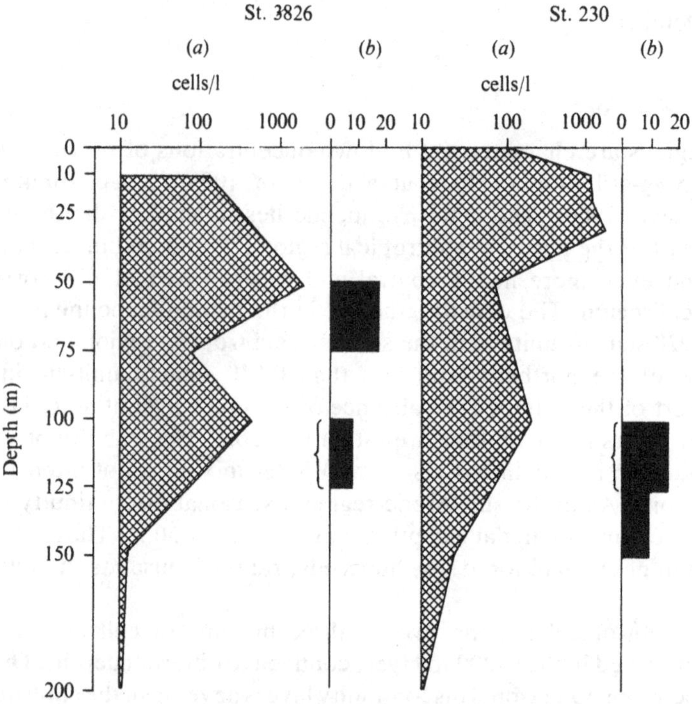

Fig. 8.4. Vertical distribution of phytoplankton in the southern subtropical region at stations with a seasonal discontinuity layer (st. 3826, 27° 15′ S, 175° 40′ W), and with no seasonal discontinuity layer (st. 230, 12° 29′ S, 83° 13′ W). Symbols as in Fig. 8.2.

the presence of a seasonal discontinuity layer is different from that observed in its absence (Fig. 8.4, st. 3826), the main difference consisting of the less frequent occurrence of maxima in the main pycnocline. Maxima in the seasonal discontinuity layer were observed at 60% of the stations, but only 30% of the stations had a maximum of cell numbers in the main pycnocline (Table 8.2). Seventy-five per cent of the stations also had maxima situated above the seasonal discontinuity layer, while the maxima in the main pycnocline were about equal to the maxima in the seasonal discontinuity layer. Thus in subtropical regions phytoplankton is more uniformly distributed within the biotope even in the presence of a seasonal discontinuity layer than it is in the sub-Arctic region, when a discontinuity layer is well-developed in spring and in summer.

Stations poor in plankton were found to have a more uniform distribution of phytoplankton than richer stations; the unevenness of distribution increasing with increasing abundance of plankton (over 1000 cells/l). The same phenomenon could be traced in the vertical distribution of diatoms (Marumo, 1957). Marumo, who conducted seasonal observations at station 'T', situated to the south of Japan at 29° N, 135° E, and in a section extending from 33° N to about 30° N, found that diatoms concentrated in the seasonal discontinuity layer only when they occurred in great quantities (more than 50000 cells/l); when present in small quantities they were far more evenly distributed through the whole 0–150 m layer, irrespective of whether the layer was isothermic or contained a seasonal discontinuity layer. The frequency of occurrence of maxima of diatom numbers through the water column was rather uniform and even more so during the period of the existence of a seasonal discontinuity layer.

Equatorial regions

Eastern equatorial region

In this region high concentrations of nutrients are observed (phosphate more than 1.5 μg-at/l at a depth of 100 m; Bogoyavlensky & Shishkina, 1971). The main pycnocline lies at a shallow depth (10–50 m), its density gradient is large, varying from 0.030 to 0.100 sigma-t units/m. In this region some of the stations are characterized by a low stability of water layers in the biotope. Some other stations are characterized by a high stability, but only in the layer next to the main pycnocline, with a thickness of merely 10–20 m. Below the main pycnocline stability is also usually low. No seasonal discontinuity layers occur in this region, and no other temporary discontinuity layers were observed at the stations above the main pycnocline. The development of phytoplankton is not limited by light; quantities in this region reach tens and hundreds of cells per litre.

Table 8.3. *Frequency of occurrence of maxima of cell numbers (%) in the main pycnocline in the equatorial regions*

Region	Number of stations	Frequency of maxima in the main pycnocline (%)
Eastern equatorial region	17	82
Western equatorial region	18	73

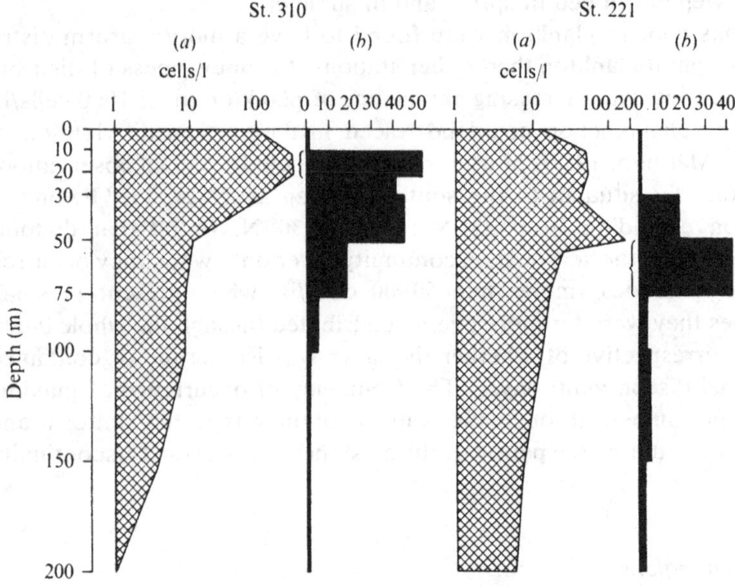

Fig. 8.5. Vertical distribution of phytoplankton in the eastern equatorial region at stations 310 (1° 59′ S, 87° 29′ W) and 221 (2° 0′ S, 84° 88′ W). Symbols as in Fig. 8.2.

A characteristic feature of this region is the impoverishment of the surface layers. One or two maxima occur in the biotope, one of them usually at the depth of the main pycnocline (Table 8.3), and the other slightly higher, both approximately equal to one another in cell numbers (Fig. 8.5).

Central equatorial region

Here the concentration of nutrients is lower than in the eastern equatorial region (phosphate less than 1 μg-at/l at a depth of 100 m; Sapozhnikov & Mokievskawa, 1966). The main pycnocline is deep-seated (its upper

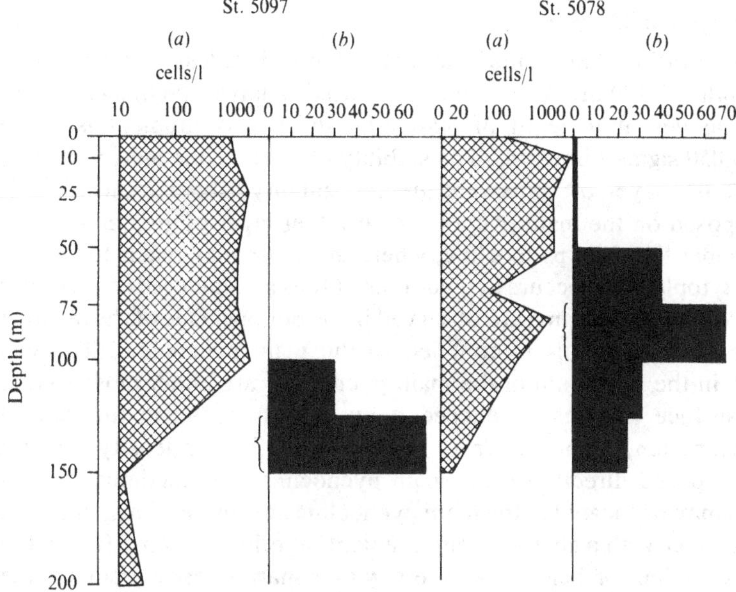

Fig. 8.6. Vertical distribution of phytoplankton in the central equatorial regions at stations 5097 (3° 30′ S, 140° 0′ W) and 5078 (8° 58′ N, 139° 52′ W). Symbols as in Fig. 8.2.

boundary lies at 75–200 m), the density gradient large (0.020–0.070 sigma-t units/m). The stability of water layers in the biotope is low at most stations, although at some of them a layer of greater stability occurs above the main pycnocline, 25–50 m thick, and another one, of the same thickness, under the main pycnocline.

The region is rich in phytoplankton (hundreds, thousands and tens of thousands of cells per litre). Two or three not very sharp maxima occur in the biotope. We were unable to estimate the frequency of their occurrence in the main pycnocline which lies here at a great depth, because our samples were widely spaced (50 m) and not always extended down to 200 m. Nevertheless the vertical distribution of phytoplankton in the biotope may be characterized as rather uniform. No impoverishment of surface layers is observed in this region.

Maxima of cell numbers may be confined both to the main pycnocline and to the higher layer of greater stability; below the latter the quantity of phytoplankton diminishes, not always increasing again in the main pycnocline (Fig. 8.6).

Western equatorial region

The concentration of nutrients is rather low (phosphate less than 0.5 μg-at/l at a depth of 100 m; Sapozhnikov & Mokievskaya, 1966). The main pycnocline lies at a depth of 50–125 m, its density gradient is sharp (0.030–0.050 sigma-t units/m). The stability of water in the biotope is low. Most often a layer of somewhat greater stability, about 25 m thick, is superimposed on the main pycnocline. Another layer of greater stability occurs under the main pycnocline, where thickness may reach 100 m and more. Phytoplankton occurs in quantities of tens and hundreds of cells per litre. Two or three maxima are observed in the biotope. One of the maxima at some of the stations is confined to the main pycnocline. The cell numbers in the maximum in the main pycnocline are either close to that of the surface maxima or exceed it by 2–3 times. As in the central equatorial region, when a layer (or layers) with a sharper density gradient is superimposed directly on the main pycnocline, the maximum of cell numbers may be located in the main pycnocline at some stations, or above it, in the layer with a sharp density gradient, at other stations. Below the main pycnocline (or below the layers with a sharp density gradient) the amount of phytoplankton decreases markedly.

In this region some stations had two discontinuity layers, but which of the two is the main pycnocline is as yet not clear. Maxima of cell numbers were discovered in both of them.

Vertical distribution of phytoplankton in a region of the Atlantic Ocean

The region investigated in the Atlantic Ocean lies between the coast of Africa and 4° W, and between 5° S and 23° S (Fig. 8.7). The region is characterized by a high concentration of nutrients; the concentration of phosphate exceeds 0.5–1 μg-at/l. The main pycnocline occurs in shallow depths (varying from 25–50 to 0–10 m in the north of the region and nearshore) and deepens in its south-western part (from 50–75 to 75–100 m). Phytoplankton is abundant at the coasts (hundreds and tens of thousands of cells per litre) decreasing offshore, towards the open part of the region, especially in a south-western direction (thousands of cells on 10° W and hundreds and tens of cells in the south-west) (Truong Ngoc An, 1971; Semina & Truong Ngoc An, 1973).

The density gradient in the main pycnocline is sharp in the north and nearshore (more than 0.050 sigma-t units/m) and small in the south-west (0.0150–0.020 sigma-t units/m). In the north and near the coast the main pycnocline is situated in the middle of a layer of high stability. In the south-west the stability of water in the biotope is usually low. In this region

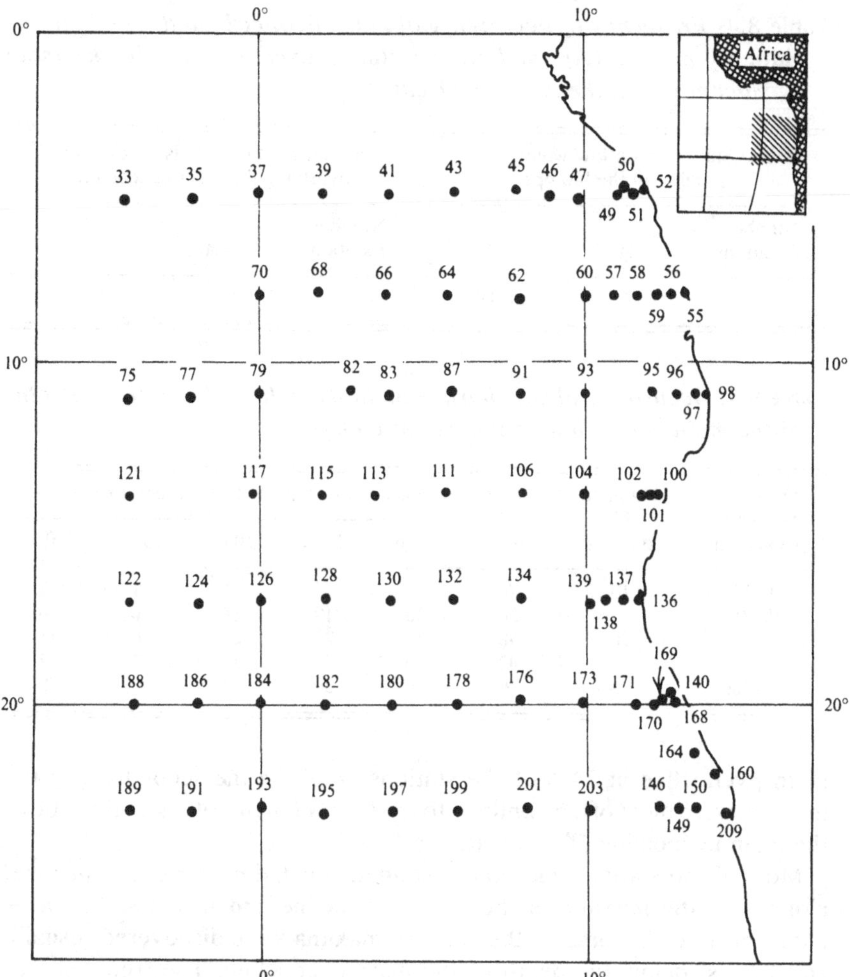

Fig. 8.7. Map of stations in the Atlantic Ocean, off the west coast of Africa (see insert). Stations are designated by points.

one or two maxima of cells occur in the biotope, one of them very often at the depth of the main pycnocline.

All the stations were divided into two groups. To one group we assigned the stations with low water stability above the main pycnocline (and also, conditionally, the stations where the main pycnocline was situated in the 0–10 m layer). The other group contained those stations at which the nearest layers above the main pycnocline had sharp density gradients. It was found that in the first group maxima of cell numbers occurred in the

Table 8.4. *Frequency of occurrence of maxima of cell numbers (%) in the main pycnocline (A), and in the stable layer adjacent to the main pycnocline (B) in the Atlantic Ocean*

Stations with low water stability in the biotope			Stations with stable layer directly overlying the main pycnocline		
Number of stations	*A*	*B*	Number of stations	*A*	*B*
58	74	no layer	14	29	71

Table 8.5. *Occurrence of maxima at standard depths and at different depths of the main pycnocline in the Atlantic Ocean*

Depth of the main pycnocline	Number of stations	Occurrence of maxima at standard depths (%)					
		0 m	10 m	25 m	50 m	75 m	100 m
0–10	10	70	2	2	1	1	2
10–25	16	50	38	19	19	0	0
25–50	26	46	34	23	27	0	11
50–75	14	42	0	13	35	42	7
75–100	5	60	0	20	20	40	0

main pycnocline at 74% of the stations, while in the second group the maxima were most often confined to the layer of high water stability above the main pycnocline (Table 8.4).

Most of the stations had other maxima, situated nearer to the surface, along with the maxima in the main pycnocline and in the stable layer adjacent to it. At some of the stations maxima were discovered, usually small ones, below the depth of the main pycnocline. The frequency of occurrence of maxima at standard depths along with varying depth of the main pycnocline is shown in Table 8.5. It is apparent that the maxima in the main pycnocline are most often confined to its upper boundary.

The frequency of occurrence of maxima above the depth of the main pycnocline is greatest at the surface, where they occurred at more than 40% of the stations. Maxima are usually less frequent in the main pycnocline than at the surface, although more frequent than in the layers situated below the surface but above the main pycnocline. Below the main pycnocline a marked decrease is observed in the frequency of occurrence. The frequency of occurrence of maxima extends to greater depths with increasing depth of the main pycnocline, i.e. with increasing thickness of the biotope. This is true for the Pacific Ocean also.

Table 8.6. *Mean thickness of the trophogenic layer and mean depth of the upper and lower boundaries of the main pycnocline in the Pacific and in the region investigated in the Atlantic Ocean*

Region	Number of stations	Thickness of trophogenic layer (m)	Upper and lower boundaries of the main pycnocline (m)
Pacific Ocean			
Sub-Arctic region (spring)	23	Not less than 80[a]	91–122
Northern sub-tropical region (central part)	8	147	106–122
Southern subtropical region	20	97	96–125
Eastern equatorial region	15	50	33–52
Central equatorial region	24	87	111–156
Atlantic Ocean	67	34	21–44

[a] In this region samples were taken only down to 100 m.

General patterns of vertical distribution of phytoplankton in the ocean

Although the vertical distribution of phytoplankton varies in different regions of the ocean, one of the maxima of cell numbers is usually confined to the main pycnocline. Maxima in the main pycnocline were found in different regions of the Pacific Ocean at 70–80% of the stations (see Tables 8.1, 8.2, 8.3), and more than 70% of the stations in the Atlantic Ocean, within the region investigated (Table 8.4). Hence it may be inferred that the vertical distribution of phytoplankton depends to a considerable degree on the main pycnocline. A fact common to all the regions of the ocean is a marked decrease in phytoplankton concentrations at depths below that of the main pycnocline.

A trophogenic layer (a layer with high phytoplankton concentration: Naumann, 1931) could be recognized at all our stations except the poorest ones, which had to be excluded from examination. The average thickness of this layer approaches the depth of the main pycnocline, increasing with the increase of the latter (Table 8.6).

In the Pacific Ocean, the trophogenic layer is thin in the eastern equatorial region, and thickest in the northern subtropical region. In the peripheral parts of the northern subtropical region, rich in plankton, the trophogenic layer may be limited by the layer of seasonal discontinuity (see above). In the subtropical region, the equatorial regions, and in the investigated region of the Atlantic Ocean the trophogenic layer may be bounded by a stable layer adjoining the main pycnocline from above (Fig. 8.6).

In the sub-Arctic region, in the spring, before the development of a well-expressed seasonal discontinuity layer, as well as during the early period of its development, the thickness of the trophogenic layer depends on the depth of the main pycnocline, while in the summer and autumn the trophogenic layer is limited by the depth of the seasonal discontinuity layer. Thus, according to our data, in the Bering Sea the trophogenic layer is about 100 m deep in spring and only 35 m deep in autumn (Semina, 1957).

For the sub-Antarctic and the Antarctic the thickness of the trophogenic layer in summer is indicated by the decrease in phytoplankton concentration, as may be seen in the figures given by Hasle (1969). The mean thickness of the trophogenic layer was about 90 m at the stations with no seasonal discontinuity layer. The depth of the main pycnocline is indicated by the increase of water stability, as shown in the same figures, and it is close to the depth of the lower boundary of the trophogenic layer. As may be seen in other figures of the same author (Hasle, 1969) the depth of the trophogenic layer was close to the depth of the seasonal discontinuity layer at the Antarctic stations that had a well-developed seasonal discontinuity layer. The vertical distribution of phytoplankton within the biotope also follows a definite pattern: it is more even when water mixing is intensive and the main pycnocline deep-seated, and less uniform in the presence of a seasonal discontinuity layer and a shallow main pycnocline.

The sizes of maxima situated in the main pycnocline differ but little from the sizes of maxima nearer to the surface, although occasionally the surface maxima may happen to be far greater than those in the main pycnocline and vice versa. Maxima situated nearer to the surface (above the upper boundary of the main pycnocline) may be found everywhere in the ocean, the frequency of their occurrence depending on the depth of the main pycnocline; the greater this depth, the farther down the occurrence of maxima extends through the standard depths to that of the main pycnocline. Below it the frequency of maxima markedly decreases (see Table 8.5).

It is noteworthy that the maxima in the main pycnocline are as a rule larger than the small maxima occurring at greater depths. This is illustrated in Table 8.7 showing the ratios of maximum cell numbers in the main pycnocline and in the depths above and below it, in the region investigated in the Atlantic Ocean.

Maxima of cell numbers occurring in different parts of the biotope are produced by different causes. Those situated nearer to the surface are formed by cell proliferation under favourable light conditions (Steele, 1957). A portion of these cells inevitably sink to greater depths, so that appreciable maxima may occur within the euphotic zone and below the euphotic zone, at all depths down to the depth of the main pycnocline or

Table 8.7. *Mean ratio of maximum cell numbers in the main pycnocline (A) and maximum cell numbers in the layers above (B) and below (C) the main pycnocline*

Depth of the main pycnocline (m)	Number of stations	$A:B$	$A:C$
0–10	10	—	100:1
10–25	16	—	20:1
25–50	26	1:7	17:1
50–75	14	6:1	30:1
75–100	5	1:3	—

of a seasonal discontinuity layer. Here the greater part of the cells are retained due to the larger density gradient.* The cells concentrate in the discontinuity layers not only because their further descent is mechanically impeded by the greater density of the water, but also because conditions in depths affect their physiology. In discontinuity layers (including the main pycnocline) situated above the compensation depth, active photosynthesis is possible. Conditions of mineral nutrition are far better here than in the upper layers, because the waters below the seasonal discontinuities (in high and temperate latitudes), and everywhere below the main pycnocline are always richer in nutrients. Therefore when a seasonal discontinuity layer or the main pycnocline occurs above the compensation depth, reproduction in these layers is possible. The aggregations encountered here are formed both by cells sinking from upper layers and by their reproduction within discontinuity layers.

When seasonal discontinuities or the main pycnocline are situated in depths greater than the depth of the compensation point the aggregations are accounted for mainly by the sinking of cells and the decrease in the rate of sinking in the discontinuity layers. Another factor contributing to the retention of cells within these layers is their greater buoyancy resulting from the dark synthesis in these waters rich in nutrients (Steele & Yentsch, 1960). Some slow reproduction seems possible in these deep-seated discontinuity layers (Venrick, McGowan & Mantyla, 1973).

The phytoplankton concentrating in discontinuity layers contributes to an increase in their content of chlorophyll *a* (although aggregations of cells do not always coincide in depth with high chlorophyll concentrations). The ratio of chlorophyll *a* to phaeophytin (a pigment inactive in photosynthesis) decreases with increasing depth (Yentsch, 1965). According to Yentsch

* An uneven distribution of phytoplankton within the biotope may depend on other causes too, such as uneven grazing by zooplankton.

the ratio is always smaller in deep-seated discontinuity layers (close to the depth of the compensation point), which means that the cells contain much inactive pigment, or, in other words, that photosynthesis is very weak or entirely absent. In discontinuity layers situated in lesser depths (above the depth of the compensation point) the ratio of chlorophyll *a* to phaeophytin is high, indicating active photosynthesis.

Let us compare the depths of the main pycnocline, of seasonal discontinuity layers and of compensation depth in the Pacific Ocean. In the northern Pacific and in the sub-Antarctic and the Antarctic, during the early vegetation period before the development of a seasonal discontinuity layer, the compensation point (1% light level) lies in a depth corresponding approximately to the middle of the biotope, as may be inferred from the figure of Hasle (1969) and Taniguchi (1969). Here, therefore, in the absence of any well-expressed seasonal discontinuity layer, some part of the population will be confined to the shadowed part of the biotope. In these latitudes the reproduction of algae at the depth of the main pycnocline is prevented by poor conditions. The aggregations at the depth of the main pycnocline are formed by the sinking of cells from the upper part of the biotope and their retention by the main pycnocline.

During the period of formation of the seasonal discontinuity layer when it is as yet weakly expressed, the compensation point is deeper (deeper than the seasonal discontinuity, but usually higher than prior to its appearance) (Hasle, 1969). At this time phytoplankton aggregations occur both within the seasonal discontinuity layer and beneath it, the latter consisting of cells sunk before its formation.

Later on, with the increase of the density gradient in the seasonal layer, phytoplankton, as a rule, concentrates in this layer and above, but not below it. At this time the depth of the compensation point is situated at a slightly lesser depth than the discontinuity layer when the phytoplankton is abundant. The compensation point is slightly deeper than the seasonal discontinuity layer when phytoplankton concentrates only within this layer while the upper layers are impoverished (Taniguchi, 1969). In the first case the cells aggregating in the discontinuity layer seem to be incapable of active photosynthesis, while in the second case active photosynthesis does take place, as has been recorded by Taniguchi (1969) by the increase of the assimilation number* at the depth of the seasonal discontinuity layer.

The absence of cell aggregations in the layer between the seasonal discontinuity and the main pycnocline, observed in the sub-Arctic region in summer, is accounted for by the low light intensity and small density gradient in the main pycnocline. Illumination is poor even directly beneath

* The assimilation number is the amount of carbon-14 fixed per mg of chlorophyll *a* in 1 hour.

the seasonal discontinuity layer, and poorer still at the depth of the main pycnocline. As compared with the light intensity in the north-eastern Pacific (which, in summer, at the depth of the seasonal discontinuity layer is near to compensation intensity), in the sub-Arctic region at the depth of the main pycnocline illumination does not exceed 0.1% of that at the surface (Anderson, 1969). Under such conditions and in the absence of a sharp density gradient in the main pycnocline, intensive reproduction is hardly possible. This explains the absence in summer of appreciable aggregations of phytoplankton below the seasonal discontinuity layer and in the main pycnocline despite the high concentrations of nutrients.

In the north-eastern Pacific a maximum of chlorophyll is often observed in the layer between the seasonal discontinuity and the main pycnocline (Anderson, 1969). According to Anderson, this maximum is the result of an increased amount of chlorophyll in the process of adaptation of the cells to low light intensity, rather than of an increase in cell numbers (Jørgensen, 1964; Steemann Nielsen & Jørgensen, 1968).

In the eastern equatorial region the depth of the compensation point is either equal to, or greater than the depth of the main pycnocline (by 20–50 m).* Active reproduction seems to take place at the depth of the main pycnocline. The high concentrations of cells observed in the main pycnocline are formed partly by the reproduction of cells at this depth and partly by their sinking from upper layers (as may be inferred from the presence of maxima in the layers above the main pycnocline).

In subtropical regions the compensation point occurs at the depth of the main pycnocline nearshore, while, in the central parts of the regions, it occurs some 50–100 m above the main pycnocline. In the western equatorial region the compensation point is situated in a lesser depth than the main pycnocline in some areas, and close to it in others. Lately some new data have appeared on an increase in production observed in the concentrations of phytoplankton occurring below the depth of the compensation point (Vinogradov, Gitelzon & Sorokin, 1970, 1971; Venrick, McGowan & Mantyla, 1973). It may be assumed therefore that the aggregations within the main pycnocline situated below the compensation point are formed both by cells sinking from upper layers and by slow reproduction at the depth of the main pycnocline.

In the central parts of the subtropical regions, very poor in nutrients and phytoplankton, the seasonal discontinuity layer has little effect on the position of the compensation point, because under the good light conditions of these regions the relatively small aggregations of cells within the seasonal layer are too thin to overshadow the underlying layers as they do in temperate and high latitudes. Thus aggregations of phytoplankton

* For the tropical zone of the ocean (from 30° N to 30° S) we are using the data on the depth of the compensation point given by E. N. Khalemsky (1971).

may occur below the seasonal discontinuity layer where illumination is sufficient for their formation, providing the phytoplankton in the seasonal layer is scarce. In the peripheral parts of the subtropical regions rich in phytoplankton, large aggregations of cells in the seasonal discontinuity layer can most probably overshadow the lower layers of the biotope, which explains the scarcity of phytoplankton below the seasonal discontinuity layers under these conditions.

Our data on the vertical distribution of phytoplankton are based on the collections taken from discrete depths. Recently, methods have been proposed for continuous recording of chlorophyll (Strickland, 1968; Venrick, McGowan & Mantyla, 1973) and of luminescent plankton (Vinogradov, Gitelzon & Sorokin, 1970, 1971), but none of them reflects the distribution of cell numbers of phytoplankton. We know that diatoms, coccolithophorids and blue-green algae which constitute the bulk of the phytoplankton are not luminescent, and although variations in chlorophyll content follow in a general way the quantitative changes in phytoplankton, conflicting results may arise from the use of these two indices, caused by differences in the size of cells with different chlorophyll concentrations, different chlorophyll content of algae belonging to different groups, such as for instance diatoms and peridineans, as well as by increased chlorophyll production by cells in the process of their adaptation to low light intensities. Chlorophyll content may vary also with vertical distribution of cells (Ohwada, 1971; Taniguchi & Kawamura, 1972). We have compared data on chlorophyll content for the eastern equatorial region (Vedernikov & Starodubtsev, 1971) with our own data on cell numbers and discovered differences in the position of chlorophyll maxima and maxima of cell numbers.

In our view, therefore, no other method can replace at present the method of estimating the vertical distribution of cell numbers by their enumeration in collections taken from discrete depths. It is true that these collections may, in some cases, fail to disclose the presence of narrow layers of phytoplankton concentrations, as has been pointed out by Strickland (1968), Vinogradov, Gitelzon & Sorokin (1970, 1971) and by Venrick, McGowan & Mantyla (1973), but our data are evidence that sampling in discrete depths makes it possible to elucidate the general patterns of vertical distribution of phytoplankton in the ocean.

Conclusions

The following patterns of vertical distribution of phytoplankton are common to all the regions of the ocean.

(i) One to three maxima of cell numbers always occur within the biotope. As a rule they are larger than the increases in cell numbers encountered below the lower boundary of the main pycnocline. Nearly 70–80% of the stations had maxima of cell numbers in the main pycnocline, either at its upper or lower boundary or in its central parts.

(ii) The aggregations of phytoplankton in the main pycnocline are always formed of cells sunk from upper layers and delayed in their descent by the main pycnocline. They may also reproduce at this depth. Their reproduction is intensive in the eastern equatorial region where the main pycnocline lies at the depth of the compensation point or above it, but some slow reproduction also seems possible in those parts of the subtropical and western equatorial regions where the main pycnocline occurs in depths greater than the depth of the compensation point.

(iii) The concentrations of phytoplankton are high only within the biotope, decreasing markedly at the depth of the main pycnocline (below the depth of a maximum frequently confined to the main pycnocline). Therefore the lower boundary of the trophogenic layer in the ocean lies close to the average depth of the main pycnocline.

The vertical distribution in the upper 200 m layer of the ocean has its own peculiarities dependent on the peculiarities of the biotope. We recognize seven regions differing in the characteristics of their biotopes and the vertical distribution of phytoplankton within the biotopes (Fig. 8.1).

(i) A characteristic feature of the sub-Arctic region is a great seasonal variability of conditions in the biotope such as stability of water, illumination, and concentration of nutrients. The vertical distribution of phytoplankton varies in accordance with the changes of these conditions. Phytoplankton is most uniformly distributed in early spring, less uniformly with increasing water stability and least uniformly in summer, when, with the development of a well-expressed discontinuity layer, phytoplankton tends to concentrate within this layer or above it. In the sub-Arctic region and in the Antarctic the trophogenic layer extends to the depth of the main pycnocline in early spring (prior to the formation and at the appearance of the seasonal discontinuity layer), in summer and in autumn when the seasonal discontinuity layer is well-developed the trophogenic layer is bounded by its depth and does not reach the main pycnocline.

(ii) The biotopes of subtropical regions are characterised by more stable conditions than those of the sub-Arctic region. The mixing of water in the biotopes is less intensive (due to a less intensive convection), illumination is higher and more constant. The concentration of nutrients is very low so that phytoplankton is scarce. Its vertical distribution is more uniform and the maxima of cell numbers are usually less sharply expressed than in the sub-Arctic region. The seasonal discontinuity layer only occasionally influences the vertical distribution of phytoplankton. Here the trophogenic layer is seldom bounded by the depth of the seasonal discontinuity (only in the presence of great quantities of phytoplankton), otherwise (with small quantities of phytoplankton) the trophogenic layer extends to the main pycnocline irrespective of the presence or absence of a seasonal discontinuity layer.

(iii) The biotopes of equatorial regions are characterized by large density gradients in the main pycnocline and high light intensity. In some other respects the biotopes differ markedly in the eastern, central and western equatorial regions. In the eastern equatorial region the main pycnocline is less deep (depths of 50 m or less), while in the central and western regions it is situated in greater depths (75–200 m). The concentration of nutrients decreases from east to west. Owing to the stable conditions in the biotope the vertical distribution of phytoplankton varies little from season to season. It is more uniform in the biotopes of the western and central parts, and least uniform in the eastern equatorial region. The trophogenic layer, as a rule, is limited by the depth of the main pycnocline, or, occasionally, by the depth of the layer of high stability superimposed on the main pycnocline.

References

Anderson, G. C. (1969). Subsurface chlorophyll maximum in the north-east Pacific Ocean. *Limnology and Oceanography*, **14**(3), 386–91.

Arseniev, V. S. (1967). *Currents and water masses of the Bering Sea.* Moscow: Nauka. (In Russian.)

Bogoyavlensky, A. N. & Shishkina, O. V. (1971). On hydrochemistry of the Peru–Chile region. *Trudy Instituta Okeanologii Akademii Nauk SSSR*, **89**, 96–105. (In Russian.)

Burkov, V. A. (1972). *General water circulation in the Pacific Ocean.* Moscow: Nauka. (In Russian.)

Dobrovolsky, A. D. & Arseniev, V. S. (1961). A hydrological characteristic of the Bering Sea. *Trudy Instituta Okeanologii Akademii Nauk SSSR*, **38**, 64–96. (In Russian.)

Fleming, R. H. (1958). Notes concerning the halocline in the north-eastern Pacific Ocean. *Journal of Marine Research*, **17**, 158–73.

Hasle, G. R. (1969). An analysis of the phytoplankton of the Pacific Southern Ocean. *Hvalrådets Skrifter*, **52**, 1–168.

Hentschel, E. (1936). Allgemeine Biologie des Südatlantischen Oceans. *Meteor Forschungsergebnisse*, **9**(11), 1–344.

Jørgensen, E. G. (1964). Adaptation to different light intensities in the diatom *Cyclotella meneghiniana* Kutz. *Physiologia Plantarum*, **17**, 136–45.

Khalemsky, E. N. (1971). Depth of position of the compensation point in the Pacific Ocean. *Dolkadў Akademii Nauk SSSR*, **196**(2), 445–7. (In Russian.)

Kozlova, O. G. (1964). *Diatoms of the Indian and Pacific sectors of the Antarctic*, pp. 1–168. Moscow: Nauka. (In Russian.)

Kuzmina, A. I. (1959). Some data on the vernal-summer phytoplankton of the North-Kurile region. *Trudy Instituta Okeanologii Akademii Nauk SSSR*, **36**, 215–29. (In Russian.)

Lohmann, H. (1920). Die Bevölkerung des Ozeans mit Plankton. *Archiv für Biontologie*, **4**, 1–617.

Marumo, R. (1957). Vertical distribution of microplankton in the open seas. In Suisangaku Shusei (Collection of oceanic publications), pp. 385–91. (In Japanese.) University of Tokyo.

Marumo, R., Kitou, M. & Ohwada, M. (1958). Vertical distribution of plankton at 40° N, 150° E in the Oyashio water. *Oceanographical Magazine*, **10**(2), 179–84.

Naumann, E. (1931). Limnologische Terminologie. In *Handbuch der biologischen Arbeitsmethoden*, vol. 9.

Ohwada, M. (1960). Vertical distribution of living and dead diatoms to one thousand metres off Sanriku, Northern Japan. *Memoirs Kobe Marine Observer*, **14**, 1–5.

Ohwada, M. (1971). Distribution of chlorophyll and phaeophytin in the sea of Japan. *Oceanographical Magazine*, **23**(1), 234.

Ohwada, M. (1972). Vertical distribution of diatoms in the sea of Japan. In *Biology and Oceanography of the northern North Pacific Ocean*. Tokyo: Idemitsu shoten.

Riley, G. A., Stommel, H. & Bumpus, D. F. (1949). Quantitative ecology of the plankton of the western North Atlantic. *Bulletin of the Bingham Oceanographic Collection*, **12**(3), 1–169.

Ryther, J. H. & Hulburt, E. M. (1960). On winter mixing and the vertical distribution of phytoplankton. *Limnology and Oceanography*, **5**(3), 327–30.

Sapozhnikov, V. V. & Mokievskaya, V. V. (1966). Organic and inorganic phosphorus. In *Pacific Ocean. Chemistry of the waters of the Pacific Ocean*, ed. V. G. Kort, vol. 3, pp. 116–67. Moscow: Nauka. (In Russian.)

Seckel, G. R. (1962). Atlas of the oceanographic climate of the Hawaiian Islands region. *Fishery Bulletin. Fish and Wildlife Service*, **61**.

Semina, H. J. (1955). On the vertical distribution of phytoplankton in the Bering Sea. *Dokladȳ Akademii Nauk SSSR*, **101**(5), 947–9. (In Russian.)

Semina, H. J. (1957). Factors affecting the vertical distribution of phytoplankton *in the sea. Trudy Vsesoyuznogo Hydrobiologicheskogo Obshchestva*, **8**, 119–29.

Semina, H. J. (1966). Biotope and the quantity of phytoplankton in the Oceans. *Uspekhi Sovremennoy Biologii*, **62**(2), 289–306. (In Russian.)

Semina, H. J. & Truong Ngoc An (1973). Phytoplankton of the coastal cyclonic gyre in the tropical Atlantic. *Trudy Vsesoyuznogo Hydrobiologicheskogo Obshchestva*, **20**, (In Russian.)

Steele, J. H. (1957). Production studies in the northern North Sea. *Rapport du Conseil pour l'Exploration de la Mer*, **144**, 79–83.

Steele, J. H. & Yentsch, C. S. (1960). The vertical distribution of chlorophyll. *Journal of the Marine Biological Association of the United Kingdom*, **39**, 217–26.

Steemann Nielsen, E. & Jørgensen, E. G. (1968). The adaptation of plankton algae. I. General part. *Physiologia Plantarum*, **21**, 401–13.

Strickland, J. D. H. (1968). A comparison of profiles of nutrient and chlorophyll concentration taken from discrete depths and by continuous recording. *Limnology and Oceanography*, **13**, 388–91.

Taniguchi, A. (1969). Regional variations of surface primary production in the Bering Sea in summer and vertical stability of water affecting the production. *Bulletin of the Faculty of Fisheries Hokkaido University*, **20**(3), 169–79.

Taniguchi, A. & Kawamura, T. (1972). Primary production in the Oyashio region with special reference to the subsurface chlorophyll maximum layer and phytoplankton-zooplankton relationships. In *Biology and oceanography of the northern North Pacific Ocean*, pp. 232, 243. Tokyo: Idemitsu shoten.
Truong Ngoc An (1971). Phytoplankton in the south-eastern Atlantic Ocean. PhD thesis. Institut Okeanologii Akademii Nauk SSSR, Moscow.
Tully, J. P. (1964). Oceanographic regions and assessment of temperature structure in the seasonal zone of the North Pacific Ocean. *Journal of the Fisheries Research Board of Canada*, **21**(5), 941–70.
Tully, J. P. & Giovando, L. F. (1964). Seasonal temperature structure in the eastern sub-Arctic Pacific Ocean. In *Marine distributions*, ed. M. J. Dunbar, pp. 10–36. Toronto: University of Toronto Press.
Uda, M. (1963). Oceanography of the sub-Arctic Pacific Ocean. *Journal of the Fisheries Research Board of Canada*, **20**(1), 119–79.
Vedernikov, V. I. & Starodubtsev, E. G. (1971). Primary production and chlorophyll in the South-eastern Pacific. *Trudy Instituta Okeanologii Akademii Nauk SSSR*, **89**, 33–42. (In Russian.)
Venrick, E. L., McGowan, J. A. & Mantyla, A. W. (1973). Deep maxima of photosynthetic chlorophyll in the Pacific Ocean. *Fish. Bulletin*, **71**(1), 41–52.
Vinogradov, M. E., Gitelzon, I. I. & Sorokin, Yu. I. (1970). The vertical structure of a pelagic community in the tropical ocean. *Marine Biology*, **6**(3), 187–94.
Vinogradov, M. E., Gitelzon, I. I. & Sorokin, Yu. I. (1971). On the spatial structure of communities in the euphotic zone of the tropical waters of the ocean. In *Funkts. pelagich. commun. trop. reg. of the ocean*, pp. 255–64. Moscow: Nauka. (In Russian.)
Wyrtki, K. (1964). The thermal structure of the eastern Pacific Ocean. *Dt. hydrogr. Z.*, (*Erg.H*), A**6**, 6–84.
Yentsch, C. S. (1965). Distribution of chlorophyll and phaeophytin in the open ocean. *Deep-Sea Research*, **12**, 653–66.

9. The Dutch Wadden Sea

M. VAN DER EIJK

9.1. Chemical and physical environment

Introduction

Biological production is regulated mainly by such basic factors as available light, nutrient supply and temperature.

Compared to the land, the open sea offers an attractive opportunity for the measurement of basic productivity since organic matter is chiefly associated with phytoplankton, which in general is more evenly distributed than land vegetation.

Much more complicated conditions are found in coastal waters. The mixing of fresh and salt water creates brackish water bodies which, depending on the duration of their existence, may or may not have specific plankton populations, whilst fast rates of change of salinity and high turbidity may depress productivity. Nutrient supply from the land, by rivers and pipelines, may cause an increase in fertility over wide areas, and may also change the ionic ratios of the nutrient salts. Transportation and mixing by the tides will also play a role. The primary production of benthic organisms and sessile plants must also be considered in addition to that of phytoplankton.

The hydrography and material transport of the coastal water and the Wadden Sea

The shallow mean depth of the coastal zone (25 m) and strong tidal currents lead to intensive vertical mixing, and a thermocline cannot develop; thus the water column is homothermal all the year round.

The oceanic water inflow (salinity 35‰) through the Straits of Dover amounts to 3000–7000 km^3/year and gives rise to a residual water transport along the coast of the Netherlands in a northerly direction. Calculations from two salinity situations in two successive years gave an estimated residence time of three to four months for the water bounded by the 53° N parallel.

Freshwater from the highly polluted Rhine, Meuse and Scheldt rivers enters the Southern Bight in the delta area with a mean annual discharge of about 70 km^3 (see Fig. 9.1). Another 14 km^3 is discharged via Lake IJssel and the Dutch Wadden Sea. Mixing of saline and freshwater and residual transport results in a pattern of isohalines almost parallel to the coast.

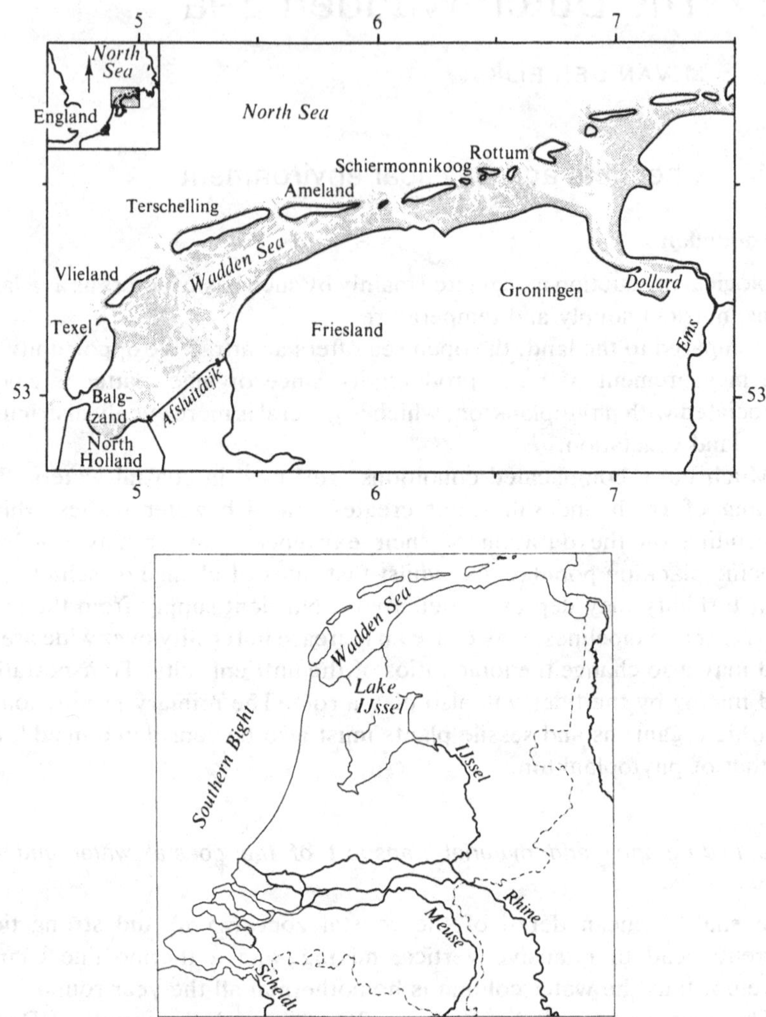

Fig. 9.1. The Wadden Sea and Lake IJssel.

Huge quantities of soluble and suspended nutrients are discharged by the west European rivers into the coastal seas. The composition of the discharged Rhine water has changed drastically over the last decades. In the ten-year period 1959–68 the annual phosphate load of the river Rhine increased from 4000 to 15000 tons phosphorus, and ammonia and nitrate increased from 190000 to 340000 tons nitrogen. The annual dissolved silicate load in the period 1967–71 has been estimated to be 400000 tons (SiO_2).

Table 9.1. *Nutrient concentrations in μM; the values refer to total phosphorus and nitrogen, and reactive silicate; the Wadden Sea mixture (S = 29.75‰) is based on 85% North Sea, 7.5% Rhine and 7.5% IJsselmeer water (in parentheses the ratio in a mixture with 5% Rhine and 10% IJsselmeer water)*

	Winter			Summer		
	P	N	Si	P	N	Si
Rhine	18.0	380.0	130.0	18.0	380.0	50.0
Lake IJssel	3.5	250.0	40.0	2.0	100.0	7.0
North Sea	0.4	5.6	2.8	0.1	1.4	0.7
Wadden Sea calculated	2.0	52.0	15.1	1.6	37.2	4.9
Ratio in mixture		1:26:7.6 (1:31:8)			1:33:3 (1:25:3.2)	

Phosphorus and nitrogen are contained in human, agricultural and industrial waste products. The increase in phosphorus is most spectacular and is partly due to the introduction of detergents in the mid-1960s. By contrast reactive silicate which is not a common waste product has not increased greatly. In the Southern Bight in summer 50–70% of the soluble phosphorus is suppplied directly or indirectly (from mineralization of suspended phosphorus) by the rivers.

Phytoplankton growth during August is limited in high salinity water by nitrogen, but over a wide area near to the coast the limiting factor for the abundant diatoms is silicon. During spring and summer, concentrations of dissolved silica in the coastal zone may be below 0.5 μg at/l and nitrate+ammonia values exceed 15 μ at/l.

The Wadden Sea consists of a mixture of water from the North Sea, the river Rhine and Lake IJssel. A useful method of indentifying these three sources is by measuring differences in atomic ratios of phosphorus nitrogen and silicon (P:N:Si). In coastal areas, organic detritus may be partly preserved at the bottom. This can lead to a shift in the atomic ratio. Moreover, rivers may introduce water with a different composition. Both processes are at work along the Dutch coast and in the Wadden Sea. Therefore care has to be taken in calculating atomic ratios. Differences in the kinetics of mineralization and long residence time cause each nitrient element to have its own seasonal cycle. The maxima for the different substances do not coincide, nor do the minima (see Table 9.1). Nevertheless the data on atomic ratios, given in this table, do show general trends.

The Rhine itself has a P:N:Si ratio of 1:21:7 in winter and 1:21:3 in summer, with an annual average of 1:21:5. The values given are for total phosphorus and nitrogen and for reactive silicon on the assumption that most organic phosphorus and nitrogen from the river is quickly mineralized in the sea, although this may not always be the case, particularly in winter. Clearly, there is a continuous surplus of nitrogen and a pronounced shortage in silica in summer in relation to the normal planktonic ratio. In Lake IJssel plankton converts a considerable part of the nutrients into bottom material, at the same time changing the ratio.

The winter ratio for the water leaving the lake is 1:70:11 and in the summer a ratio of 1:50:3.5, with an annual average of 1:60:7. The nitrogen excess has greatly increased compared with the original Rhine value. The same holds for silica in winter. The variations can be explained by changes in supply combined with fixation by a plankton population in the lake containing relatively few diatoms.

According to the table, the composition of Wadden Sea water is not very sensitive to the ratio of Lake IJssel and Rhine water. In each case there is a large excess of nitrogen throughout the year, a slight excess of reactive silicate in winter and a severe shortage in summer. These calculations indicate that the Wadden Sea receives a water mixture with a nutrient ratio quite different from the planktonic ratio. The *actual* ratios found for inorganic phosphorus, nitrogen and silicon are not quite the same; in winter inorganic P:N:Si = 1:20:13 and in summer 1:15:3.

Studies on the nutrient budgets

A number of surveys have been conducted of the nutrient distribution and primary production in the Wadden Sea. The central theme of these is the influence of the river Rhine on the state, production and mineralization of organic matter and the distribution of nutrients. Eutrophication effects have also been taken into account and an investigation made of the distribution of a number of trace metals (Duinker, Eck & Nolting, 1974). The influence of freshwater from Lake IJssel and the river Ems has been studied and particular attention has been paid to processes on tidal flats (Bennekom *et al.*, 1974; Cadée & Hegeman, 1974*a*, *b*; Helder, 1974; Jonge & Postma, 1974; Manuels & Postma, 1974).

In the entire western Wadden Sea the average freshwater content is about 15 %. One-half to two-thirds of this freshwater comes from Lake IJssel, the rest is Rhine water flowing north along the Dutch coast. Part of Lake IJssel's water comes from the Rhine, but its composition has been changed by losses of material in the Lake IJssel-basin. Going east, the Wadden Sea contains less and less Rhine water until, finally, near the German border the river Ems provides the freshwater component. However, this does not mean that the Rhine influence is restricted to the western part. Suspended matter from the Rhine enters the Wadden Sea in relatively large quantities. The actual amount is unknown, but possibly up to 50% of the silt discharge of the Rhine enters the western Wadden Sea. This material is carried east as far as the Ems estuary and it may provide a very large percentage of the fine grained material accumulated in small channels, mud-flats and salt-marshes. The Wadden Sea thus acts as a trap for sediments from the North Sea.

The concentrations of nutrients in the Wadden Sea are determined by:

(i) The amount of freshwater (outflow from the Rhine, as well as the discharge from Lake IJssel).

(ii) Decomposition of organic material, formed in the adjacent North Sea, which by a process of tidal accumulation is brought into the Wadden Sea.

(iii) Production and subsequent decomposition of organic matter in the area itself.

The seasonal cycles of reactive silicate and suspended diatoms

In estuaries and coastal zones reactive silicate is fixed by diatoms. Since no evidence can be found that diatom frustules are recycled completely in these shallow areas it is generally assumed that reactive silicate is removed. Research on the various processes was undertaken in order to get a general impression of the budget of this element.

Silica and the processes involved

The concentration of reactive silicate in the western Wadden Sea is determined by three processes, each having a seasonal cycle:

(i) Inflow of reactive silicate, into the Wadden Sea area. This shows a maximum in winter and is negligible in May and September.

(ii) Diatom growth, which peaks twice and sometimes three times a year.

(iii) Mineralization and transport from interstitial water.

The two most important water sources contributing reactive silicate to the Wadden Sea are:

(1) Dutch coastal water, present at the inlets, with a salinity of about 30‰ which carries the fresh water discharged from the rivers Rhine, Meuse and Scheldt. The silica values showed maxima of 20–25 μg at/l in January and February. In spring and summer the values drop to 1–6 μg at/l, partly because of lower concentrations of dissolved silica in the Rhine and partly because of periodic depletions by diatoms along the Dutch coast (Bennekom *et al.*, 1974).

(2) Lake IJssel water, discharged intermittently through sluices (during

1970 18.5 km^3 freshwater, containing 31 000 tons SiO_2). Seasonal variations are much more pronounced than in the river Rhine, maxima of about 80 μg at/l being attained in late March, compared with values of 1–10 μg at/l from May to November. During this latter period 45% of the water but only 11% of the silicate is discharged.

Periodic diatom production affects the reactive silicate concentration. The amount of silica that is fixed in suspended diatoms reaches maxima two or three times a year coinciding with minima of about 1 μg at/l in dissolved silicon. This effect is a function of water depth. In deeper water depletion occurs in April, to a lesser extent in June and is not detectable in September.

Despite the correspondence between diatom maxima and silicon minima, the limitation by silicon is not proven conclusively because nitrogen compounds also reached very low values, nitrate being exhausted long before silica.

The role of silicon can only be understood in relation to other nutrients. In the Wadden Sea the nitrate concentration is high during the spring diatom bloom, but in summer total dissolved nitrogen compounds may in some regions decrease to less than 1 μM for short periods (Helder, 1974). Likewise phosphate–phosphorus values as low as 0.2 μM have been measured (Jonge & Postma, 1974) and in these cases primary production is just limited by nitrogen and phosphorus. Limitation by silicon is found during longer periods and is more widespread (Gieskes & Bennekom, 1973), but will only influence the diatoms. In the Walden Sea and the coastal waters the spring diatom bloom is usually followed by a large bloom of *Phaecocystis poucheti* which can consume the phosphorus left over by the diatoms, and *Skeletonema costatum*.

Light, grazing pressure and turbulence are also important factors in the development of phytoplankton blooms, but the fact that in temperate regions these blooms usually occur in a series of peaks (Cadée & Hegemans, 1974*a*; Postma & Rommerts, 1970) means that primary production cannot be attributed to the same limiting factors throughout the whole year. It seems that light and grazing pressure determine whether growth occurs or not. Once it occurs, the exponentially increasing nutrient demand rapidly outgrows the nutrient supply.

Diatom growth

The standing stock of diatoms in the Wadden Sea is the net result of growth, exchange with Dutch coastal water and addition from Lake IJssel on the one hand, and on the other hand the effect of direct sedimentation and grazing.

Estimates of grazing can be obtained from the density of filter feeders and zooplankton. The total volume of water filtered is calculated to be 0.42 km^3/day by the mussel population and 0.04 km^3/day by the cockles (Verwey, 1952, 1966). Each week 55% of the water is filtered, leading to dilution of plankton, and 43% of the original amount is converted into faecal pellets and pseudo-faeces.

Assuming a dry weight of zooplankton of 1 mg/l in June and 0.1 mg/l in August and October, a daily food intake which equals zooplankton body weight and an average suspended load taken of 1–2 mg organic carbon, or 4 mg organic matter/l (Manuels & Postma, 1974), it can be calculated that in a week 15–100% of the diatoms are converted into faecal pellets. Net removal of material is probably small compared with faecal pellet formation. It is thought that 16000 tons of SiO_2 is added to the sediment as well as 12500 tons from fresh water species. The benthic diatoms, which are responsible for a large part of the primary production (Cadée & Hegemans, 1974*a*) and whose biomass remains essentially constant during the year, obtain the silicon for their production from the interstitial water. Because of the time-lag between production and dissolution, part of the dissolved silicon will come from benthic diatoms.

The main source of silicon produced *in situ* is the mineralization of diatom debris in the surface sediments. These sediments in muddy, shallow regions contribute much more to mineralization than does the suspended matter.

It may be concluded that diatomaceous silica accumulated in the sediment of the Wadden Sea is only partly dissolved (Straaten, 1954). But it is uncertain whether only diatom frustules take part in this process and therefore the word dissolution is preferred rather than mineralization. Only coarse diatom skeletons are preserved in the sediments of the Wadden Sea.

The transport of silicon from interstitial water is caused by turbulent exchange between the surface sediment and overlying water rather than by diffusion. The steepest gradient in silicon was found at 15 cm below the sediment surface, while silicon concentration in the upper 10 cm was constant (Rutgers van der Loeff, 1974). The average rate of release from the bottom of the Wadden Sea is estimated to be 100 mg SiO_2/m^2/day. This large contribution is thought to be increased not only by temperature but also with wind force, as wind determines the effectiveness of the subtidal pump. In this context it might be mentioned that also copper and zinc are introduced from the interstitial water to the Wadden Sea at a tidal watershed, particularly during periods of strong winds (Duinker *et al.*, 1974).

In the light of this it is not surprising that higher concentrations of dissolved silica are found in the eastern part of the Dutch Wadden Sea. The areas of tidal flats are proportionally greater and the silt and clay fractions in the sediment are more important. Inflow plays a minor role here. The production of silicon *in situ* stops in January, possibly due to the low temperatures.

The amounts of silicon show marked seasonal variations, not only in the Wadden Sea itself but also in the waters transported into the area. Maxima are found in January for the tidal inlets and in March for Lake IJssel. The minima coincide with diatom blooms. From the beginning of March a zone in which silicon is being removed moves gradually to the inner parts of the Wadden Sea, while the sum of silicon and silica in diatoms remains about constant.

Large quantities of water discharged from Lake IJssel during March and April ensure further growth of diatoms. By the end of April, the zone in which silicon is almost completely removed covers the entire Wadden Sea. After the spring bloom the silicon values are higher in the inner parts of the Wadden Sea than in the adjacent North Sea and Lake IJssel. This is a result of the production of silicon *in situ*, which starts in mid-May but is interrupted by the second diatom bloom in June, when part of this silicon is consumed. The production reaches a maximum in August–September and lasts till December.

In short, in summer the growth of diatoms and the solution of silica are the most important processes, whilst in autumn the solution of silica and the addition of organic matter from the river Rhine and Lake IJssel determine the distribution.

Silicon budget

From May to November concentrations of dissolved silica in the innermost parts of the area are higher than those in the tidal inlets and Lake IJssel. During this period the amount of silica dissolved from the sediment is many times higher than the amount discharged from Lake IJssel. The increase in dissolved silica concentrations occurs after almost all diatoms have disappeared from the water column. This process of dissolution from the upper sediment layers seems to be the most important factor.

From the seasonal cycles of the reactive silicate and the diatoms a silicon budget can be derived, based on the freshwater budget, the grazing rates of diatoms and the mineralization.

In the western Wadden Sea, where an important contribution from mineralization can be expected, budget calculations show a net removal of reactive silicate via diatoms. The amount of reactive silicate removed is comparable with the amount of diatomaceous silica, the estimation of which is based on the number of diatoms.

Of the 140000 tons SiO_2 transported annually into the western part, only 80000 tons leave the area. Of the remainder, part is sedimented in the

Wadden Sea, and part is transported by the residual current in suspension and is again subjected to the cycle of accumulation and mineralization in the eastern part of the Wadden Sea and the North Sea. Because of the long residence time, slow solution of diatom debris in suspension probably cannot be neglected.

The greater part of the silicon from the Rhine does not enter the Wadden Sea again and will be converted into diatoms; the productive season lasts longer offshore (Gieskes, 1974*b*). About half of the reactive silicate, present in the two inflow waters (coastal water and Lake IJssel) is converted into diatoms, either in the Wadden Sea itself or during transport to the Wadden Sea. Less than a third of the silica from these diatoms is mineralized. Production of diatoms is therefore responsible for a considerable net removal of reactive silicate in the estuary of the Rhine.

From data on the density of filter feeders and zooplankton it is inferred that removal of diatoms by mussels is the most important factor; in early summer the contribution of zooplankton cannot be neglected.

The cycle of dissolved inorganic nitrogen compounds (Helder, 1974)

A study has been made of the origins of ammonia, nitrite and nitrate in the western Wadden Sea, where the concentrations are influenced by the conditions in Lake IJssel and the adjacent North Sea.

In order to understand this cycle it is essential to realize that nitrate has to be reduced to amino-nitrogen before incorporation into cellular proteins. This reduction is carried out by nitrate reductase, an enzyme inhibited by ammonia. Most marine species prefer the uptake of ammonia to that of nitrate and only when ammonia is seriously depleted ($< 2 \mu$g at/l) will organisms shift to nitrate uptake (Eppley, Coatsworth & Solorzano, 1969).

Western Wadden Sea

Lake IJssel, which acts as a storage basin, receives about 60% of its water from the river IJssel, a branch of the river Rhine. Over the period 1960–72 the concentration of nitrate in the Rhine increased considerably, but the increase in the concentration of nitrogen compounds in the river IJssel over this period did not apparently result in any increase in concentration in the lake.

In the western Wadden Sea, in the area south of the islands of Texel, Vlieland and Terschelling, the use of ammonia was possible almost all the year round. Only in June, at some stations, did ammonia concentrations fall below 1 μg at/l, and to ensure maximum growth, nitrate had to be used.

A situation in which the total dissolved inorganic nitrogen concentration

(NH_4^+. NO_3^-) was responsible for growth inhibition (< 1 μg at/l) was detected at very few stations. This decrease was not a consequence of nitrate depletion by primary production; a dramatic fall of nitrate concentration occurred when the amount of nitrate sluiced into the area from Lake IJssel was decreased seriously by diminishing the introduced freshwater flow.

Thus the nitrate concentrations in the western part of the Wadden Sea are determined mainly by the amount of water sluiced into the area from Lake IJssel.

Over the period 1960–72 the ammonia concentration increased by about 100 %, equivalent to a maximum increase of 1.5 μg at/l. This is partly the result of changes in the Rhine and IJssel, but some of the increase may have its origin in the western Wadden Sea itself. The annual average nitrite concentration for the period 1961–72, especially in areas of lower salinity, appears to have doubled (Helder, 1974). At stations with high salinities, as well as in Lake IJssel, no significant change has occurred in the 10-year period. These data indicate that the increase of nitrite and also of ammonia (precursor of nitrite in the mineralization process) must have its origin in the Wadden Sea itself. Primary production in the Wadden Sea seems to have remained unchanged in the period 1963–73 (Postma & Rommerts, 1970; Cadée & Hogemans, 1974*a*), implying that the increases in ammonia and nitrite concentrations are probably associated with a greater input of organic material from the adjacent North Sea, where primary production increased during that decade (Postma, 1973). This is also suggested by the fact that greater amounts of ammonia and nitrite are found at places where organic material is deposited by tidal accumulation.

Eastern Wadden Sea

The eastern part of the Wadden Sea, south of the islands of Ameland and Schiermonnikoog, is shallow and, lacking a large freshwater input, has a somewhat higher salinity of 31.4‰, which is about 1.5‰ higher than in the western part.

Ammonia plays a greater role throughout the year in the eastern Wadden Sea where in contrast to the western part the concentrations of dissolved nitrogen compounds are not strongly influenced by the condition of Lake IJssel and the North Sea and where depletion can occur more readily.

Times at which total dissolved organic nitrogen concentrations are less than 1 μg at/l, occur in June in the whole area, with the exception of some stations situated near a watershed, where mineralization is at a maximum because of the deposits of organic material. Maximum nitrate concentrations are reached in March–April and are found in the inner parts of the

Table 9.2. *Ammonia, nitrate and nitrite: annual average concentrations*

	Lake IJssel		Western Wadden Sea		Eastern Wadden
	1960–2	1971–2	1960–2	1971–2	1971–2
Ammonia	5.8	4.5	6.1	7.7	8.3
Nitrite	1.3	1.0	1.4	1.5	1.2
Nitrate	57.0	47.7	17.6	12.2	8.0

Data for 1960–2 from Postma (1956).
Data for 1971–2 from Helder (1974).

Wadden Sea. Minima occur during June, July and August with concentrations of 1–2 μg at/l.

Maximum concentrations of ammonia are found in April and May, and may be associated with spring blooms in the adjacent North Sea. Maxima are also reached from September to December (20–30 μg at/l), which are obviously connected with mineralization of organic matter produced in the preceding summer. Minimum values in the range of 1–5 μg at/l are found in the months June, July and August, when consumption of ammonia by plankton organisms sometimes exceeds production of ammonia by mineralization. This mineralization is the result not only of the process of bacterial proteolysis, but is also caused by the activities of detritus feeders (e.g. *Arenicola marina*) which by the digestion of organic matter contribute to the total amount of ammonia.

In the eastern Wadden Sea the freshwater of the river Ems brings a large amount of nitrate into the estuary. In addition a large amount of proteinaceous material derived from sugar factories and potato mills is sluiced into this area, particularly in winter.

Conclusions

In the Wadden Sea as a whole a gradient exists with high concentrations of ammonia and nitrate on the landward side and low concentrations in the tidal inlets between the islands. Most organic material is accumulated at the shallowest places, the so-called 'wantijen' (watersheds). The annual average ammonia concentrations are high in these areas (10.4 μg at/l). Table 9.2 gives the annual average concentrations, calculated for two different periods.

Phosphorus compounds

The Rhine, which in 1970 carried about twice as much phosphorus to the sea as in 1950, is responsible for the significant increase in the supply of this element to the Dutch coastal water (Jonge & Postma, 1974).

In most shallow temperate regions phosphate concentrations vary with the seasons, the highest values being found in winter, when phytoplankton production is low, and the lowest in spring or summer, when the production is high. Generally the concentrations in the Wadden Sea are higher than in the adjacent North Sea, especially during summer. This difference appears to be attributable to the transport of particulate organic matter from the North Sea into the Wadden Sea together with a relatively rapid decomposition of this material. There is a net flow of phosphate from decomposing organic matter (particulate phosphorus) to the Wadden Sea and of dissolved phosphate seaward. The same appears to apply to dissolved organic phosphorus. The net influx from the North Sea was estimated in 1950 to be 190000 tons of particulate organic matter per tide and the total loss of dissolved phosphorus from the tidal area was 1900 kg P per tide. This value is of the same order as the organic matter produced *in situ* by primary production (Postma & Rommerts, 1970); hence it forms an essential part of the food supply of the Wadden Sea.

Phosphate and organic matter

An interesting observation is the fact that high concentrations of phosphate are found on the tidal watersheds in summer. There, the water masses are renewed and contain relatively large amounts of particulate organic matter. The release of phosphate from the bottom in these shallow depths adds to the total concentration. In summer biological activity is high, which holds both for microbiological activity and for animals in the water and on the bottom, and at the same time the concentration of particulate organic matter mainly from phytoplankton is high, so that much organic matter is decomposed and phosphate liberated.

Phosphate distribution is more homogeneous in winter than in summer, when steep gradients are built up. The highest phosphate concentrations are generated near the shores on the tidal watersheds, behind the islands and in the Ems–Dollard estuary. Most of the decomposition processes take place in these very shallow and turbid areas.

The Ems–Dollard estuary shows an elevated concentration, because of the extra supply of phosphate. This is due to the freshwater inlet from the river Ems which contains an average of about 3 μg at/l and another freshwater outlet from Groningen which contains much higher amounts of phosphate. Additional to this is the breakdown of the large amount of particulate organic matter suspended in this Dollard water.

Particulate organic phosporus and phosphate release

As stated earlier, the amount of particulate organic phosphorus in the Wadden Sea has risen by a factor of two between 1950 and 1970 (to 0.9 kg P/sec in 1970). This increase in supply has probably also led to an increase of primary production in the North Sea over a widening zone along the Dutch coast. This would cause an increase in the concentration of suspended organic matter, which in turn would lead to a considerable increase of dissolved phosphorus in the Wadden Sea. Particulate phosphorus concentration doubled from 2.4 and 5.1 μg at P/100 mg carbon-free suspended matter (Jonge & Postma, 1974).

The gradient of dissolved phosphorus between the Wadden Sea and the North Sea increased by a factor of three, and so did the net transport of phosphorus to the North Sea. Thus the release of phosphate has risen by a factor of three from 0.5 to 1.7 μg at/l average (Jonge & Postma, 1974). The explanation of this phenomenon could be found either in differences in primary production or in the changed rate of mineralization.

Primary production is assumed to have remained unchanged (Cadée & Hegemans, 1974*a*; Postma & Rommerts, 1970). On this assumption, the total amount to be mineralized has doubled. This is in proportion to the amount of particulate organic matter in the water, which has also increased by a factor of two. Consumption and subsequent release of phosphorus from organic matter in the Wadden Sea takes place by the agency of bacteria, benthos, zooplankton, fish, other free swimming animals and birds. Only the first three consume organic detritus directly. A calculation of the contribution of each of these groups to the total mineralization will show which group has developed over the last 20 years.

Mineralization budget

The amount of organic matter consumed by benthos, excluding mussels, is about 100 g/m^2/year and has not increased, and mineralization of this amount releases 1 g P/m^2/year. Due to increased cultivation, the mussel population of the Wadden Sea has increased from 46×10^6 kg in 1949 to 475×10^6 kg in 1966 (Verwey, 1952, 1966). Assuming that mussels contain 4% of dry organic matter and that they consume ten times their own weight annually, the figures for consumption are 12 and 122 g organic matter/m^2/year respectively, or roughly 0.1 and 1.2 g P/m^2/year. The role of zooplankton may have been widely underestimated previously and may actually be of the same order as benthos. This seems to be confirmed by other work (Manuels & Postma, 1974). It will therefore be assumed that zooplankton mineralizes 1 g P/m^2/year (Jonge & Postma, 1974).

This calculation involved only consumption of organic matter in the second step of the food-chain, the first step being the supply of basic

Table 9.3. *Balance of production and mineralization of organic phosphorus of the western Wadden Sea in 1950 and in 1970*

	Organic particulate phosphorus produced ($g/m^2/year$)			Organic phosphorus mineralized ($g/m^2/year$	
	1950	1970		1950	1970
Supplied by North Sea	2	6	Benthos, except for mussels	1.0	1.0
Primary production *in situ*	2	2	Mussels	0.1	1.2
			Zooplankton	1.0	1.0
			Bacteria	1.9	4.8
To be mineralized:	4	8	Mineralized:	4	8

organic matter itself. It does take into account that part of the phosphate is released only after benthos is consumed in subsequent steps. Moreover, bacteria and zooplankton are consumed by benthos and fish. For this and other reasons no great importance should be attached to the actual numbers presented. However, they illustrate that in the process of mineralization, bacterial decomposition may be predominant, whereas the role of the mussel population has increased (Table 9.3).

The amount of particulate phosphorus supplied by the North Sea (Postma, 1954) is equal to the escape of dissolved phosphate. This amount must be three times higher in 1970 than in 1950.

Primary production *in situ* is assumed to have remained unchanged, because phosphate did not decrease to zero in spring and was therefore not a limiting factor for the growth of phytoplankton (Cadée & Hegemans, 1974*a*; Postma & Rommerts, 1970).

Sulphate in water and sediment

The considerable amount of organic matter present in the Wadden Sea is derived from primary production (Postma & Rommerts, 1970; Cadée & Hegemans, 1974*a*, *b*), secondary production (Beukema, 1970) and from the adjoining North Sea (Postma, 1954, 1961). The greater part of it is mineralized by micro-organisms in the sediment. When oxygen is present mineralization takes place by aerobic respiration, but when the oxygen is depleted, further decomposition is effected by fermentation and anaerobic respiration. The black colour of the sediments in the Wadden Sea is characteristic and indicates that some of the mineralization takes place by sulphate reduction.

Experiments have been carried out in order to estimate the rate of sulphate production and to study the factors which have a bearing on this process. The rate of sulphate reduction can be derived from changes in the SO_4:Cl ratio. Sulphate was determined by turbidimetry and the chlorinity by Mohr's method.

In a few places, where a great quantity of organic matter is discharged into the Wadden Sea, very low values were found for the SO_4:Cl ratio, clearly indicating a reduction of sulphate. This reduction took place in the anaerobic water column which accordingly smelt strongly of sulphide. The SO_4:Cl ratios at various depths in different cores showed a very clear decrease of sulphate downwards, but in some cores the SO_4:Cl ratio at the surface is unexpectedly high. This may be explained by an upwards diffusion of sulphide and its subsequent oxidation to sulphate by organisms such as thiobacilli. The decrease in sulphate is caused by sulphate-reducing bacteria. Counts of samples from anaerobic sediment showed that 5700 to 70000 of such bacteria were present per cubic centimetre. In samples taken from mud-flats, sulphate reducers had maxima of 20000/ml in the oxidation layer and 200000/ml in the reduction layer (Sandkvist, 1968).

A relative measure for the rate of sulphate reduction was obtained by means of ^{35}S-labelled sulphate, which was added to a natural sediment. These experiments showed that the rate of reduction in an anaerobic mud-flat sample, taken in winter, was increased considerably by the addition of lactate, from which it may be inferred that the limiting factor for the growth of bacteria was the amount of organic matter. Further experiments showed that temperature also had a considerable effect on the sulphate reduction. Although an intensive consumption of sulphate may be observed in the bottom of the Wadden Sea, there is generally no decrease of sulphate to be traced in the water near the bottom.

On the behaviour of copper, zinc, iron and manganese, and evidence for mobilization processes

A number of metals associated with suspended matter in the freshwater environment are mobilized when the suspended matter is brought into salt water (Groot, 1963, 1966, 1973). This process starts in the intermediate region between freshwater and seawater. Therefore, the suspended matter from the Rhine, when it enters the Wadden Sea, has lost a large fraction of a number of metals.

But other processes may lead to the reverse effect. In sea water, dissolved metals can form complexes with a wide range of organic and inorganic complexing agents, depending on pH, oxidation state and relative abundance of the complexing agents. There are three types of processes by which metals can be removed from the dissolved state (Krauskopf, 1956).

(i) Adsorption on suspended matter.

(ii) Chemical precipitation or in reaction with compounds of more abundant elements.

(iii) Uptake by marine organisms.

Bearing those processes in mind, it is clear that the ecological significance of metals in the Wadden Sea, for example in uptake by bottom organisms, may be underestimated if only the metal concentrations in the bottom material are considered and not the concentrations in suspension as well. The ecological significance of trace metals in the marine environment appears to depend on their chemical state, so that the determination of the total amount of any particular metal within a sample, without information about its chemical or physical state, is inadequate, if it is to be interpreted in terms of ecological significance.

The processes of the metal balance

In order to get an impression of the processes which take place in the Wadden Sea and of the factors which are involved, samples of water, suspended matter and bottom sediment from a tidal watershed in the middle of the Wadden Sea have been analysed for their metal content. Research has been carried out on copper and zinc specifically; their supply to the marine environment may have ecological significance and their chemistry in sea water is relatively well known. Moreover, both elements contribute significantly to the metal load of the Rhine. The elements iron and manganese have been included because of the scavenging properties of their hydrous oxides and the usefulness of manganese as a tracer for sediment transport studies.

It became apparent that several factors play an important role in the metal balance. From previous work (Postma, 1954, 1961; Straaten & Kuenen, 1957) it is known that continuous exchange of bottom and suspended sediments takes place. The levels of metal concentration in suspended matter vary considerably with time and place, and exchange processes of metals between water, suspended sediment, and bottom sediment complicate the course of events.

In this study (Duinker, van Eck & Nolting, 1974), it was found that the level of metals in interstitial water reaches high values by uptake of metals mobilized in bottom sediment, and not by uptake of metals from the overlying water. Part of the suspended matter that is introduced from the North Sea into the Wadden Sea will settle and lose part of its metal content under the reducing conditions which exist in the bottom layer. The metals thus become more readily available to the environment.

Exchange of metals between interstitial water and water near the bottom takes place only under particular hydrographic and meteorological conditions. Sharp and strong maxima are found in the concentrations of copper and zinc in the water phase, and high concentrations of iron, manganese, copper and zinc are found in the interstitial water sampled. A small exchange between interstitial water and bottom water may cause considerable enrichment of the latter. Tidal and wind action increase the efficiency of this process. The metals are either taken up from the water phase as soluble complexes or they are attached to suspended matter (i) in colloidal form, (ii) by adsorption (clay minerals) or (iii) by uptake in marine organisms.

The existence of a very fine fraction, probably consisting of colloidal iron and manganese oxides, and including trace elements, has been demonstrated. This fraction does not settle under the prevailing local conditions and will probably settle elsewhere, perhaps in the deep ocean. It has a relatively high metal content.

The processes discussed here take place under natural conditions; they are of a general character and applicable to any estuarine environment. The most efficient exchange of metals between interstitial water and overlying water takes place during periods of maximum current velocities. Estuarine sediments are influenced more by one short period of extreme wind conditions than by a long period of average conditions. Tidal action and reducing conditions are partly responsible for the mobilization processes that occur. The absence of tidal action from an area may cause high metal concentrations in interstitial water, with great ecological significance.

9.2. Biological production

Primary production

Phytoplankton composition

The species composition was fairly uniform at all stations in the western and eastern Wadden Sea and in the Ems estuary, and the seasonal cycle consisted of the spring bloom and a lesser mid-summer peak followed by a small peak in the late summer. The spring bloom, however, was not formed of the same species each year. Thus in 1969 the bloom was caused exclusively by *Chaetoceros radians*; in 1970 it consisted of *Skeletonema costatum*, *Chaetoceros radians* and *Biddulphia aurita*; and in 1970 the diatoms were poorly represented and the bloom consisted almost entirely of the flagellate *Phaeocystis poucheti*.

Diatom development starts in February with the freshwater species *Diatoma elongata*, *Asterionella formosa* and *Melosira* sp. Marine species appear early in March. The spring maximum occurs everywhere in mid-April. These freshwater species are also found in the Wadden Sea. At some places in the Wadden Sea they make up the bulk of the diatoms both in numbers and in amounts of silica, althouth their relative importance decreases in April. *Skeletonema costatum* has the highest numbers (2000–5000 cells/ml); *Plagiogramma brockmannii*, *Asterionella japonica* and *A. kariana* reach 400–1400 cells/ml. At the entrance of the Wadden Sea, *Chaetoceros radians* is important with 4000 cells/ml, but in the inner parts *Biddulphia aurita* may be present up to 1000 cells/ml. Together *Biddulphia regia* and *B. aurita* contain more than half of the silica in the marine diatoms. A second diatom bloom, occurring at the end of June, is most pronounced in the Marsdiep–Den Oever area. Highest numbers of up to 5500 cells/ml are found for *Chaetoceros radians*. *Eucampia zodiacus*, *Cerataulina bergonii* and *Skeletonema costatum* reach up to 400 cells/ml. In the Marsdiep–Den Oever region a third relatively small bloom is found in September, consisting of *Thallassiosira condensata*, *Ditylum brightwellii*, *Eucampia zodiacus* and *Chaetoceros radians*.

It is thought that the spring diatom bloom has its origin in the North Sea; large quantities of silicon from Lake IJssel ensure further growth in the Wadden Sea while at the same time freshwater species are advected into that area. Sometimes a small peak is found at the end of the year, consisting of freshwater algae, mostly *Scenedesmus* spp. This is the result of a high discharge of water from Lake IJssel. Experiments have shown that the phytoplankton of Lake IJssel is not killed immediately when it enters the Wadden Sea, but remains photosynthesizing, although at a lower rate than in freshwater.

Pennate benthic diatoms form on the average 3% of the population in most of the Wadden Sea. Only in the shallow eastern part do they occasionally form up to 35% of the cell numbers of the diatoms. In the Dollard – the shallowest part of the Wadden Sea with extensive tidal-flats – they may form up to 100%. For the Danish Wadden Sea varying amounts (1–70%) of benthic diatoms are reported (Grøntved, 1960). Comparison of the numbers of benthic diatoms present per cm^2 on the total flat with those in the water column show that only a small proportion of the benthic diatoms is suspended in the water (0.1–0.3%).

Primary production inshore and offshore

The seasonal cycle of primary production in shallow temperate zones differs significantly from that in offshore waters of the same latitude (Ryther, 1963). The offshore seasonal cycle in our general region is bimodal, with a summer minimum. This is shown on the Fladen ground (Steele, 1956) and off Plymouth (Harvey *et al.*, 1935). The annual primary production offshore is probably of the same order of magnitude as inshore, in spite of the large differences in nutrient content between the two areas.

Probably the full inshore production potential is not realised because of the high turbidity, also because the annual production cycle in the turbid inshore waters is slower to develop in spring and stops earlier in the autumn than in offshore regions (Gieskes, 1974*a*). In shallow waters high production normally continues throughout the summer. This has been observed for Danish coastal waters (Steemann Nielsen, 1937, 1951, 1958), Long Island Sound (Riley, 1956), embayments in the Woods Hole area (Ryther, 1963) and now for the Dutch Wadden Sea (Postma & Rommerts, 1970; Cadée & Hegeman, 1974*a*).

In offshore waters the low production in summer is caused by the low nutrient content of the euphotic layer, shut off by a thermocline from the deeper waters, which are richer in nutrients. Sometimes the opposite is true in periods of very calm weather, which cause a haline stratification of the water column with surface salinities below 30‰. This has been observed in the Dutch offshore waters where freshwater from the Rhine remains at the surface. It is thus relatively less mixed with the saline water of the North Sea than under usual wind conditions. Sedimentation of diatoms may also take place in this stratified water (Gieskes, 1974*a*). The chlorophyll concentrations at the bottom of the euphotic zone may be up to twice as high as at the surface. In shallow water no thermocline is formed, resulting in a better supply of nutrients from the bottom. Moreover summer temperatures will increase the rate of mineralization of nutrients during the production season. Mineralization in the Wadden Sea takes place especially on the tidal flats. As these flats are most extensive in the watershed areas, mineralization is most important near the tidal watersheds (see part 1 of this chapter).

In the present studies estimations of primary production were made by two different approaches: (i) by the determination of functional chlorophyll *a*, as a measure for living phytoplankton (Cadée & Hegeman, 1974), and (ii) by the determination of adenosine triphosphate (ATP) as a measure of total living material (Manuels & Postma, 1974).

The estimation of primary production based on the functional chlorophyll a method

The seasonal cycle found for the western and eastern parts of the Dutch Wadden Sea is similar to the cycle found for Isefjord, the Sound off Helsingör and for the Great Belt (Steemann Nielsen, 1937, 1951, 1958), namely a relatively short spring peak and a broad summer peak. Previous studies, however, have shown differences in these cycles and great care must be taken not to assume an identical pattern. The geographical distribution of the estimated production *in situ* indicates that spring values are highest in the Marsdiep inlet area and decrease regularly towards the inner parts of the Wadden Sea. This suggests that the spring peak is a North Sea phenomenon, with a large effect in the outer, but a smaller effect in the inner parts of the Wadden Sea. The spring bloom development rate is most rapid from February to early March in offshore waters and slowest in turbid Dutch coastal waters, where it does not occur before the end of March. The broad summer peak may be typical of the whole of the Wadden Sea. The same seasonal phenomenon exists in the benthic microflora; the bimodal seasonal curve applies to both the planktonic and the benthic flora.

As stated earlier it may be concluded that nitrogen does not seem to limit primary production in the Wadden Sea; phosphate may be limiting during the spring bloom, whereas silica limits the diatom production during the peaks. When the spring bloom consists almost exclusively of the flagellate *Phaeocystis*, however, silica can be excluded as a limiting factor. Phosphate measurements indicate that phosphate may have been limiting in the inflow areas, where 0.2 μg at P/l was found (see part I of this chapter). In the inner part of the Wadden Sea, south of Terschelling, phosphate values were higher but production values were lower, and nitrogen and phosphate values were too high to limit the production (Jonge & Postma, 1974). An important factor limiting production *in situ* appears to be grazing. Seasonal summer production showed remarkably low values in an area of extensive mussel culture. As these mussels filter large quantities of water (Verwey, 1952), it is suggested that the low chlorophyll value is a result of this filtration by mussels. The effect of this grazing was largest in summer, during the growing season of the mussels.

In the western part of the Wadden Sea and at inlets from the North Sea, chlorophyll values are relatively low. At the sluices of Lake IJssel freshwater with a high chlorophyll content enters the Wadden Sea. Chlorophyll values in the inner part of the western Wadden Sea and in the eastern Wadden Sea are higher than in North Sea water in the inlets. The Ems estuary has chlorophyll values similar to those of the western Wadden Sea, whereas higher values are found in the Dollard. In Lake IJssel the chlorophyll values are the highest, about four times the values of the western Wadden Sea. Chlorophyll *a* has to be regarded as a crude index of the phytoplankton cell concentrations. The highly variable chlorophyll to cell volume ratio depends above all upon the size of the algae. Factors that influence the amount of pigment in the same species are light intensity and the physiological condition of the cells.

There is a much closer relationship, more independent of species composition, between chlorophyll *a* and measured primary production, although the variation of the production per unit chlorophyll *a* is a well known phenomenon, as appears from a review of the literature for marine phytoplankton (Curl & Small, 1965). Factors influencing this ratio include

Table 9.4. *Annual mean values for functional chlorophyll a, potential primary production, potential production per unit functional chlorophyll a, Secchi disc visibility and estimated* in situ *production for different areas studied (the data of the cruise of April 1973, when during the spring bloom only the western Wadden Sea was studied, are omitted to obtain comparable data)*

Area	Functional chl. *a* (mg/m³)	Potential production (mg/C/m³ hour)	Potential production/ chl. *a* unit	Secchi disc (m)	Production *in situ* (g/C/m² year)
Western Wadden Sea	6.3	22.7	3.6	1.25	100
Eastern Wadden Sea	10.5	35.9	3.4	0.75	120
Ems estuary	6.4	21.0	3.3	0.70	55
Dollard	9.1	11.9	1.3	0.17	13
Northern Lake IJssel	25.5	160.0	6.3	0.75	400

season, nutrients, physiological state and light adaptation of the phytoplankton. The estimated annual production for the western Wadden Sea, amounting to 100 g C/m², is very similar to the primary production measured *in situ* in earlier studies (Postma & Rommerts, 1970), which, after correction (Cadée & Hegeman, 1974), amounts to 92.5 g C/m². The estimated primary production, *in situ*, for the eastern part of the Wadden Sea gave a figure of 120 g C/m²/year. This value is of the same order of magnitude as the values found for the adjacent North Sea. The values for the Ems estuary and Dollard were lower, 55 and 13 g C/m²/year respectively. For the northern part of Lake IJssel a higher value of 400 g C/m²/year was found (see Table 9.4).

Light is probably the most important limiting factor for phytoplankton primary production in the Wadden Sea. Conversion of solar energy into organic matter is least efficient in turbid zones. Although functional chlorophyll *a* and potential primary production in the eastern Wadden Sea are 1.6 to 1.7 times as high as in the western part, production *in situ* is only 1.2 times as high, due to the higher turbidity. The very high turbidity in the Dollard, partly caused by pollution, but mostly due to its shallowness, results in a very low production *in situ*.

M. van der Eijk

The estimation of primary plankton production by measurements of ATP and organic carbon

Particulate organic matter in the Wadden Sea originates from two sources: primary production *in situ* and input from the adjacent North Sea (Postma, 1954). The latter amounts to about three times the former (Jonge & Postma, 1974) and is therefore the origin of the major part of organic particles in suspension. The accumulation of this material in the Wadden Sea depends on settling velocity (Steemann Nielsen *et al.*, 1957).

A comparatively large fraction of non-living organic matter can be expected to be present and it was the purpose of this investigation to measure this amount. The dead particulate organic matter was determined as the total organic carbon minus living carbon, as estimated by the ATP method. For the separation of phyto- and zooplankton a mesh size of 150 μm was assumed to be satisfactory. The ratio of ATP to organic carbon in living cells is fairly constant. Therefore it may be expected that the seasonal variation of ATP shows the same pattern as the seasonal variation of production *in situ* in the surface water. This seems to be true – ATP values were found to be 700 μg/l (0.18 mg/l of carbon) in summer and 5 μg/l (0.0013 mg/l of carbon) in winter. Primary production figures (expressed as carbon) were 600 mg/m^3/day in summer and 5 mg/m^3/day in winter (Postma & Rommerts, 1970).

The horizontal distribution of ATP shows large differences from one part of the Wadden Sea to the other. There is a tendency for the eastern part to have higher values in winter than the western part, and the reverse usually, but not always, occurs in summer. In mid-summer the values on the watersheds are distinctly lower than on the tidal inlets. The seasonal variation of organic matter in the Wadden Sea is much less than that of ATP, ranging from a maximum of roughly 15% (for 150 μm) and 7% (for total C) of suspended matter in summer to about 3% in winter, indicating that a large part consists of non-living material. The Ems estuary shows a small peak of organic carbon of about 7% in spring. Most of the year the values are about 4%; most of this particulate matter originates from the river Ems. In the inner part of the Dollard the average carbon content is 15%, as a result of the large amounts of organic sewage sluiced into this area. Lake IJssel has a very distinct seasonal distribution of organic carbon with maximum values of 30% and even of 70% in summer and minima of 10–15% in winter. Yet this variation is still much smaller than that of ATP. These percentages are relative amounts.

The absolute concentration of organic carbon in suspended material smaller than 150 μm varies between 0.3 and 5 mg/l for the Wadden Sea, 1.5 and 10 mg/l for the Ems estuary and 0.8 and 7 mg/l for the northern part of Lake IJssel. For the Wadden Sea, values of organic carbon in the

Table 9.5. *Average concentrations of plankton and percentages of living matter in relation to organic C in the Wadden Sea*

Group	Living matter		
	(mg C/l)	(% of C fraction)[a]	(% of total C)
Western Wadden Sea			
Phytoplankton	0.054	5.4	4.1
Zooplankton	0.041	13.7	3.2
Total	0.095		7.3
Eastern Wadden Sea			
Phytoplankton	0.051	2.7	1.7
Zooplankton	0.037	3.4	1.2
Total	0.088		2.9

[a] The percentage of C fraction refers to either the phytoplankton (< 150 μm) or the zooplankton (> 150 μm) fraction of the sample.

inner parts of the area and in the eastern part are higher than the values for the tidal inlets in the western parts (about twice as high in the eastern part). This is mainly caused by locally higher concentrations of total particulate matter due to the supply of organic matter; the values in the Dollard may even go up to 20 mg/l. Maximum carbon assimilation in summer is of the order of 1 g/m²/day, and production in winter is of the order of 0.001 g/m²/day. Taking the average depth of the area as 4 m, we find a standing crop of 0.8 g C/m² of phytoplankton in summer, and 0.003 g C/m² in winter. These values give a 120–30% daily growth of the population, with the highest increment in summer.

The amount of functional chlorophyll *a* in summer was estimated at 9.7 mg/m³ (Cadée & Hegeman, 1974*a*). Assuming that phytoplankton contains about 5% chlorophyll the average concentration is 0.2 mg C/l, which is about equal to an average summer value of 0.17 mg C/l, calculated from ATP values. In winter, an average of 3.6 mg chlorophyll was found, which corresponds to 0.07 mg C/l phytoplankton. Using the ATP method, only 0.0008 mg C/l was found. Another difference between the two methods is that the ATP measurements yield practically the same phytoplankton values for the eastern and western Wadden Sea, whereas chlorophyll values indicated amounts twice as high in the eastern part. The reason for these discrepancies is unclear. One possibility is that the ATP content of cells in winter is much lower than in summer. Another possibility is that the chlorophyll measurements include at least part of the chlorophyll in dead cells.

For the Wadden Sea as a whole, phytoplankton concentrations are

Table 9.6. *Data on the percentage of phytoplankton in organic matter in different areas*

Southern North Sea (Krey, 1956)	15–17%
Bay of Naples (Korringa & Postma, 1957)	27%
Long Island Sound (Riley, 1959)	12%

slightly higher than zooplankton values (see Table 9.5) and only about 5% of the total organic carbon is contained in living organisms, the remainder being organic detritus. This is in accordance with the assumption that large amounts of organic detritus are carried into the Wadden Sea from the North Sea by the tides. The percentage of living matter goes up to 20% in summer, of which roughly half is phytoplankton and half zooplankton. These values are of the same magnitude as those for more open ocean areas, as can be seen in Table 9.6.

The primary production in the brackish part of the estuary is very low; the plankton concentrations are five to ten times lower than those found elsewhere. The reason for this difference is that in an environment where freshwater and salt-water rapidly mix, such as the Ems estuary, most of the phytoplankton species from the freshwater and seawater perish, while the residence time of the plankton is too short for the establishment of an autochthonous plankton population. High ATP concentrations are recorded in the inner part of the Dollard, where water with a very high load of organic sewage is discharged; but since the river is permanently anaerobic, the plankton content is insignificant and the ATP values reflect mainly high bacterial concentrations. This also explains the high ATP values in winter. In the northern part of Lake IJssel ATP values and phyto-and zooplankton concentrations are lower than in the Wadden Sea. Earlier measurements (Postma & Rommerts, 1970; Kloet, 1971; Cadée & Hegeman, 1974*a*), however, indicate that in the northern part of Lake IJssel primary production is higher than in the Wadden Sea. The reason for this discrepancy might be that the ATP measurements are too low, because of the difficulties in extraction from the genus *Scenedesmus* which is dominant in the lake.

In summary: the seasonal variation of ATP lies between less than 5 μg/l in winter and more than 5000 μg/l in summer. The average amount of phytoplankton is about 0.05 mg C/l with a peak in summer of 0.2 mg C/l. For zooplankton these values are 0.04 mg C/l and 0.15 mg C/l respectively. Particulate organic carbon shows very little seasonal change. The percentage in suspended matter is slightly higher in summer than in winter (about 7% compared with 3% in winter) but total suspended matter concentrations are higher in winter. In summer sometimes 80% of carbon may be represented by living material locally, but the average in summer is only 20% and in winter less than 0.1%, with an overall average of 5%. On the watersheds in nearshore areas these values are lower.

Primary Production of the benthic microflora (Cadée & Hegeman, 1974a; Cadée, 1977)

Tidal-flats comprise approximately 50% of the Dutch Wadden Sea. They are inhabited by a large and varied flora of microscopic algae among which diatoms play an important role. In this study the importance of this microflora as primary producers was estimated by the ^{14}C method. Minimum production values are found in winter, with 50–100 mg C/m²/day. In April–May a sharp increase in production is found, leading to the highest production values in the period of June–August, after which a decrease to winter values follows. This seasonal variation was found for functional chlorophyll *a*, phaeopigment, organic carbon, ATP and primary production of the benthic microflora. The annual primary production of the benthic microflora showed a good correlation with the annual average functional chlorophyll *a* content of the sediment.

No explanation can be found for the large differences found between the data from different years. The summer values varied from 500 to 600 mg C/m²/day in 1968 and 1969, 100 to 500 mg C/m²/day in 1970 and 1971, and 500 to 1100 mg C/m²/day in 1972. The mean annual production of the microflora living on the tidal-flats in the western part of the Wadden Sea amounts to 100±40 g C/m²/year, based on 5 years of measurements. The average value for 14 Balgzand stations was 85 g C/m²/year. The annual primary production of benthic microflora for the eastern part was estimated at 80 g C/m², and in the Dollard values of 100 g C/m² were found. For primary production work it is clear that longer-range studies are essential. In general, year-to-year variation in phytoplankton production is less than in benthic microflora production. Different stations showed in the same year annual productions of 65.80 and 145 g C/m². The explanation for these differences can be found in the relation between primary production and tidal level. The high tidal-flats along the coast have a higher phaeopigment–chlorophyll ratio and a higher content of dead organic carbon than the lower flats in the northern part of the area. The increase in organic carbon content of the sediment on the high tidal-flats, from low values in winter to a summer peak, is much larger than can be accounted for by the production of benthic microflora and phytoplankton. This combination of facts indicates that the amount of organic carbon present for consumption by secondary producers and bacteria, is supplied not only by the primary production, but also by accumulation of allochthonous organic matter. High tidal-flats receive more allochthonous organic carbon than the other flats, and the main source of allochthonous organic matter in the Wadden Sea is the North Sea.

Annual production varied from 29 g C/m^2 on the lowest tidal-flat station to 188 g C/m^2 on the highest tidal-flat station near the coast. The primary production of the phytoplankton in the water above the tidal-flats was estimated at 20 g C/m^2/year. The production of the phytobenthos is thus more important than the phytoplankton production. The opposite was true for the shallow continuously-submerged part of the Wadden Sea, with a water depth of 2–4 m, where the estimated annual production is 70 g C/m^2 for the phytoplankton and 10 g C/m^2 for the phytobenthos.

Measurements of the loss of labelled organic matter showed that the excretion rate was low, varying from 0.1 to 9.5% of the primary production. When production was low, as in winter and spring, these percentages were higher. Excretion varied between 0.1 and to 4.5 mg C/m^2/day. The annual production of the excretion products amounted to less than 1 g C/m^2, or approximately 1% of the primary production of the benthic microalgae. These figures are low compared with the excretion found in marine phytoplankton, e.g. 7% in Georgia estuaries and 13–21% in coastal water (Thomas, 1971). Primary production measurements showed that the free-living microflora contained in the suspendible fraction was in general less important than the microflora attached to sand grains, contained in the sand fraction.

Functional chlorophyll *a* data were collected to provide an estimate of the standing stock of microalgae living on the tidal-flats, which showed a seasonal cycle similar to that found in the phytoplankton and similar to the cycle described for an estuarine mud-flat near Aberdeen, Scotland, where higher values of functional chlorophyll *a* were found in summer (35 μg/g sand) than in winter (25 μg/g sand) (Leach, 1970).

Measurements of phaeopigments showed that these values were on the average one-third of the amount of functional chlorophyll *a*, which illustrates the necessity to distinguish between functional chlorophyll *a* and phaeopigments in tidal-flat sediments. Data on chlorophyll depth distribution on the tidal-flats showed the highest values in the top centimeter of the sediment and a regular decrease with depth. Below 10 cm the quantities found were negligible. Fluctuations of the chlorophyll *a* content in the deeper layers were smaller than the fluctuations in the surface layer. The algae sampled in the sediment below the euphotic zone were found to photosynthesize when exposed to light; they therefore constitute an important stock of potential primary producers that take over photosynthesis when gales remove the upper few centimeters of the sediment. The processes which play a role in the depth distribution of algae were thought to be sediment transport during gales and an active migration of the algae (Fenchel & Staarup, 1971). The influence of the burrowing activity of *Arenicola* on the algae could not be demonstrated.

Table 9.7. *Comparison of available annual production values of microflora living on tidal-flats.*

Area	Annual production (g C/m²/year)	Source
Saltmarsh, Georgia	200	Pomeroy (1959)
Danish fjords	116	Grøntved (1960)
Danish Wadden Sea	115–178	Grøntved (1962)
False Bay	143–226	Pamatmat (1968)
Ythan estuary	31	Leach (1970)
Southern New England	81	Marshall *et al.* (1965)
Western Wadden Sea	100±40 this study	

Studies on the influence of herbivore grazing on the primary production of benthic microflora showed significant changes in production with only small changes in herbivore biomass (Hargrave, 1970; Cooper, 1973). Year-to-year variation in grazing may effect year-to-year variations in primary production (standing crop).

The annual production of microflora living on the tidal-flats amounts to 100±40 g C/m². Phytoplankton production in the water on the tidal-flat during submersion adds some 20 g C/m²/year. The total production thus amounts to approximately 120 g C/m²/year. The phytoplankton production in the inlets estimated from potential production measurements gave values of 100 g C/m²/year (Cadée & Hegeman, 1974). Phytoplankton production in the deeper waters of the western Wadden Sea in thus very similar to the microflora production on the tidal-flats. In Table 9.7 a comparison is made of available annual primary production values of microflora living on tidal-flats in different areas.

Tidal-flats comprise 40–50% of the Dutch Wadden Sea. As production of the benthic microflora on the flats is of the same magnitude as the phytoplankton production in deeper waters, the benthic microflora is as important as the phytoplankton for the primary production in the area. When the supply of organic matter from the North Sea is added to the Wadden Sea production, the total amount of organic matter available for secondary production will amount to approximately 360 g C/m²/year. Primary production in the western Wadden Sea is higher than in the Mediterranean, offshore in the Black Sea and the Pacific, at stations of comparable latitude. The value of the total amount of organic matter is less than the production found in the eutrophic waters of the Black Sea, Pacific Ocean and Long Island Sound. Thus the Wadden Sea does not stand out as an exceptionally fertile area. Although a relatively high amount of the primary production in a shallow area such as the Wadden Sea becomes available to the benthic fauna (the biomass of the bottom fauna in the Wadden Sea is three to nine times higher than in the North Sea) the zooplankton consumption in deeper waters must make up the difference.

References

Becker, H. B. & Postuma, K. H. (1974). Enige voorlopige resultaten van vijf jaar 'Waddenzeeprojekt'. *Visserij*, **27**(2), 69–79.

Bennekom, A. J. van, Krijgsman-van Hartingsveld, E., Veer, G. C. M. van der & Voorts, H. F. J. van (1974). The Seasonal cycles of reactive silicate and suspended diatoms in the Dutch Wadden Sea. *Netherlands Journal of Sea Research*, **8**(2–3), 174–209.

Beukema, J. J. (1970). De Waddenzee, een bijzonder milieu. *Chem. Weekbl.*, **66**, 62–6.

Beukema, J. J. (1974). Seasonal changes in the biomass of the macro-benthos of a tidal flat area in the Dutch Wadden Sea. *Netherlands Journal of Sea Research*, **8**(1), 94–107.

Beukema, J. J. (1976). Biomass and species richness of the macro-benthic animals living on the tidal of the Dutch Wadden Sea. *Netherlands Journal of Sea Research*, **10**(2), 236–61.

Beukema, J. J. & Bruin, W. de (1977). Seasonal changes in dry weight and chemical composition of the soft parts of the tellinid *Macoma balthica* in the Dutch Wadden Sea. *Netherlands Journal of Sea Research*, **11**(1), 42–55.

Boddeke, R. (1963). Donkere wolken boven de kustvisserij. *Visserij Nieuws*, **16**(8), 194–7.

Boddeke, R. (1967). Visserij-biologische veranderingen in de westelijke Waddenzee. *Visserij Nieuws*, **20**(9), 213–22.

Boddeke, R. (1972). De garnalenstand langs de Nederlandse kust van 1947 tot 1972. *Visserij*, **25**(3), 189–96.

Boddeke, R. & Daan, N. (1971). Waar zijn de garnalen gebleven? *Visserij*, **24**(6), 323–35.

Boer, P. (1971). Vogeltellingen op de Noordzee in augustus 1970. *Limosa*, **44**, 23–8.

Braber, L. & Groot, S. J. de (1973). The food of five flatfish species (*Pleuronectiformes*) in the southern North Sea. *Netherlands Journal of Sea Research*, **6**(12), 163–72.

Bregnballe, F. (1961). Plaice and flounders as consumers of microscopic bottom fauna. *Meddelelser fra Kommissionen for Danmarks Fiskeri – og Havund – ersøgelser*, **3**, 133–82.

Cadée, G. C. (1976). Sediment reworking by *Arenicola marina* on tidal flats the Dutch Wadden Sea. *Netherlands Journal of Sea Research*, **10**(4), 440–60.

Cadée, G. C. (1977). Distribution of primary production of the benthic microflora and accumulation of organic matter on a tidal flat area Balgzand, Dutch Wadden Sea. *Netherlands Journal of Sea Research*, **11**(1), 24–41.

Cadée, G. C. & Hegeman, J. (1974*a*). Primary production of phytoplankton in the Dutch Wadden Sea. *Netherlands Journal of Sea Research*, **8**(2), 240–59.

Cadée, G. C. & Hegeman, J. (1974*b*). Primary production of the benthic microflora living on tidal flats in the Dutch Wadden Sea. *Netherlands Journal of Sea Research*, **8**(2), 260–91.

Cooper, D. C. (1973). Enhancement of net primary productivity by herbivore grazing in aquatic laboratory microcosms. *Limnology and Oceanography* **18**(1), 31–7.

Curl, H. J. & Small, L. F. (1965). Variations in photosynthetic assimilation ratios in natural marine phytoplankton communities. *Limnology and Oceanography*, **10**(suppl.), 67–73.

Duinker, J. C., Eck, G. T. M. van & Nolting, R. F. (1974). On the behaviour of copper, zinc, iron and manganese, and evidence for mobilization processes in the Dutch Wadden Sea. *Netherlands Journal of Sea Research*, **8**(2), 214–39.

Eppley, R. W., Coatsworth, J. L. & Solorzano, L. (1969). Studies of nitrate reductase in marine phytoplankton. *Limnology and Oceanography*, **14**, 194–205.

Essink, K. (1972). Onderzoek naar de gevolgen van lozing van ongezuiverd industrieel afvalwater door middel van een persleiding van Hoogkerk naar de Waddenzee. *Interim-rapport over 1971, Zoölogisch Laboratorium der Rijksuniversiteit te Groningen*, Z 72/122, 1–24.

Fenchel, R. & Staarup, B. J. (1971). Vertical distribution of photosynthetic pigments and the penetration of light in marine sediments. *Oikos*, **22**, 172–82.

Gieskes, W. W. C. (1974*a*). Eutrophication and primary productivity studies in the Dutch coastal waters. In *Netherlands contribution to the International Biological Programme.* Final report 1966–1971, pp. 65–7. Amsterdam: North-Holland Publishing Company.

Gieskes, W. W. C. (1974*b*). Phytoplankton and primary productivity studies in the Southern Bight of the North Sea, eastern part, in 1972. *Annales Biologiques*, **29**, 54–61.

Gieskes, W. W. C. & Bennekom, A. J. van (1973). Unreliability of the ^{14}C method for estimating primary productivity in eutrophic Dutch coastal waters. *Limnology and Oceanography*, **18**, 494–5.

Gieskes, W. W. C. & Kraay, G. W. (1975). The phytoplankton spring bloom in Dutch coastal waters of the North Sea. *Netherlands Journal of Sea Research*, **9**(2), 166–96.

Grøntved, J. (1960). On the productivity of microbenthos and phytoplankton in some Danish fjords. *Meddelelser fra Kommissionen for Danmarks Fiskeri – og Havundersøgelser*, **3**(3), 55–92.

Grøntved, J. (1962). Preliminary report on the productivity of microbenthos and phytoplankton in the Danish Wadden Sea. *Meddelelser fra Kommissionen for Danmarks Fiskeri – og Havundersøgelser*, **3**(12), 347–78.

Groot, A. J. de (1963). Mangaantoestand van Nederlandse en Duitse holocene sedimenten. *Verslag van het Landbouwkundige Onderzoek in Nederland*, **69**, 1–164.

Groot, A. J. de (1966). Mobility of trace elements in deltas I. *Trans. Comm.* II *&* IV *int. Soc. Soil Sci. Aberdeen*, 267–79.

Groot, A. J. de (1973). Occurrence and behaviour of heavy metals in river deltas, with special reference to the Rhine and Ems rivers. In *North Sea science*, ed. E. D. Goldberg, pp. 308–25. Cambridge, Mass.: MIT Press.

Hancock, D. A. & Franklin, A. (1972). Seasonal changes in the condition of the edible cocke (*Cardium edule* L.). *Journal of Applied Ecology*, **9**, 567–79.

Hargrave, B. T. (1970). The effect of a deposit-feeding amphipod on the metabolism of benthic microflora. *Limnology and Oceanography*, **15**(1), 21–30.

Harvey, H. W., Cooper, L. H. N., Lebour, M. V. & Russell, F. S. (1935).

Plankton production and its control. *Journal of the Marine Biological Association, Plymouth*, **20**, 407–41.

Hauser, B. (1973). Bestandsänderungen der Makrofauna an einer Station im ostfriesischen Watt. *Jber. Forsch. St. Norderney*, **24**, 171–203.

Helder, W. (1974). The cycle of dissolved inorganic nitrogen compounds in the Dutch Wadden Sea. *Netherlands Journal of Sea Research*, **8**(2), 154–73.

Hempel, G. (1954). The abundance of young plaice in the Wadden Sea area between Weser and Lange-oog. *Annales Biologiques, Copenhagen*, **10**, 116–17.

Hempel, G. (1958). Zur Beziehung zwischen Bestandsdichte und Wachstum in der Schollenbevölkerung de Deutschen Bucht. *Berichte der Kommission zur Wissenschaftliche Meeresuntersuchungen*, **15**, 132–45.

Hughes, R. N. (1970). An energy budget for a tidal-flat population of the bivalve *Scrobicularia plana* (Da Costa). *Journal of Animal Ecology*, **39**, 357–81.

Jespersen, P. (1929). On the frequency of birds over the high Atlantic Ocean. *Verhandlungen 6 Internationalen Ornithologischen Kongress Kopenhagen*, 163–72.

Jonge, V. N. de & Postma, H. (1974). Phosphorus compounds in the Dutch Wadden Sea. *Netherlands Journal of Sea Research*, **8**(2), 139–53.

Kay, D. G. & Knights, R. D. (1975). The macro-invertebrate fauna of the soft sediments of south-east England. *Journal of the Marine Biological Association, Plymouth*, **55**, 811-32.

Kloet, W. A. de (1971). Het eutrofieringsproces van het IJsselmeergebied. *Meded. Hydrobiol. Ver.* **5**(1), 23–38.

Korringa, P. & Postma, H. (1957). Investigations in the fertility of the Gulf of Naples and adjacent salt water lakes, with special reference to shellfish cultivation. *Pubbl. Staz. zool. Napoli*, **29**, 229–84.

Krauskopf, K. B. (1956). Factors controlling the concentrations of thirteen rare metals in sea water. *Geochimica et Cosmochimica Acta*, **9**, 1-32B.

Krey, J. (1956). Die Trophie küstennaher Meeresgebiete. *Kieler Meeresforchungen*, **12**, 46–71.

Kuipers, B. (1973). On the tidal migration of young plaice (*Pleuronectes platessa*) in the Wadden Sea. *Netherlands Journal of Sea Research*, **6**(3), 376–88.

Leach, J. H. (1970). Epibenthic algal production in an intertidal mudflat. *Limnology and Oceanography*, **15**(4), 514–21.

Longbottom, M. R. (1970). The distribution of *Arenicola marina* (L.) with particular reference to the effects of particle size and organic matter of the sediments. *Journal of Experimental Marine Biology and Ecology*, **5**, 138–57.

McIntyre, A. D. (1970). The range of biomass in intertidal sand, with special reference to the bivalve *Tellina tenuis*. *Journal of the Marine Biological Association, Plymouth*, **50**, 561–75.

Manuels, M. W. & Postma, H. (1974). Measurements of ATP and organic carbon in suspended matter of the Dutch Wadden Sea. *Netherlands Journal of Sea Research*, **8**(2–3), 292–311.

Marshall, N., Oviatt, C. A. & Skauen, D. M. (1971). Productivity of the benthic microflora of coastal estuarine environments in southern New England. *Internationale Revue der Gesamten Hydriobiologie*, **56**(6), 947–56.

Meyer Waarden, P. F. & Tiews, K. (1965). Der Beifang in den Fängen der deutschen Garnelenfischerei in den Jahren 1954–1960. *Berichte der Kommission zur Wissenschaftliche Meeresuntersuchungen*, **18**(1), 13–78.

Moore, H. B., Davies, L. T., Fraser, T. H., Gore, R. H. & Lopez, N. R. (1968). Some biomass figures from a tidal flat in Biscayne Bay, Florida. *Bulletin of Marine Science Gulf Caribb*, **18**, 261–79.

Mörzer Bruijns, M. F. & Braaksma, S. (1954). Vogeltellingen in het staatsnatuurreservaat Boschplaat van 1951 t/m 1953. *Ardea*, **42**, 175–210.

Pamatmat, M. M. (1968). Ecology and metabolism of a benthic community on an intertidal sandflat. *Internationale Revue der Gesamten Hydriobiologie*, **53**(2), 211–98.

Pomeroy, L. R. (1959). Algal productivity in salt marshes of Georgia. *Limnology and Oceanography*, **4**, 386–97.

Postma, H. (1954). Hydrography of the Dutch Wadden Sea. *Archives Néerlandaises de Zoologie*, **12**, 319–49.

Postma, H. (1954). Hydrography of the Dutch Wadden Sea. *Archives Néerlandaises de Zoologie*, **10**(4), 405–511.

Postma, H. (1961). Transport and accumulation of suspended matter in the Dutch Wadden Sea. *Netherlands Journal of Sea Research*, **1**, 148–90.

Postma, H. (1966). The cycle of nitrogen in the Wadden Sea and adjacent areas. *Netherlands Journal of Sea Research*, **3**, 186–221.

Postma, H. (1973). Transport and budget of organic matter in the North Sea. In *North Sea science*, ed. E. D Goldberg, pp. 326–34. Cambridge Mass.: MIT Press.

Postma, H. & Rommerts, J. W. (1970). Primary production in the Wadden Sea. *Netherlands Journal of Sea Research*, **4**(4), 470–93.

Reynders, P. J. H. (1976). The harbour Seal (*Phoca vitulina*) population in the Dutch Wadden Sea: size and composition. *Netherlands Journal of Sea Research*, **10**(2), 223–35.

Rommets, J. W., Postma, H. & Bennekom, A. J. van (1974). Budget aspects of biologically important chemical compounds in the Dutch Wadden Sea. *Netherlands Journal of Sea Research*, **8**(2–3), 318–21.

Riley, G. A. (1956). Oceanography of Long Island Sound 1952–54: IX. Production and utilization of organic matter. *Bulletin of the Bingham Oceanographic Collection*, **15**, 324–44.

Riley, G. A. (1959). Note on particulate matter in Long Island Sound. *Bulletin of the Bingham Oceanographic Collection*, **17**, 83–6.

Rooth, J. (1960). Vogeltellingen op Vlieland 1953–1956. *Limosa*, 134–59.

Rooth, J. (1966). Vogeltellingen in het gehele Nederlandse waddengebied, augustus 1963.

Rutgers van der Loeff, M. M. (1974). Transport van reactief silikaat uit Waddenzee sediment naar het bovenstaande water. Intern Verslag Nederlands Instituut voor Onderzoek der Zee, Texel 1974–11 (mimeo).

Ryther, J. H. (1963). Geographic variations in productivity. In *The sea*, ed. R. Hill, vol. 2, pp. 347–80. New York: Wiley-Interscience.

Ryther, J. H. (1969). Photosynthesis and fish production in the sea. *Science*, **166**, 72–6.

Sanders, H. L., Goudsmit, E. M., Mills, E. L. & Hampson, G. E. (1962). A study of the intertidal fauna of Barnstable Harbor, Massachusetts. *Limnology and Oceanography*, **7**, 63–79.

Sandkvist, A. (1968). Microbiological investigation of modern Dutch tidal sediments. *Stockholm Contributions in Geology*, **15**, 67–113.

Spaans, A. L. (1967). Wadvogeltellingen in het gehele Nederlandse waddengebied in December 1966. *Limosa*, **40**, 206–15.

Steele, J. H. (1956). Plant production on the Fladen Ground. *Journal of the Marine Biological Association, Plymouth, Cambridge*, **35**, 1–33.

Steele, J. H. & Baird, I. E. (1968). Production ecology of a sandy beach. *Limnology and Oceanography*, **13**(1), 14–25.

Steemann Nielsen, E. (1937). The annual amount of organic matter produced by the phytoplankton in the Sound off Helsingor. *Meddelelser fra Kommissionen for Danmarks Fiskeri – og Havundersøgelser (Plankton)*, **3**(3), 1–37.

Steemann Nielsen, E. (1951). The marine vegetation of the Isefjord. A study of ecology and production. *Meddelelser fra Kommissionen for Danmarks Fiskeri – og Havundersøgelser (Plankton)*, **5**(4), 1–114.

Steemann Nielsen, E. (1958). A survey of recent Danish measurements of the organic productivity in the sea. *Rapport et Procès-Verbaux des Réunions (Conseil Permanent International pour l'Exploration de la Mer)*, **144**, 92–5.

Steemann Nielsen, E., Straaten, L. M. J. U. van & Kuenen, P. H. (1957). Accumulation of fine grained sediments in the Dutch Wadden Sea. *Geologie en Mijnbouw*, **19**(8), 329–54.

Straaten, L. M. J. U. van (1954). Composition and structure of recent marine sediments in the Netherlands. *Leidsche Geologische Mededeelingen*, **19**, 1–108.

Straaten, L. M. J. U. van (1958). Tidal action as a cause of clay accumulation. *Journal of Sedimentary Petrology*, **28**, 406–13.

Straaten, L. M. J. U. van & Kuenen, P. H. (1957). Accumulation of fine grained sediment in the Dutch Wadden Sea. *Geologie en Mijnbouw*, **19**, 329–54.

Swennen, C. (1971). Het voedsel van de Groenpootruiter (*Tringa nebularia*) tijdens het verblijf in het Nederlandse waddengebied. *Limosa*, **44**, 71–83.

Swennen, C. (1975). Aspecten van voedselproductie in Waddenzee en aangrenzende zeegebieden in relatie met de vogelrijkdom. *Het Vogeljaar*, **23**(4), 141–56.

Thomas, J. P. (1971). Release of dissolved organic matter from natural populations of marine phytoplankton. *Marine Biology*, **11**, 311–23.

Thijssen, R., Lever, A. J. & Lever, G. (1974). Food composition and feeding periodicity of o-group plaice (*Pleuronectes platessa*) in the tidal area of a sand beach. *Netherlands Journal of Sea Research*, **8**(4), 369–77.

Vaupel-Klein, J. C. van & Weber, R. E. (1975). Distribution of *Eurytemora affinis* (Copepoda; Calanoida) in relation to salinity: field and laboratory observations. *Netherlands Journal of Sea Research*, **9**(3–4), 297–310.

Verwey, J. (1952). On the ecology of distribution of cockles and mussels in the Dutch Waddensea, their role in sedimentation and the source of their food supply. *Archives Néelandaises de Zoologie*, **10**, 171–239.

Verwey, J. (1966). De rijke Waddenzee. *Natuur Landsch.* **19**, 129–52.

Vosjan, J. H. (1974). Sulphate in water and sediment of the Dutch Wadden Sea. *Netherlands Journal of Sea Research*, **8**(2–3), 208–13.

Warwick, R. M. & Price, R. (1975). Macrofauna production in an estuarine mudflat. *Journal of the Marine Biological Association, Plymouth*, **55**, 1–18.

Zijlstra, J. J. (1972). On the importance of the Waddensea as a nursery area in relation to conservation of the southern North Sea fishery resources. *Symposia of the Zoological Society of London*, **29**, 233–58.

Zimmerman, J. T. F. (1974). Circulation and water exchange near a tidal watershed in the Dutch Wadden Sea. *Netherlands Journal of Sea Research*, **8**(2), 126–38.

10. Seaweed utilization in the Philippines

G. T. VELASQUEZ

The Republic of the Philippines is composed of 7100 islands, only 400 of which are populated. The rest are for the most part rocky or mountainous with hardly any fertile soil suitable for agriculture where people might settle indefinitely. Some of the islands may serve only as temporary resting places where fishermen may spend a day to a week to catch fish. Fresh water is also not readily available and often is not potable.

The climate of the country is characteristically tropical, located within latitude 4°–20° N and longitude 116°–120° E. The Pacific Ocean limits the country in the east while the great expanse of the South China Sea borders the coastline to the west. The China Sea converges with the Pacific Ocean in the north towards the mainland of Taiwan. In the south are the Sulu Sea towards Borneo and the Celebes Sea towards Indonesia. Such a geographical arrangement has exposed the country to the typhoons which come during the rainy season. Between 15 and 18 typhoons annually, originating over the Pacific Ocean, pass through the Philippines area, generally with destructive effects upon the country.

The unusual number of islands separated by channels of varying sizes produces a luxuriant growth of marine algae (seaweeds). Many species of seaweeds have formed part of the diet of the natives who have been living along the sea coasts since primitive times. These people, during the Spanish regime, were never aided by any of the economic and social improvement schemes found in the society which was developing in many inland municipalities. Accordingly, the primitive way of utilizing the seaweeds has persisted among them up to the present.

Much later, the growing volume of trade in the provinces caused by the development of easily passable roads and other communications leading to the cities and large towns started to increase the role of the marine algae as part of the people's economy. For example, the Chinese established small factories for crude agar in many densely populated cities and municipalities. In Manila alone there were three small factories for crude agar. This agar, better known as 'gulaman', became very important as part of the diet of many people. The collection of the edible seaweeds received much encouragement throughout the country, and the Chinese used to send collectors of edible algae to distant rural areas to supply the raw materials for the agar factories. Among the important marine algae which are especially utilized in the manufacture of agar are *Gracillaria*

Table 10.1. *Food preparation of edible marine seaweeds in Ilocos Norte*

Scientific name	Common name	Method of preparation
Enteromorpha plumosa Kuetz. *Enteromorpha compressa*. L.	Lumot	Always eaten raw in the form of salad mixed with sliced tomatoes, onions and a dash of salt, after thorough washing in fresh water
Ulva lactuca L.	Gamgamet	Eaten in the form of salad after thorough washing in fresh water, prepared as above; can be cooked with vegetables or meat
Caulerpa racemosa (Forskaal)	Ar-arucip	Always eaten raw and fresh in the form of salad, garnished with sliced tomatoes and onions and a dash of salt
Codium papillatum Tseng & Gilbert *Codium intricatum* Okamurai *Codium* sp.	Pocpoclo	Always eaten raw and fresh in the form of salad, garnished with sliced tomatoes and onions, with a dash of salt
Hydroclathrus clathratus (Bory) Howe	Balbalulang	Always eaten raw in the form of salad prepared as above, after blanching in boiling water for 2 minutes and rendered free from small crabs that breed in the thallus
Porphyra crispata Kjellman	Gamet	Eaten raw and dried in the form of salad prepared as above; dried sheets of *Porphyra* are similarly made into soup or omelet, or broiled over live charcoal and mixed with sliced tomatoes and onions; sometimes with sliced cucumber and radish and a little vinegar to taste
Gelidiella acerosa (Forskaal) Feldmann et Hamel *Laurencia papillosa* (Forskaal) Greville *Laurencia okamurai* Yamada *Acanthophora spicifera* (Vahl.) Boergesen	Culot	Eaten fresh or in dried form; fresh – used in salads prepared as above; dried – used for future consumption and reconstituted in boiling water. Blanched for two minutes and mixed with vegetables or meat

confervoides, *Gelidiella acerosa*, *Acanthophora spicifera*, and two or three species of the genera *Hypnea*, *Eucheuma*, and *Gelidium*.

At the same time, the household utilization of the seaweeds collected for general use by the people received similar encouragement. The building of more highways made the freshly-collected seaweeds accessible to the inland provinces where many cities and large municipalities are located. Families, especially of the poor and middle classes, go to open markets to buy the seaweeds.

How the raw seaweeds are used as food by the people is shown in Table 10.1, quoted from the publication 'Taxonomy, distribution and seasonal

Halymenia durvillei Bory	Gayong-gayong	Always eaten raw and fresh; prepared as salad, with meat
Eucheuma muricatum (Gmelin) Weber van Bosse	Canot-canot	Always eaten fresh in the form of salad; can be added to vegetables or meat
Hypnea charoides Lamouroux	Culot ti pusa	Eaten raw or dried as salad with garnishings as above; if dried, it is similarly reconstituted to nearly fresh condition and prepared as salad
Gracilaria calicornia (C. Ag.) Dawson *Gracilaria verrucosa* (Huds.) Papenfuss. *Gracilaria coronopifolia* J. Ag.	Caocaoayan	Always eaten raw in the form of salad, prepared as above, or cooked with vegetables or meat. Crude agar is extracted from these three species and *Eucheuma muricatum* by boiling the blanched thalli for a few hours until the gelose is extracted. The digested portion is separated from the fibrous residue by straining while hot through sinaway cloth stretched over a large vat where the solution filters. Sugar and food colouring are added; when cooled, it is cut into small cubes and served with ice cream

occurrence of edible marine algae in Ilocos Norte, Philippines' by Galutira & Velasquez.*

Although the common names of the edible seaweeds may differ in many localities, their methods of preparation are similar. Other species also grow in these localities, easily accessible to the natives. The people are beginning to realize the role of marine algae in their diet, especially those who live in rural areas far from public markets but provided with transportation. In addition to the seaweeds which the Chinese use in their agar manufacturing, the many species are generally used as sources of salad and vegetable preparations, such as species of *Caulerpa*, *Codium*, *Ulva*, *Enteromorpha*, *Porphyra*, *Halymenia*, *Hydroclathrus*, and *Sargassum* (as shown in Table 10.1). Many people find the collection of seaweeds to be a good source of business during the months of December to May when the algae are in season.

In the mid-1960s, the culture of *Eucheuma cottonii* and *E. striatum* was introduced by Maxwell S. Doty, and towards the end of the decade the success of the culture was beginning to be realized. Harvested *Eucheuma* are cured and exported, mostly to Marine Colloids Inc., in the

* Galurita, E. & Velasquez, G. T. (1963). *Philippine Journal of Science*, **92**(4).

United States. By April of 1973, Doty reported the success of *Eucheuma* culture in the Philippines in 'Eucheuma farming for carrageenan' (Doty, 1973).* The culture, first in the southern part of Mindoro and later in Sulu province and Zamboanga in Mindanao, started to attract the attention of dwellers along the coasts.

During this time, G. C. Trono, Jr also published an interesting article on the culture of *Eucheuma*.† The Marine Fisheries Research Division of the Bureau of Fisheries and Aquatic Resources became interested and issued mimeographed brochures describing in detail the method of culturing *Eucheuma*. This same method was later found to be applicable to other red algae such as *Gracillaria confervoides* and *Hypnea* species. The culture of *Eucheuma* was done successfully in southern Mindoro, Sulu province, and Zamboanga. *Eucheuma* farming in these places enjoyed a boom of production during the late 1960s to the early 1970s when the price of carrageenan was very high. After the boom export of the dried algae, the cost of carrageenan became unusually low and would not cover the expense of farming.

Recently the price of carrageenan, as reported by Doty (1973), caused loss of enthusiasm among the residents in continuing the culture. After some time several residents along the shores where the culture had become a part of the principal source of income stopped the seaweed farming and returned to their original work. But some people are still operating their seaweed farms. These daring workers hope to recover the losses due to the low price of carrageenan caused by world recession. Perhaps the export of *Eucheuma* may become a boom once again. Farming of at least *E. cottonii* and *E. striatum* may be reactivated as a lucrative source of income in the people's economy.

The overall utilization of the seaweeds in the country has always been highly commended. It has been a part of the diet of the great majority of Filipino families, especially those who have made their living along the seashore. Many of them learned to adapt their family life to a dependence on edible seaweeds during the season when they are available. One of the sources of income for the natives during December to May every year is collecting seaweed, which sells well in the markets. The people of many inland towns also rely upon seaweeds for much of their vegetables and salads. There is need for a standard recipe book for seaweed utilization, available to the people. The use of seaweeds is a sure way to improve the quality of living and the economy of the Filipino nation.

* This research is sponsored by NOAA Office of Sea Grant, Department of Commerce and U.S. Atomic Energy Commission.

† In the 'Challenge to biologists in the 70's: The escalation of food production', pp. 28–41. Alpha Chi Chapter, Phi Sigma Society, sponsored a series of lectures to honour Gregorio T. Velasquez on his retirement from active government service.

11. Trophic relationships in communities and the functioning of marine ecosystems: I

Studies in trophic relationships in pelagic communities of the southern seas of the USSR and in the tropical Pacific

T. S. PETIPA

Studies of the relationship between structure and function of marine ecosystems have become one of the principal tasks of marine biology. Trophic relationships between species basically determine the functioning of pelagic ecosystems and the metabolism of organic matter in the sea. Organic matter produced by one organism is transferred and utilised by the others through the food webs of the communities. A discrete water mass, with all its inhabitants and trophic processes, may be regarded as a monolithic functioning system producing organic matter in one form or another. Thus the analysis of the trophic relationships between organisms and populations is considered to be the most important element in studies of the functioning of pelagic ecosystems. Creation of a scheme of food relationships within communities and of the transfer of matter and energy through them, makes possible the mathematical simulation of the whole complex of trophodynamic processes and the construction of models of pelagic ecosystems. Mathematical models of communities and of the functioning of ecosystems may be further applied as a step towards the understanding of the maintenance of communities and of the production of organic matter by their separate components.

Two types of trophic interrelationship may be distinguished in the pelagic ecosystem: (1) predator–prey relationships, (2) non-predatory relationships based on the external metabolites excreted into the environment by one organism and consumed by the others. The first type, comprising direct relationships between organisms, prevails at higher trophic levels; the second dominates at lower levels (bacteria, algae, some lower invertebrates) (Khailov, 1971). Predator–prey relationships have recently been given much attention, while studies on the non-predatory relationships are just beginning.

To understand the effects of trophic interrelationships on communities and on ecosystem structure, we must examine the two types of trophic relationship within them.

In the first part of this work we shall discuss the results of studies on trophic-web schemes and on rates and efficiencies of matter and energy flow through them in neritic and oceanic plankton communities of the south seas, using examples from the Black Sea and the tropical Pacific. The main emphasis here is given to predator–prey relationships. The following chapter (chapter 12) contains some original data on the functioning of the pelagic tropical Pacific ecosystem.

Most of the material for the studies on trophic webs in the neritic Black Sea communities has been obtained by intensive research lasting many days at anchored stations in the Black Sea (1959–63) on board the RV *Academ. A. Kovalewsky* of the Institute of Biology of South Seas, Academy of Sciences, UkSSR. Analogous data have been obtained in the equatorial Pacific on the cruises of the RV *Vityaz* aimed at studying the functioning of pelagic tropical ecosystems. The cruises were carried out by the Institute of Oceanology of the USSR Academy of Sciences.

The description of the actual trophic chains or webs of marine communities with calculated data on energy transmission requires complex and extensive research on nutrition and energy balance. In particular, to create a scheme of trophic relations between separate organisms, or between their groupings in the model community, one must know not only the contents of the animals' intestines, one must also evaluate the animals' potential abilities to consume food of different shape, size and value, etc. The latter is especially important because of seasonal and daily fluctuations in the number and composition of marine organisms. The potential abilities of heterotrophic animals and some plants to consume different food depends upon the characteristics of oral apparatus, methods of food capture, composition, distribution and abundance of food objects in the sea, and specificity of an animal to a certain type of food.

To obtain all this information, intensive research has been carried out in the Black Sea (beginning in 1959) and in the tropical Pacific (1968–71). The principle nutritional characteristics of the common planktonic species, representing different ecological groups, have been studied. The description of the whole scheme of trophic relationships in the seas is simplified by the differentiation of plankton communities into ecological groups or life-forms (Petipa, 1967*a*). The patterns of quantitative nutrition processes of large ecological groups of organisms may be evaluated by those of their typical representatives.

Quantitative values for the trophic relationships and for the matter and energy transfer through food chains have been determined from the magnitude of the daily food intake of planktonic organisms and by the amount of food utilised.

All the elements of the energy balance of the main representatives of the groups have been calculated by the radiocarbon method (basically developed by Yu. Sorokin, 1966; Petipa, Pavlova & Sorokin, 1971), or by direct measurement of the amount of food eaten, of body weight increment, and of respiration. The ration composition of the main representatives of 10–15 groups has been evaluated by the amount of food actually eaten. For the other organisms, not included in these groups, the ratio of food objects in the sea and in the rations has tentatively been considered as unity.

The methods and calculations used in the study of nutrition process characteristics and of the matter and energy balance of separate planktonic species and life forms are found in the following articles: Mironov, 1954, 1960, 1967; Lebedeva, 1959; Pavlova, 1959*a*, *b*, 1964, 1967; Petipa, 1959*a*, *b*, 1965, 1966*a*, *b*, 1967*a*, *b*; Delalo, 1961, 1964; Arashkevich, 1969; Pavlova & Sorokin, 1970; Pavlowskaya, 1970; Sorokin, Petipa & Pavlova, 1970; Zaika & Pavlowskaya, 1970.

To calculate average quantitative indices of biomass of ecological groups within the communities and systems studied, the results of observations and catches taken in the water layers of maximum and minimum concentration of the different ecosystem elements have been used. Migrations of organisms are not intensive in the tropical oceanic zones (very few large species undertake regular daily vertical migration; Vinogradov, 1968). Nevertheless diurnal and nocturnal plankton catches were made, so as to take into account the role of migrating organisms of the tropical oceans and in the Black Sea, in the trophodynamics of the communities of the euphotic zones. The average daily standing stock of organisms within ecosystems was then determined.

Special attention has been paid to the role of smaller zooplankton forms (up to 0.5 mm) which are impossible to sample adequately by the use of nets alone. Organisms have been counted directly in non-fixed measured samples (data of V. E. Zaika & T. V. Pavlowskaya, personal communication).

The nutrition of the major plankton forms in the Black Sea and in the tropical Pacific has been determined by summarising and analysing all the known literature in addition to new original material obtained. Ecological groups of planktonic organisms, regarded as the initial points for construction of trophic web schemes, have been distinguished in the ocean areas investigated. Not only nutrition types, but also data on behaviour and distribution of the marine plankton were taken into consideration.

Trophic webs in neritic and oceanic ecosystems

Pycnoclines, and other spatial changes in physico-chemical characteristics caused by various factors, divide water masses vertically into natural, more or less independent layers which are inhabited by specific communities of organisms. Two communities are marked out in the euphotic zone of the Black Sea: the epiplankton, occupying the surface strata above the thermocline (12–25 m layer) and the bathyplankton, inhabiting the water below the thermocline in a layer from 12–25 m to 100 m and deeper (Petipa, Pavlova & Mironov, 1970). One community may be distinguished in the euphotic zone of the equatorial Pacific approximately down to 70–100 m. It is separated from deeper-living communities by a small thermo- or pycnocline or by the layer of dynamic equilibrium of two life-limiting factors, namely the amount of light penetrating from above and the supply of nutrients from deeper layers (Sorokin, 1959; Vinogradov, Gitelzon & Sorokin, 1971; Vinogradov, Menshutkin & Shushkina, 1972), or by a combination of these two factors. With the stable stratification of water masses, aggregations of living and dead organisms occur at the boundary zone. This boundary zone plays a most important role in the metabolism of organic matter within the ecosystem.

Copepoda constitute the main mass of zooplankton observed in marine communities (40–80% of the total zooplankton biomass). Most investigators regard the Calanoidea (*Calanus*, *Eucalanus*, *Rhincalanus*, *Pseudocalanus*, *Acartia*, etc.) as filter-feeders, the others (*Candacia*, *Euchaeta*, Pontellidae, *Haloptilus*, *Oncaea*, etc.) as grasping predators, but studies on the functional morphology of oral appendages and methods of food capture indicate that copepods, especially Calanoidea, show great versatility in the movements of their oral appendages, so that their methods of catching food and the type of food taken are highly diverse. Thus typical filter-feeders at all stages, combining different movements of appendages and abdomen, are capable of short, frequent leaps, as well as rapid and longer leaps. Crustaceans catch their food either by filtering out the smaller food objects, or by scooping up medium-sized material, or by actively grasping the largest food.

Combinations of behaviour patterns in food capture are characteristic of copepod predators. For example, Calanoidea, *Candacia* and *Euchaeta* leap and grasp such animals as large as *Sagitta*, but also glide slowly in the water to filter algae and even bacteria. Experimental studies on nutrition and food selectivity of copepods in the Black Sea, Indian and Pacific Oceans confirm their wide nutrition spectrum (Mullin, 1966; Petipa, 1959*a*, *b*; Petipa *et al.*, 1974).

On the other hand the movements and methods of food capture of certain other copepods are determined by their structural characteristics and the main functions of their oral appendages, and indicate a certain specialisation. By directly observing crustacean behaviour when different kinds of food are abundant, we have estabished that some genera (*Acartia*, *Euchaeta*, *Oithona*) prefer one method of movement and food capture (leaping and grasping) while the others (*Pseudocalanus*, *Paracalanus*) use

Table 11.1. *Biomass, daily ration and composition of food in main ecological groupings from the epiplankton community in the Black Sea*

Groups	Biomass (cal/m³)	Daily ration (cal/m³)	Daily ration (% of body weight)	Composition of food: Animals (% of the ration)	Composition of food: Algae (% of the ration)	Composition of food: Bact. Detr. (% of the ration)	Assimilation (%)
1. Detritus	1903	—	—	—	—	—	—
2. Phytoplankton							
small-sized forms (17–30 μm)	6.97	—	—	—	—	—	—
medium-sized forms (30–50 μm)	10.11	—	—	—	—	—	—
large-sized forms (50 μm)	5.98	—	—	—	—	—	—
3. *Noctiluca* (0.5–0.7 mm)	210.24	186	89	—	10	90	87.0
Producers and saprophages	233.30	186	89	—	10	90	87.0
4. Nauplii, I–VI (0.1–0.32 mm)	30.8	11.07[a]	45	—	100	—	81.2
5. Copepodites, I–III (0.23–1.6 mm)	25.0	26.2	105	—	100	—	86.8
6. *Paracalanus*, IV–VI (0.5–0.7 mm)	44.1	20.6	48	—	8	92	84.4
7. Appendicularia (0.15–0.58 mm)	7.19	4.3	60	—	6	94	76.7
8. Larvae of Mollusca and polychaetes (0.1–1.4 mm)	0.22	0.024	11	—	14	86	80.0
9. IV, VI *Pseudocalanus* and *Calanus* (0.7–3.3 mm)	3.26	0.16[b]	5	—	10	—	91.7
Herbivores	110.57	62.354	45.7	—	54.7	45.3	83.5
10. *Acartia*, IV–V,	—	—	—	—	—	—	—
Oithona, IV–V and	—	—	—	—	—	—	—
Acartia, ♀, ♂, (0.38–1.2 mm)	22.2	18.4	83	34	29	37	91.1
Mixed food consumers	22.2	18.4	83	34	29	37	91.1
11. *Oithona minuta*, *O. similis*, ♀, ♂ (0.5–0.67 mm)	24.0	39.6	165	87	—	13	93.9
Primary carnivores	24.0	39.6	165	87	—	13	93.9
12. *Sagitta* (1–10 mm, av. 3.96 mm)	8.33	7.0	85	100	—	—	77.7
13. Pleurobrachia, Medusae (0.3–3 mm, av. 1 mm)	1.05	3.5	323	100	—	—	43.3
Secondary and tertiary carnivores	9.38	10.5	204	100	—	—	60.5

[a] For 80% feeding nauplii. Stages I and II do not feed.
[b] Data are given on half an hour feeding while present in the epiplankton system.

quite different methods (gliding and filtration). But it has also been found that environmental conditions such as the composition, size and quantity of food, as well as light and temperature, have an important effect upon the animals' preference of movement and nutrition. Copepods usually select food objects, and adopt movements and manners of food capture which quickly meet their nutrition requirements with the least expenditure

Table 11.2. *Biomass, daily ration and composition of food in main ecological groupings from the bathyplankton community of the Black Sea*

Groups	Biomass (cal/m³)	Daily ration (cal/m³)	Daily ration (% of body weight)	Composition of food: Animals (% of the ration)	Algae (% of the ration)	Bact. Detr. (% of the ration)	Assimilation (%)
1. Detritus	288.9	—	—	—	—	—	—
2. Phytoplankton,							
small-sized forms (17–30 μm)	1.04	—	—	—	—	—	—
medium-sized forms (30–50 μm)	1.18	—	—	—	—	—	—
large-sized forms (50 μm)	6.77	—	—	—	—	—	—
3. *Noctiluca* (0.5–0.7 mm)	81.3	70.14	86	0.5	0.6	98.9	79.9
Producers and saprophages	90.29	70.14	86	0.5	0.6	98.9	79.9
4. Nauplii, I–VI (0.1–0.32 mm)	21.0	3.78[a]	21	—	100	—	88.3
5. Copepodites, I–III (0.23–1.6 mm)	6.06	4.06	67	—	47	53	78.8
6. *Paracalanus*, IV–VI (0.5–0.7 mm)	2.56	0.56	22	—	3	97	86.4
7. Appendicularia (0.15–0.58 mm)	1.42	0.43	31	—	5	95	80.0
8. Larvae of Mollusca and polychaetes (0.1–1.4 mm)	0.43	0.03	8	—	38	62	78.6
9. IV, VI *Pseudocalanus* and *Calanus* (0.7–3.3 mm)	34.54	42.14	122	—	88	12	92.5
Herbivores	66.01	51.00	45.2	—	46.8	53.2	84.1
10. *Acartia*, IV–V, *Oithona*, IV–V and *Acartia*, ♀, ♂, (0.38–1.2 mm)	1.94	0.97	50	59	25	16	78.4
Mixed food consumers	1.94	0.97	50	59	25	16	78.4
11. *Oithona minuta*, *O. similis* (0.5–0.67 mm)	3.51	3.05	89	80	—	20	90.2
Primary carnivores	3.51	3.05	89	80	—	20	90.2
12. *Sagitta* (1–10 mm, av. 3.96 mm)	3.2	2.10	64	100	—	—	33.0
13. Pleurobrachia, Medusae (0.3–3 mm, av. 1 mm)	10.24	8.20	80	100	—	—	32.8
Secondary and tertiary carnivores	13.44	10.30	77	100	—	—	32.9

[a] For 86% feeding nauplii.

of energy. Animals often apply one method of movement and food capture after another, depending upon the food available to them. Thus, taking into consideration the observed peculiarities of Copepoda and other planktonic groups, certain ecological nutritional types have been distinguished.

The ecological nutritional type is, in our opinion, predetermined by a whole complex of functional and morphological properties of the mouths and digestive apparatus of the consumers, which in turn is determined by

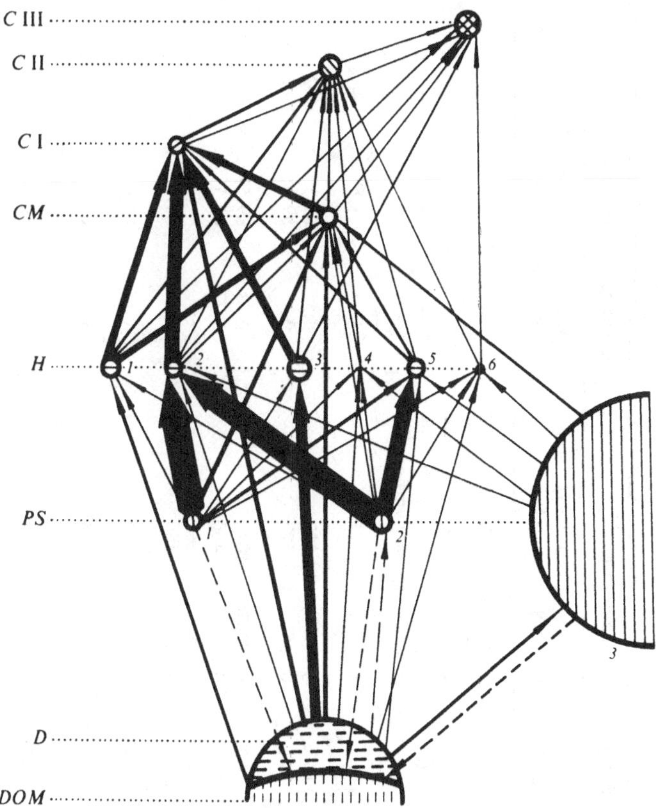

Fig. 11.1. Food webs in the epiplankton community of the Black Sea. C III, tertiary carnivores; C II, secondary carnivores; C I, primary carnivores; CM, mixed-food (plant and animal) consumers; H, herbivorous organisms; *PS*, producers and saprophagous organisms (phytoplankton); *D*, detritus; *DOM*, dissolved organic matter; *1*, *2*, *3*, etc., written near the circles, are the ecological groups of the trophic level. The level of producers: *1*, small-sized forms; *2*, medium-sized forms; *3*, large-sized forms. The level of herbivores: *1*, *Oikopleura*; *2*, I–III copepodites; *3*, IV–VI copepodites of *Paracalanus*; *4*, larvae of molluscs and polychaetes; *5*, nauplii of *Pseudocalanus* and *Calanus*; *6*, IV–V copepodites of *Pseudocalanus* and *Calanus*. Areas of circles are proportional to the average daily standing stock of the ecological groups. Thickness of solid arrows is proportional to daily specific rate of food consumption. Dotted arrows indicate the processes of dissolved organic matter excretion and the processes of detritus consumption, where no quantitative data were obtained.

environmental variables (food abundance, its quality, temperature and light). The principle ecological groups and the ration composition of their main representatives are shown in Table 11.1–11.3. Schemes of trophic relations between these groups of organisms are shown in Figs. 11.1–11.3.

The numbers of groups (g) in the communities investigated are not equal. It is known that the plankton of tropical waters has a higher species

Table 11.3. *Biomass, daily ration and composition of food in the main ecological groupings in the tropical planktonic community of the Pacific Ocean (oceanic zone)*

Trophic level	Groups	Biomass (cal/m³)	Daily ration (cal/m³)	Daily ration (% of body weight)	Composition of food: Animals	Algae (% of the ration)	Detritus (% of the ration)	Bacteria	Assimilation (%)
Producers and	1. Dissolved organic matter	15000.0	—	—	—	—	—	—	—
saprophagous	2. Detritus	279.0	—	—	—	—	—	—	—
consumers	3. Phytoplankton,	108.0	94.0	90	—	—	—	—	—
(phytoplankton)	small-sized forms (0.010 mm)	1.6	—	—	—	—	—	—	—
	medium-sized forms (0.04 mm)	9.4	94.0	(photosynthesis)					
	large-sized forms (0.309 mm)	97.0	—	—	—	—	—	—	—
Mainly	4. Bacteria	32.0	71.0	221	—	—	1.8[a]	—	—
phytophagous	5. Ciliates and								
and	Radiolaria (0.2 mm)	0.07	0.159	227	—	0.5	89.3	10.2	45
detritophagous	Infusoria (0.057 mm)	0.06	0.136	227	—	0.5	89.3	10.2	45
consumers	6. Appendicularia,								
	small (0.075 mm)	0.000065	0.000676	1040	0.02	0.51	89.57	9.9	60
	large (3 mm)	0.007	0.028	400	0.03	25.72	66.5	7.75	60
	7. Salpae	0.876	—	—	—	—	—	—	—
	8. Nauplii of Copepoda (0.150 mm)	3.30	4.62	140	0.1	3.4	86.6	9.9	60
	9. Copepodita of Copepoda,								
	small (0.35 mm)	4.13	6.61	160	0.1	3.4	86.6	9.9	60
	10. '*Calanus*'-type								
	small forms (up to 1.5 mm)	3.11	2.8	90	1.1	3.4	85.7	9.8	60
Consumers of	11. '*Calanus*'-type,								
plant, animal and	large forms (2–4 mm)	0.473	0.236	50	63	7	16	14	50
other food	12. '*Centropages*' type,								
	large forms (1.5–5 mm)	1.6	0.69	43	82	3	7	8	60
	13. '*Acartia*'-type,								
	small forms (up to 1.28 mm)	0.239	0.239	100	3.1	3.3	83.8	9.8	60

	14. '*Acartia*' type, large forms (2–4 mm)	0.080	0.028	35	85	3	7	5	50
	15. '*Euchaeta*'-type, large forms (1.8–5 mm)	2.69	1.103	41	91	1.8	4.2	3	60
	16. Euphausiacea (5–18 mm)	0.481	0.082	17	2.7	25.1	64.8	7.4	50
	17. Hyperiidae, Ostracoda, Amphipoda, Gastropoda, Polychaeta (0.4–6.5 mm)	0.707	0.141	20	2.7	25.1	64.8	7.4	50
	18. '*Oithona–Oncaea*'-type, small forms (0.5–1.5 mm)	3.1	3.94	127	74	6.6	15.4	4	30
	19. '*Oithona–Oncaea*'-type, large forms (2–6 mm)	0.91	0.182	20	95	1	2	2	40
Carnivores	20. Pteropoda (3.5 mm)	0.849	0.032	3.8	100	—	—	—	—
	21. Chaetognatha (*Sagitta*) (10.5 mm)	0.296	0.250	84	100	—	—	—	65
	22. Medusae, Siphonophora (3.3 mm)	0.358	0.319	89	100	—	—	—	65

[a] The remaining 98.2% consists of dissolved organic matter.

diversity than has the plankton of temperate latitudes (Hulbert, 1923, cited from Mullin, 1967; Zaika & Andrjuschenko, 1969). The number of planktonic ecological groups in the Pacific tropical community is approximately twice as large ($g3 = 24$) as in the Black Sea neritic communities ($g1, 2 = 13$). The number of trophic relations among the tropical plankton ($m3 = 164$) is also more than three times that of the Black Sea plankton communities ($m1, 2 = 50$). The Black Sea Infusoria and bacteria are not accounted for, nor are the tropical Salpae and Pteropoda. We may imagine trophic relationships as lines connecting ecological groups. The average number of trophic interrelationships per group in the tropical community is 8.2; and in the Black Sea, 3.85. The index ($f = m/g$) may be called an index of the trophic complexity of the community. The higher the index, the more complicated the community. Comparing all the observed communities by this index, it is clear that oceanic tropical communities are the most complicated.

Our studies on ecological nutritional types show that the zooplankton of neritic Black Sea communities is more or less distinctly divisible into herbivores and carnivores, with an insignificant number of omnivores (Tables 11.1, 11.2), while in the oceanic tropical communities mixed-feeding organisms predominate, with preference for predatory feeding or detritus eating (Table 11.3).

The large numbers of omnivores and predators in the tropical areas of the ocean are emphasised by some scientists (Mullin, 1967; Heinrich, 1958). According to Timonin (1969) the large numbers of predators in stable oceanic waters may be explained by the low magnitudes of the phyto- and zooplankton biomass, by high plankton diversity and by the absence of dominating forms. The research carried out in 1972 on the RV *Vityaz* by radiocarbon methods has shown that the same factors cause an increase in predatory and mixed feeding within zooplankton in stable tropical waters of the Pacific. Animal food appears to be a necessary component in the rations of the majority of organisms. With a lack of animal food in the rations, energy requirements of organisms are usually not fulfilled even at high levels of plant and bacteria assimilation (Petipa, Pavlova & Sorokin, 1971; Petipa, Monakov, Pavlyutin & Sorokin, 1974). The importance of large-sized food objects in the rations of tropical copepods is determined by (i) the active selection by consumers of large-sized phytoplankton cells and animals (ii) by the predominance of large-sized phytoplankton cells in stable oceanic waters (Semina, 1969, 1974) when the total plankton biomass is very low.

The neritic communities of temperate areas are usually characterised by a much higher plankton biomass, by a low species diversity, and by the dominance of certain types of food. This causes the narrowing of the food spectrum of planktonic animals whose energy requirements may be met

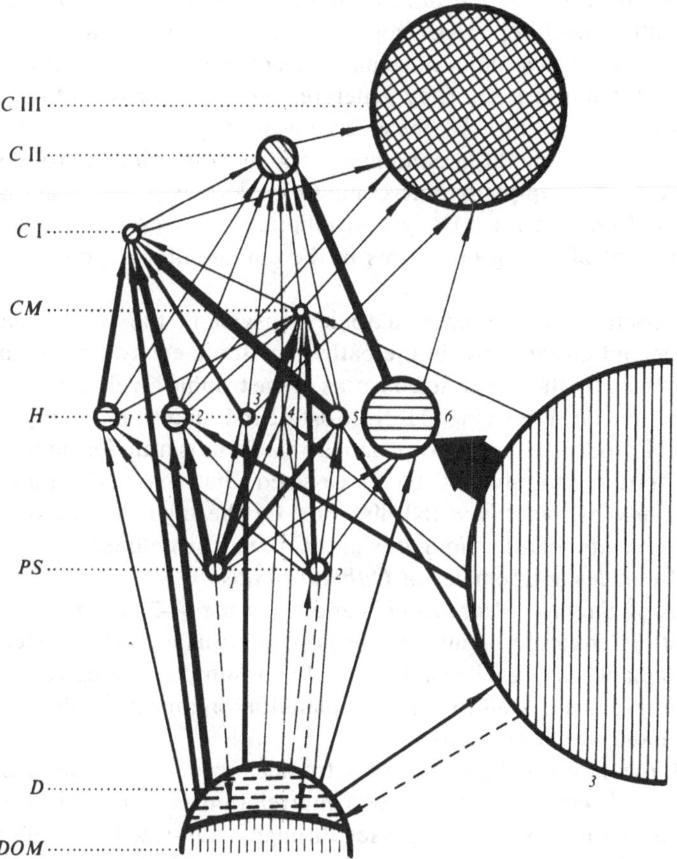

Fig. 11.2. Food webs in the bathyplankton community of the Black Sea (see Fig. 11.1. for key to symbols).

by plant or bacterial food only (Pavlova & Sorokin, 1970). The situation is similar in the neritic and oceanic communities of tropical and temperate waters: the richer the species composition of the phyto- and zooplankton and the lower their total biomass, the higher the percentage of predatory and mixed feeding among zooplankton organisms.

The flow and efficiency of energy transfer through the trophic levels of communities

The main flow of matter and energy within the epiplanktonic ecosystem of the Black Sea is through small- and medium-sized forms of life at all trophic levels, such as dinoflagellates (Pyrrhophyta) and diatoms (Bacillariophyta) up to 50 μm in size, and animals (nauplii, copepodites I–III of smaller-sized copepods, copepodites IV–VI of *Paracalanus*, *Acartia clausi* and *Oithona minuta*, small forms of *Sagitta* and *Pleurobrachia*) up to 1.5–4.0 mm in size. They generate a small biomass, but develop high rates of production. All these organisms together constitute the main volume of the plankton community and are the principle food stock for all levels. The epiplanktonic community derives a significant part of its energy from detritus. Daily consumption rates constitute 49% of the biomass of all living organisms in the community, expressed as energy units.

In contrast to this epiplanktonic ecosystem, the greater part of the matter and energy flow in the bathyplanktonic ecosystem is transmitted principally by the larger-sized forms of the trophic levels and only partially by the smaller ones (Fig. 11.2). Most of the matter and energy from the level of producers and saprophagous organisms is transferred to secondary and tertiary carnivores by large-sized diatoms and dinoflagellates, especially by *Noctiluca* (60–800 μm), by the migrating stages of large-sized *Pseudocalanus elongatus* and *Calanus helgolandicus* (1.4–3.3 mm) and by males and females of *Oithona similis* (0.7–0.9 mm), to large-sized forms of *Sagitta*, *Pleurobrachia* and *Medusae* (4–20 mm).

Bathyplanktonic organisms consume detritus almost as extensively as do epiplankton organisms. Their daily consumption rates constitute 45% of the biomass of all living bathyplankton organisms in the community, expressed as energy units.

All the ecological groups in the tropical oceanic ecosystem are closely connected with each other by food interrelationships (Fig. 11.3). The major flow of matter and energy passes through small-sized as well as through large-sized forms. In particular, large energy flows are transmitted by nauplii and copepodites of small- and large-sized copepods and small-sized species of '*Calanus*-type' to large forms of '*Euchaeta*-type' and to small forms of '*Oithona–Oncaea*-type'. Bacteria play a significant role in food relations within tropical communities. They obtain particularly large quantities of energy from dissolved organic matter and through the breakdown of detritus. This energy is then passed on to nauplii, copepodites and other ecological groups.

The daily rates of detritus consumption within tropical ecosystems equals 9% of the organisms' biomass. A comparison of the ecosystems studied has shown that in spite of the low rate of detritus consumption in a tropical community, the average percentage of detritus in the average animal's ration is the highest (53.7%) compared with 50.4% in the Black Sea bathyplanktonic system and 46% in the epiplanktonic system. The total amount of detritus per square metre in the 100 m layer of the tropical system is higher than in the Black Sea communities.

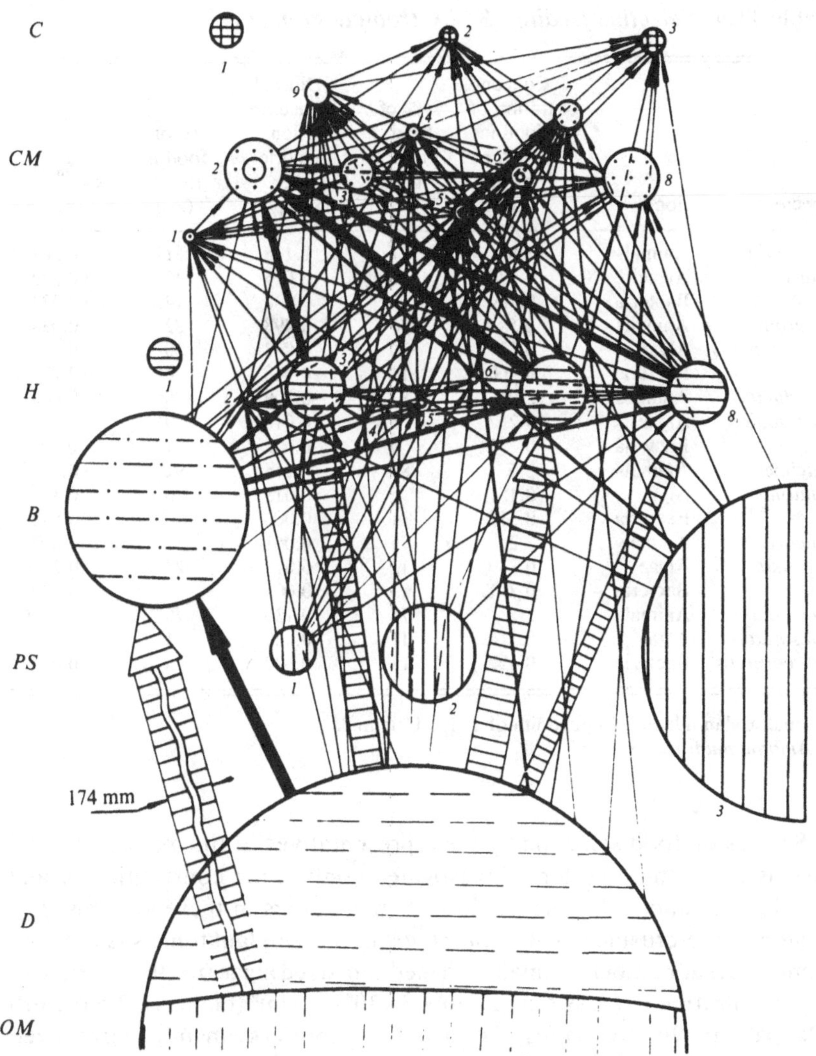

Fig. 11.3. Food webs in the tropical plankton community of the Pacific Ocean. *C*, carnivores: *1*, Pteropoda; *2*, Chaetognatha; *3*, Medusae and Ctenophora. *CM*, mixed-food (plant and animal) consumers: *1*, '*Acartia*-type', large-sized forms; *2*, '*Oithona–Oncaea*-type', small-sized forms; *3*, '*Oithona–Oncaea*-type', large-sized forms; *4*, '*Centropages*-type'; *5*, '*Acartia*-type', small-sized forms; *6*, '*Calanus*-type', large-sized forms; *7*, 'Amphipoda-type'; *8*, '*Euchaeta*-type'; *9*, '*Euphausia*-type'. *H*, mainly herbivores; *1*, Salpae; *2*, Appendicularia, large-sized forms; *3*, nauplii of Copepoda; *4*, Appendicularia, small-sized forms; *5*, Infusoria; *6*, Radiolaria; *7*, Copepodita of Copepoda; *8*, '*Calanus*-type', small-sized forms. *B*, bacteria. *PS*, producers – saprophagous organisms (phytoplankton): *1*, small-sized forms; *2*, medium-sized forms; *3*, large-sized forms. *D*, detritus. *DOM*, dissolved organic matter.

Table 11.4. *Selective feeding (S) by tropical copepods*

Species	Food	Concentration of food in environment (cal/l)	% of food in environment (A_c)	Diurnal consumption (calories per exemplar)	% of food in ration (A_n)	$s = \frac{A_n - A_c}{A_n + A_c}$
Undinula darwini	Animals[a]	1.8	77	0.108	51	−0.203
	Algae[b]	0.374	16	0.075	35	+0.372
	Bacteria	0.164	7	0.030	14	+0.333
Pleuromamma abdominalis	Animals	2.5	76	0.498	82	+0.038
	Algae	0.337	10	0.063	10	0
	Bacteria	0.460	14	0.048	8	−0.273
Candacia aethiopica	Animals	2.7	83	0.109	90	+0.040
	Algae	0.257	8	0.010	9	0
	Bacteria	0.297	9	0.002	1	−0.80
Euchaeta marina	Animals	2.5	84	0.216	92	+0.040
	Algae	0.157	5	0.015	6	+0.107
	Bacteria	0.318	11	0.006	2	−0.692
Oncaea venusta	Animals	2.2	75	0.061	74	−0.007
	Algae	0.339	12	0.018	22	+0.294
	Bacteria	0.391	13	0.003	4	−0.529
Rhincalanus nasutua a, R. cornutus	Animals	2.2	75	0.130	76	+0.0066
	Algae	0.314	11	0.020	11	0
	Bacteria	0.422	14	0.022	13	−0.037

[a] Small Calanoida and Copepodita of Large Calanoida.
[b] *Amphidinium klebsi.*

Studies on food selection by the representatives of the main ecological groups of all three systems investigated confirm the constitution of and conclusions about the energy flow characteristics in these systems. For example, the crustacean *Acartia clausi* in the epiplanktonic system consumes a great number of small- and medium-sized algae (to 20–40 μm) with a concentration not less than $2\times10^6-3\times10^6$ cells/m^3 (biomass, 2.5 mg/m^3) (Petipa, 1959*b*). *Calanus helgolandicus* prefers mainly large-sized planktonic forms (60–800 μm) which have a lower concentration ($10\times10^3-40\times10^3$ cells/m^3) but develop a higher biomass (25–1000 mg/m^3) (Petipa, 1965). Copepods of tropical oceanic ecosystems consume both small- and large-sized food though they prefer large forms (Table 11.4) (Petipa *et al.*, 1974).

Quantitative values of the efficiencies of ecological trophic levels (Slobodkin, 1962) or of the efficiency of matter and energy transfer by communities of different heterotrophic levels in the Black Sea and in the tropical Pacific are presented in Table 11.5. The energy transfer efficiency (*E*) in the epiplanktonic system of the Black Sea decreases from 59% in

Table 11.5. *Transfer efficiency of energy by the heterotrophic levels of the communities in temperate and tropical waters (E is energy transfer efficiency, G is diurnal extraction from the level and R is diurnal consumption by the level)*

Trophic level	Epiplankton community of the Black Sea (0–12 m)			Bathyplankton community of the Black Sea (12–100 m)			Plankton community of tropical areas of the Pacific		
	R (cal/m³)	*G* (cal/m³)	*E* (%)	*R* (cal/m³)	*G* (cal/m³)	*E* (%)	*R* (cal/m³)	*G* (cal/m³)	*E* (%)
Herbivores	62.35	36.57	59[c]	51.0	10.47	24	14.35	4.61	32
Mixed-food consumers	18.40	8.87	48	0.97	0.36	37	6.02	1.52	25
Primary carnivores	39.60	4.9	12	3.05	0.59	19	—	—	—
Secondary and tertiary carnivores	10.5	0.87	8	10.31	0.0	0.0[a]	0.6	0.005	0.8[b]

[a] Not grazing.
[b] Full grazing is not known.
[c] Microzooplankton is included.

the first heterotrophic level to 8% in the last one, while in bathyplanktonic and tropical ecosystems this efficiency is relatively constant for all the levels, with an average of 27–28%.

The analysis of the material collected has shown that the efficiency of energy transfer at all community trophic levels of the ecosystems studied is evidently conditioned by differences in ecosystem stability. It is well known that high species diversity within any ecosystem and consequently high complexity of food interrelationships provides higher stability of the ecosystem as a whole (Slobodkin, 1962). Hence one may suppose that the tropical oceanic ecosystem, because of the higher trophic complexity (see above), is more stable than the ecosystems of the Black Sea. However, the stabilities in the Black Sea epi- and bathyplanktonic communities studied are not the same. V. V. Menshutkin has computed their stability by the term $(\Sigma\,|K|/m)$, where $\Sigma\,|K|$ is the sum of absolute magnitudes of decreases in all ecosystem elements (ecological groups' biomass) under the experimental perturbation; m is the number of trophic relations within the system. The results obtained indicate that the bathyplanktonic community (0.318) is more stable than the epiplanktonic (0.249). As the numbers of trophic relationships and ecological groups of the Black Sea epi- and bathyplankton are practically the same, their different stabilities may be explained by the conditions of their existence, i.e. by different environmental conditions. In fact the bathyplankton community inhabits comparatively deep waters where yearly temperature, salinity and other factors are constant, while the living conditions of epiplankton are greatly influenced by seasonal changes, fluctuations in river influx, etc. As a result, the bathyplankton community has developed stable structural and physiological characteristics in the course of time. The epiplankton community, in contrast, often changes its structural and functional characteristics, which are conditioned by variable and important environmental influences.

The following conclusions may be stated. Stable temperate and tropical communities are characterised by a constant (averaging 27–28%) index of energy transfer efficiency (*E*) embracing all community heterotrophic levels. For the non-stable system, this index rapidly decreases from the first heterotrophic level (60%) to the last (8%), but the average value for all the levels is approximately the same as for the stable systems.

Inconstant, or variable values of the energy transfer efficiency of the trophic levels and high indices (*E*) for the first heterotrophic level reflect strained interrelationships between producers and consumers within the community.

References

Arashkevich, Ye. G. (1969). The food and feeding of copepods in the north-west Pacific. *Okeanologya*, **9**(5), 695–709.

Delalo, E. P. (1961). Preliminary data on the food and feeding of *Paracalanus parvus* in the Black Sea. *Trudy Sevastopolskoy biologicheskoy stantsiy*, **14**, 126–34.

Delalo, E. P. (1964). On daily rhythm in nutrition of *Pseudocalanus elongatus* (Boeck). *Trudy Sevastopolskoy biologicheskoy stantsiy*, **15**, 94–100.

Heinrich, A. K. (1958). The food and feeding of marine copepods in the tropical region. *Doklady Akademii Nauk SSSR*, **119**(5), 1028–34.

Khailov, K. M. (1971). *Ecological metabolism in the sea*. Kiev: Naukova Dumka.

Lebedeva, M. N. (1959). Bacterial threads, introduced from the hydrogen-sulphide depths of the Black Sea, as possible food for zooplankton-filtrators. *Trudy Sevastopolskoy biologicheskoy stantsiy*, **11**, 29–42.

Mironov, G. N. (1954). Nutrition of planktonic carnivores. I. *Noctiluca* nutrition. *Trudy Sevastopolskoy biologicheskoy stantsiy*, **8**, 320–40.

Mironov, G. N. (1960). Nutrition of planktonic carnivores. II. *Sagitta* nutrition. *Trudy Sevastopolskoy biologicheskoy stantsiy*, **13**, 78–91.

Mironov, G. N. (1967). Nutrition of planktonic carnivores. III. Food requirements and daily rations of *Aurelia aurita* (L.). In *Biologia i raspredelenie planktona juzhnikh morey*, pp. 124–37. Moscow: Nauka.

Mullin, M. M. (1967). On the feeding behaviour of planktonic marine copepods and the separation of their ecological niches. In *Proceedings of the Symposium on Crustacea*. Marine Biological Association of India, **2**, 956–64.

Mullin, M. M. (1966). Selective feeding by Calanoid copepods from the Indian

Ocean. In *Some contemporary studies in marine science*, ed. H. Barnes, pp. 545–54. London: Allen and Unwin.

Pavlova, E. V. (1959*a*). On *Penilia avirostris* Dana nutrition. *Trudy Sevastopolskoy biologicheskoy stantsiy*, **11**, 63–71.

Pavlova, E. V. (1959*b*). On food requirements of Cladocera *Penilia avirostris* Dana. *Trudy Sevastopolskoy biologicheskoy stantsiy*, **12**, 153–9.

Pavlova, E. V. (1964). Food requirements of the Black Sea Cladocera *Penilia avirostris* Dana and how they are covered. *Trudy Sevastopolskoy biologicheskoy stantsiy*, **15**, 446–59.

Pavlova, E. V. (1967). Food consumption and energy conversion by Cladocera populations in the Black Sea. In *Structura i dinamika vodnikh soobshchestv i populatsii, Ser. 'Biologia Morya'*, Akademiia Nauk, UkSSR, pp. 66–85. Kiev: Naukova Dumka.

Pavlova, E. V. & Sorokin, Yu. I. (1970). Bacterial feeding of the plankton copepod *Penilia avirostris* Dana from the Black Sea. In *Produktsiya i pishchevgye svyazi v soobshchestvakh planktonnykh organismov* (Production and Feeding Relationships in Plankton Organism Communities), pp. 182–99. Kiev: Naukova Dumka.

Pavlowskaya, T. V. (1970). Algae-diatoms as a food object for infusoria *Uroleptopsis viridis*. *Zoologitchesky*, **49**(12), 1775–9.

Petipa, T. S. (1959*a*). Feeding of the copepod *Acartia clausi*. *Trudy Sevastopolskoy biologicheskoy stantsiy*, **11**, 72–100.

Petipa, T. S. (1959*b*). The food and feeding of *Acartia clausi* and *A. latisetosa* in the Black Sea. *Trudy Sevastopolskoy biologicheskoy stantsiy*, **12**, 130–152.

Petipa, T. S. (1965). Selectivity in the feeding of *Calanus helgolandicus* (Claus). In *Issledovaniya planktona Chernogo i Azovskogo morey* (Plankton Studies in the Black Sea and the Sea of Azov), pp. 102–10. Kiev: Naukova Dumka.

Petipa, T. S. (1966*a*). The relationship between weight body increment, energetic metabolism and rations of *Acartia clausi* Giesbrecht. In *Fiziologiya morskikh zhivotnykh* (Marine Animals Physiology), Moscow, Oceanogr. Commn. scient. publ. Ho., pp. 82–91. Moscow: Nauka.

Petipa, T. S. (1966*b*). On the energy balance of *Calanus helgolandicus* Claus in the Black Sea. In *Fiziologiya morskikh zhivotnykh* (Marine Animals Physiology), Moscow, Oceanogr. Commn. scient. publ. Ho., pp. 60–81. Moscow: Nauka.

Petipa, T. S. (1967*a*). On the life forms of pelagic copepods and to the question of trophic levels structure. In *Structura i dinamika vodnikh soobshchestv i populatsii, Ser. 'Biologia Morya'*, Akademiia Nauk, UkSSR, pp. 108–19. Kiev: Naukova Dumka.

Petipa, T. S. (1967*b*). On the movement and food capture modes of *Calanus helgolandicus* (Claus). In *Biologya planktona i jego raspredelenie v yuzhnikh moryakh*, pp. 109–23. Kiev: Naukova Dumka.

Petipa, T. S., Monakov, A. V., Pavlyutin, A. P. & Sorokin, Yu. I. (1974). Nutrition and energetic balance of tropical Copepoda. In *Osnovy biologitcheskoy produktivnosty yuzhnikh morey*, pp. 136–52. Kiev: Naukova Dumka.

Petipa, T. S., Pavlova, E. V. & Mironov, J. N. (1970). The food web structure, utilisation and transport of energy by trophic levels in the planktonic communities. In *Marine food chains*, ed. J. H. Steele, pp. 142–67. Edinburgh: Oliver and Boyd.

Petipa, T. S., Pavlova, E. V. & Sorokin, Yu. I. (1971). Studies of feeding of mass forms of plankton in the tropical region of the Pacific Ocean by the radiocarbon method. In *Functioning of pelagic communities in the tropical ocean waters*, ed. M. E. Vinogradov, pp. 123–41. Moscow: Nauka.

Semina, G. I. (1969). The size of phytoplankton cells along Longitude 174° W in the Pacific Ocean. *Okeanologya*, **9**(3), 391–8.

Semina, G. I. (1974). *The phytoplankton of the Pacific Ocean*, pp. 237. Moscow: Nauka.

Slobodkin, L. B. (1962). *Growth and regulation of animal populations*. New York: Holt, Rinehart and Winston.

Sorokin, Yu. I. (1959). The effect of water stratification on the primary production of the sea. *Obschei Biologii*, **20**(6), 455–63.

Sorokin, Yu. I. (1966). Application of C^{14} for the study of aqueous animals. In *Plankton i bentos vnytrennikh vodoyemov* (Plankton and benthos of inland bodies of water), pp. 75–119. Moscow: Nauka.

Sorokin, Yu. I., Petipa, T. S. & Pavlova, E. V. (1970). Quantitative estimate of marine bacterioplankton as a source of food. *Okeanologya*, **10**(2), 253–60.

Timonin, A. G. (1969). The structure of pelagic associations. The quantitative relationship between different trophic groups of plankton in frontal zones of the tropical ocean. *Okeanologya*, **9**(5), 846–56.

Vinogradov, M. Ye. (1968). *Vertikal'noye raspredeleniye okeanicheskogo zooplanktona* (Vertical distribution of oceanic zooplankton), pp. 7–320. Moscow: Nauka.

Vinogradov, M. Ye., Gitelzon, I. I. & Sorokin, Yu. I. (1971). *On the space structure of euphotic zone communities in tropical oceanic waters*. Moscow: Nauka.

Vinogradov, M. Ye., Menshutkin, V. V. & Shushkina, E. A. (1972). On mathematic simulation of a pelagic ecosystem in tropical waters of the ocean. *Marine Biology*, **16**(4), 261–8.

Zaika, V. E. & Andrjuschenko, A. A. (1969). Taxonomic diversity of the Black Sea phyto- and zooplankton. *Gygrobiologitchesky*, **5**(3), 12–19.

Zaika, V. E. & Pavlowskaya, T. V. (1970). Marine infusoria grazing in one-celled algae. *Biologia Morya*, **19**, 82–96.

12. Trophic relationships in communities and the functioning of marine ecosystems: II

Some results of investigations on the pelagic ecosystem in tropical regions of the ocean

E. A. SHUSHKINA & M. E. VINOGRADOV

The production of the several levels of communities depends on the quantity of energy entering them and on the efficiency of its utilisation in the food web. Therefore the study of trophic interrelationships within a community and the study of energy flow through the system gives most important information about the functioning of communities.

Difficulty in investigating complex and variable oceanic ecosystems makes us pay special attention to modelling of the processes going on within these systems. Investigation of models of natural complex systems proves to be an effective tool in studying them and allows us not only to describe the functioning of the system and to predict system behaviour in the case of change in some of its parameters, but also to evaluate certain situations arising in a real system which may be unavailable to direct measurement.

The Institute of Oceanology of the Academy of Sciences of USSR has investigated the structure and functioning of pelagic communities in the ocean. Cruises on board the RV *Vityaz* were organized to study the biology of the western equatorial portion of the Pacific Ocean (44th cruise: November 1968–March 1969; 50th cruise: April–July 1971). Long term investigations were carried out at stations in oligotrophic and mesotrophic waters of equatorial currents and in adjacent regions (Fig. 12.1). Discussion of some of the results obtained on these cruises comprises the bulk of this chapter.

The main task of the cruises was the quantitative study of the structure of the community, the dynamics of energy exchange and recycling of organic matter in the pelagic ecosystem of the tropical ocean; from the phytoplankton, closely connected with distribution of nutrient salts and light energy, to the predatory invertebrates and fishes. In order to characterise all stages of the productive process in an ecosystem, and to try to reveal the ecological mechanisms of its realisation and the different external factors affecting it, it was necessary to carry out a number of new investigations and especially to work out and to use new methods. Only

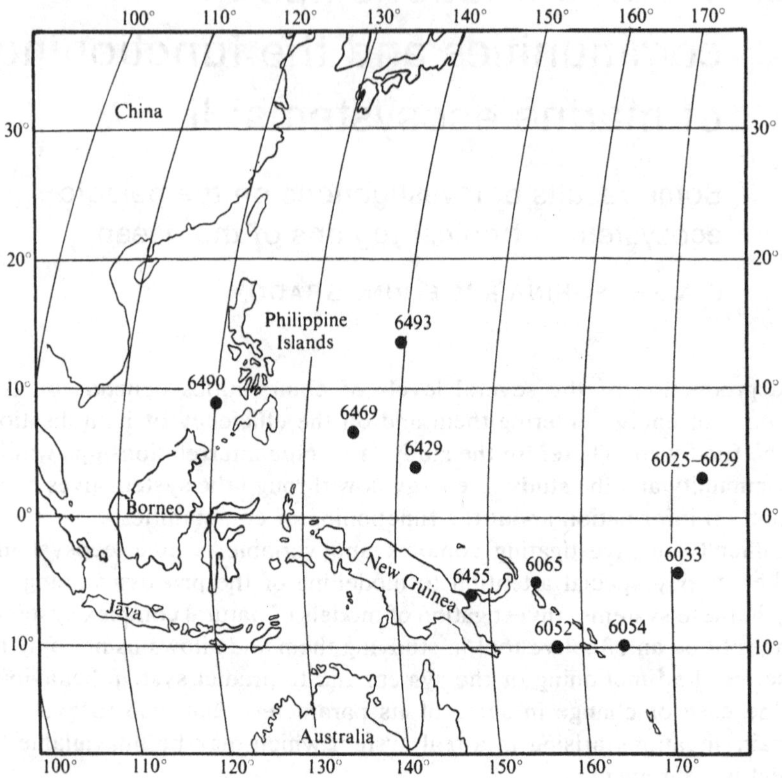

Fig. 12.1. Position of multi-day stations of RV *Vityaz* where special investigations of the pelagic ecosystem were carried out.

recently has the organisation of such combined investigations become possible due to the wide development of electronic instruments, and also isotope techniques used for studying metabolic processes and the working out of more exact methods in chemical analyses.

To have a sufficiently complete picture of the characteristics of productive processes within the community, the investigations were run in two main lines.

(i) Study of the functioning of the community: dynamics of energy and matter exchange at different trophic levels and in the trophic web as a whole. (The quantitative evaluation of these processes alone may make it possible for man to exert a rational influence upon communities at his will.)

(ii) Study of the spatial and trophic structure of the community, and of its species composition. It should be kept in mind that the existence and functioning of pelagic ecosystems in oligotrophic regions presumably are possible due to the non-homogeneous distribution of organisms that form layers and patches of increased concentrations.

In correspondence with these two main lines of enquiry the following factors were investigated.

(i) Distribution and daily dynamics of biogenous elements (phosphorus, nitrogen, silicon) as minerals and as a part of dissolved organic matter.

(ii) Quantity, distribution, activity and production of bacteria, and factors affecting their population.

(iii) Vertical and horizontal distribution of phyto-, micro-, and mesoplankton, ichthyoplankton, nekton and bio-inert components of the ecosystem. Diurnal changes in the distribution of these components.

(iv) Processes of primary production. Determinations of photosynthesis values for the surface unit and distribution with depth. Consumption and turnover of phosphates. Precise methods for the determination of primary production by ^{14}C uptake.

(v) Biochemical composition of particulate and dissolved organic matter in the water column and the characteristics of its transformation and mineralisation.

(vi) Division of the pelagic population into trophic groups *in situ*, relative to the consumption of radioactive phytoplankton, the experimental determination of rations and the assimilation of different types of food (plants, bacteria, animals) by leading species of the community.

(vii) The role of the assimilation of dissolved organic matter in the nutrition of filter-feeding invertebrates.

(viii) Change in the trophic structure of the community at successive stages.

(ix) Characteristics of the bioluminescent field and its variability. Connections between fluctuations in this field and changes in the composition and distribution of bacterio-, phyto- and zooplankton.

(x) Energetic equivalents of body mass, intensity of respiration and energetic balance in all main components of the community.

(xi) Changes in meteorological factors, optical extinction coefficients at different wavelengths and thermohaline characteristics. Measurements of currents with continuous buoy stations. Observations for the calculation of coefficients of vertical turbulence. Hydrochemical parameters of the environment.

Results

During investigations on the 44th and 50th cruises of the RV *Vityaz*, the authors followed Margalef's concept of succession in pelagic communities. If the upwelling zone is to be considered as the place of formation of a community, then considerable structural changes take place in the community with time, while the community is removed by the flow of 'aged' water from its place of origin to the convergence areas where, under oligotrophic conditions, the community attains its climactic stage. Thus successional changes in a community occur simultaneously in both time and in space.

Study of the vertical micro-distribution of plankton by the continuous recording of a bioluminescent field along with the localised collection of quantitative samples of phyto-, bacterio- and zooplankton and of hydro-chemical elements (Gitelzon, 1969; Gitelzon *et al.*, 1971) enables us to detect a sharply defined and stable stratification of the distribution of life within the upper 100–150 m euphotic layer. In meso- and oligotrophic areas with quasi-stable water stratification, i.e. in areas with a sufficiently 'ripe' community, the so-called 'two-maxima' structure in plankton distribution was found to be common. In the layers of maxima with comparably small vertical extent comprising 5–15 m, producers, consumers and reducers occur together. The lower maximum lies above the thermocline at a depth of 70–90 m while the upper maximum is observed at 20–30 m. Concentration and activity of phyto- and bacterioplankton within these maxima (especially in the lower one) may be considerably higher (sometimes about ten times) than in adjoining waters.

It is noteworthy that in oligotrophic tropical waters, the mean concentration of phyto- and bacterioplankton in the euphotic zone is too low to satisfy the food requirements of filter-feeders. It does not cover their energy expenditures derived from the data on food requirements and filtration rates in zooplankton (Jørgensen, 1962; Sushchenya, 1972). Thus the existence of zooplankton in oligotrophic regions is apparently determined by a non-homogeneous (patchy) distribution of food and by the existence of layers with high concentrations where zooplankton also actively concentrate.

The position of the 'lower' maximum layer may be explained by the fact that phytoplankton (along with bacteria and zooplankton depending on it) have optimal conditions for development in a relatively narrow layer only. This layer is limited by light deficiency (lower boundary) and by a shortage of nutrient salts (upper boundary), which come up from greater depths through the thermocline as a result of turbulent mixing (Sorokin, 1959). Phyto- and bacterioplankton consume practically all nutrient salts coming up from greater depths, and it is likely that the lowermost layer of the maximum cuts off the upper part of the community from this source of nutrients.

The population of the upper maximum lives at the expense of the horizontal transport of nutrient salts by water flow in the production-destruction cycles of the community. A certain portion of the nutrient salts always escapes recycling, sinking together with dead and migratory organisms into the deeper layers. Therefore along the time scale, while moving from the upwelling zone, the upper maximum decreases much more quickly than the lower one.

These considerations may be illustrated by a scheme of recycling nutrient salts and organic matter in the ecosystem of oceanic surface waters as proposed by Vinogradov, Gitelžon & Sorokin (1970), and later taken as the basis for the construction of a number of quantitive models of the functioning of tropical pelagic ecosystems (Vinogradov *et al.*, 1972, 1973).

Usually, when energy flow through a community is estimated, photosynthetic production is considered as the primary and main source of organic matter which maintains the existence of all other trophic links. According to Koblentz-Mishke *et al.* (1971) and Sorokin (1971*a*), its magnitude for the whole photosynthesising layer in the area under study is 100–200 mg C/m^2/day. The determination of food rations of herbivores, however, raises the suspicion that the phytoplankton production here is too low to satisfy food requirements of the higher trophic levels. Indeed, Sorokin (1971*b*) has demonstrated that more than one-third of the bacterioplankton in surface tropical waters of the ocean are represented by aggregations exceeding 3 μm in size, which may be taken directly and consumed, not only by filter-feeding copepods, but also by fish larvae, while non-aggregated bacterioplankton is unavailable even to such small-sized animals as the Calanidae. Determination of energy balance in filter-feeding copepods has shown that they consume both bacteria and algae without any marked preference when these occur in sufficient concentrations (Petipa, Pavlova & Sorokin, 1971; Sorokin, 1971*a*).

It is clear that in thoroughly warmed tropical surface waters, bacteria must utilise, apart from autochthonous organic matter, also allochthonous organic matter formed in productive cold oceanic waters, where it sinks down along with cold water and then ascends to the surface in the areas of tropical upwellings. From this point of view, the production of bacterioplankton in tropical communities may be considered as primary. Bacterial production in the upper 1000 m layer is not only commensurable with that of phytoplankton, but may also be several times higher: e.g. in the Solomon Sea (st. 6052) bacterial production comprises 1300 mg C/m^2/day and is ten times higher than that of phytoplankton, whereas at st. 6064 it is only twice as high (Sorokin, 1971*a*).

Further studies on nutrition of aquatic animals have shown that some of them consume and ingest not only bacteria but a certain portion of dissolved organic matter (Vishkwartzev & Sorokin, 1971). Consequently it appears that in tropical oceanic regions the following may be considered as essential ways of introducing organic matter into the food-chains of

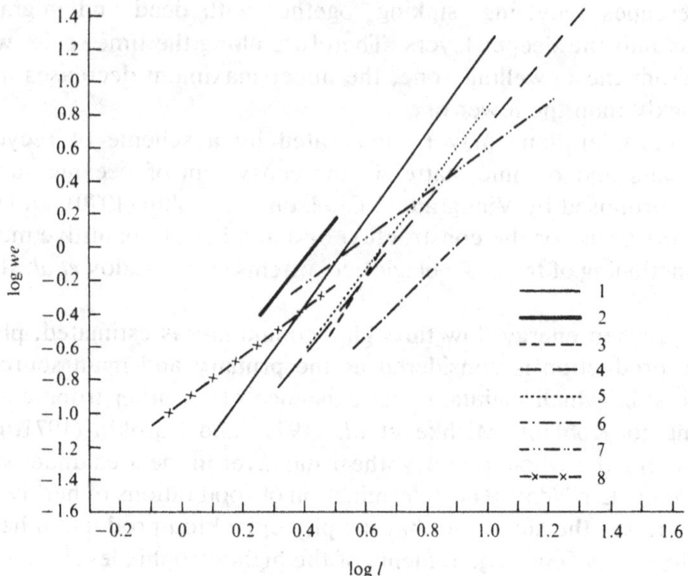

Fig. 12.2. Regression curves log $We = f(\log l)$ for planktonic crustaceans. 1, Copepoda ($l:d = 3$–3.5); 2, Copepoda ($l:d = 2$–2.5); 3, *Eucalanus* ($l:d = 4.0$); 4, Euphausiacea; 5, Mysidacea; 6, *Lucifer*; 7, Hyperiidea; 8, Ostracoda.

pelagic communities: photosynthesis – as the process based on the nutrient salt supply and solar energy; bacterial synthesis – as the process utilising dissolved autochthonous and allochthonous organic matter; direct intake of dissolved organic matter by certain multicellular organisms.

To learn more of the distribution of energy flow through the trophic web of a community, it is necessary to untangle and, when possible, evaluate quantitatively the trophic relations of community elements. These relations became evident through the analysis of intestine contents, the study of mouth morphology, and experimental observations in the feeding of the animals made in the laboratory on board the RV *Vityaz* (Arashkevich & Timonin, 1970; Pavlova, Petipa & Sorokin, 1971; Petipa *et al.*, 1970, 1971; Ponomareva, Tsikhon-Lukanina & Sorokin, 1971; and others).

The description of energy flow in the ecosystem under study was performed on the basis of the well-known equilibrium equation:

$$R = P + T + H = U(P + T) \tag{1}$$

It was necessary to determine rates of feeding (R), respiration (T), and production (P), as well as unassimilated food (H) and the efficiency of food assimilation (U) for animals belonging to different trophic groups.

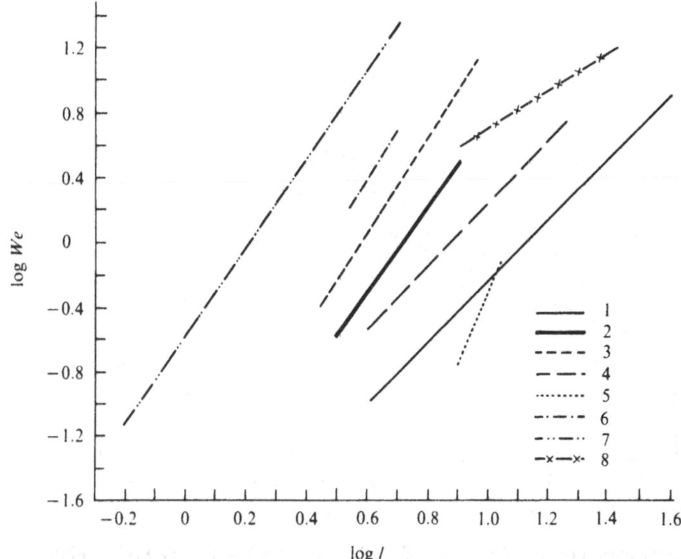

Fig. 12.3. Regression curves log $We = f(\log l)$ for certain pelagic animals. 1, Chaetognatha; 2, Medusae; 3, Ctenophora; 4, Siphonophora; 5, Appendicularia; 6, Polychaeta – *Tomopteris*; 7, Pteropoda; 8, Salpae.

Sharply differing water body contents however (for example, Medusae, Salpae and Siphonophora as compared with copepods, mysids and euphausiids) make the comparative study of these coefficients impossible when the only data available are those on wet weights of the pelagic animals belonging to the various taxonomic groups. Consequently, energy equivalents i.e. the body calorie content of individual animals – *We* (calories/individual) – of a definite size, 1 mm – appear to be more suitable equivalents of wet body weights, having the advantage of being easily defined.

Materials obtained on the cruises allowed us to carry out 700 analyses using bichromate oxidation techniques (Winberg, 1968) for calculating the caloric equivalents, *We*, of pelagic animals, belonging to 42 taxonomic groups in accordance with their size (Shushkina & Sokolova, 1972). Results are given in Figs. 12.2 and 12.3. These measurements covering representatives of practically all the pelagic animals of the south-western Pacific, were later used for determination of respiration and production rates (Shushkina, 1971; Shushkina & Pavlova, 1973).

The basic method for production estimates used was the physiological method (Winberg, 1966, 1968; Shushkina, 1968), involving the measurement of respiration rates. Rates of oxygen consumption related to body

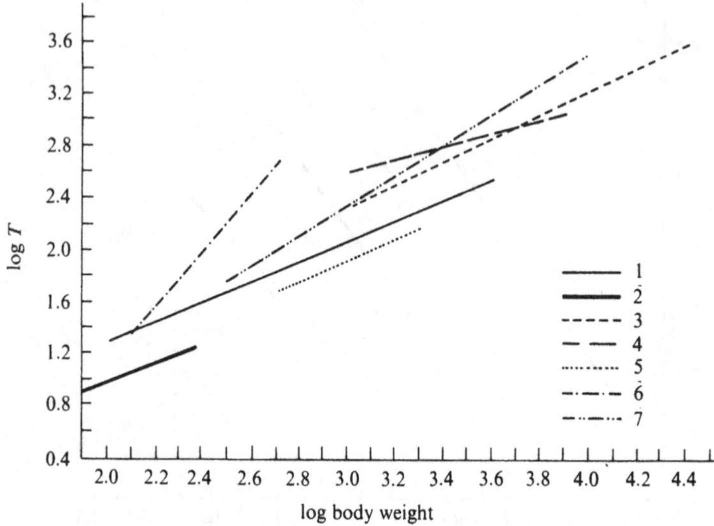

Fig. 12.4. Regression curves of the rate of exchange (*T*) related to the body weight for some planktonic tropical crustaceans. 1, Calanoidea; 2, Cyclopoida; 3, Euphausiacea; 4, Mysidacea; 5, *Lucifer*; 6, Ostracoda; 7, Hyperiidea. From Shushkina & Pavlova (1973).

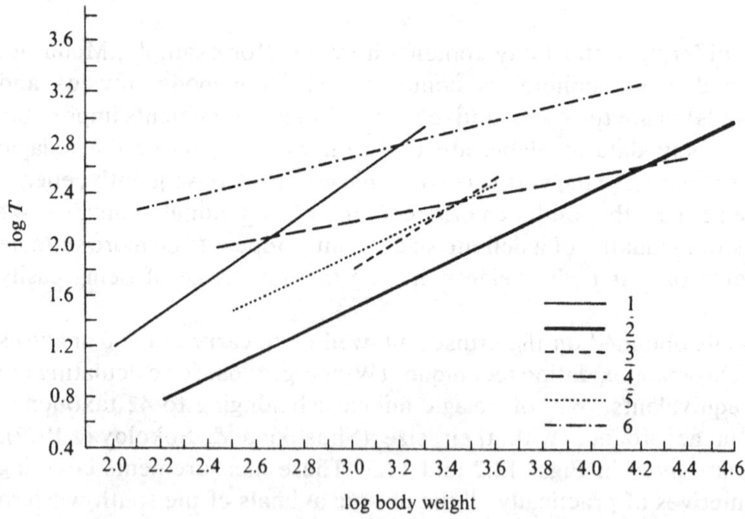

Fig. 12.5. Regression curves for the rate of exchange (*T*) related to the body weight for some tropical pelagic animals. 1, Chaetognatha; 2, Pteropoda; 3, Polychaeta; 4, Ctenophora; 5, Siphonophora; 6, Medusae. From Shushkina & Pavlova (1973).

Table 12.1. *Relationship between mean daily P/B coefficients and body lengths for certain pelagic animals (l = mm; We = millicalories/individual; T = millicalories/day/individual; $K_3 = k_2/L(1-K_2)$; $P/B = K_3 \cdot T(1)$; P/B, production per day/biomass)*

Animal groups	Range of body lengths, l (mm)	$We = f(l)$	$T = f(We)$	$P/B = f(l)$	P/B $K_2 = 0.3$	P/B $K_2 = 0.4$
Calanoidea $l:d = 3.0–3.5$	1.5–5.5	43 $l^{2.6}$	1.13 $We^{0.7}$	$0.37 \cdot K_3 l^{-0.78}$	0.12–0.04	0.18–0.07
Calanoidea $l:d = 2.0–2.5$	1.9–3.1	132 $l^{2.3}$	1.13 $We^{0.7}$	$0.26 \cdot K_3 l^{-0.69}$	0.07–0.05	0.11–0.08
Calanoidea $l:d = 4.0$	2.9–4.4	11 $l^{3.0}$	1.13 $We^{0.7}$	$0.55 \cdot K_3 l^{-0.90}$	0.09–0.06	0.14–0.10
Cyclopoidea	1.3–1.75	43 $l^{2.6}$	0.40 $We^{0.7}$	$0.13 \cdot K_3 l^{-0.73}$	0.05–0.03	0.07–0.05
Euphausiacea	5.0–14.0	27 $l^{2.3}$	0.42 $We^{0.9}$	$0.30 \cdot K_3 l^{-0.23}$	0.09–0.07	0.14–0.11
Mysidacea	4.6–8.9	24 $l^{2.5}$	13.30 $We^{0.5}$	$2.70 \cdot K_3 l^{-1.25}$	0.18–0.07	0.27–0.11
Lucifer	5.0–7.4	22 $l^{1.9}$	0.59 $We^{0.7}$	$0.23 \cdot K_3 l^{-0.57}$	0.04–0.03	0.06–0.05
Ostracoda	0.9–2.5	116 $l^{1.5}$	0.0005 $We^{2.2}$	$0.15 \cdot K_3 l^{1.80}$	0.06–0.34	0.10–0.53
Hyperiidea	2.4–16.0	240 $l^{1.4}$	0.07 $We^{1.2}$	$0.21 \cdot K_3 l^{0.28}$	0.12–0.20	0.18–0.31
Sagitta	4.6–20.0	8 $l^{1.8}$	0.04 $We^{1.3}$	$0.07 \cdot K_3 l^{0.53}$	0.07–0.15	0.11–0.23
Pteropoda	0.9–5.0	190 $l^{2.9}$	0.07 $We^{0.9}$	$0.04 \cdot K_3 l^{-0.29}$	0.02–0.01	0.03–0.01
Polychaeta	6.0–14.0	1140 $l^{0.4}$	0.02 $We^{1.2}$	$0.08 \cdot K_3 l^{0.08}$	0.04–0.0	0.06–0.07
Ctenophora	3.0–9.1	17 $l^{3.0}$	15.7 $We^{0.3}$	$2.16 \cdot K_3 l^{-2.1}$	0.09–0.01	0.15–0.01
Siphonophora	4.3–14.0	18 $l^{2.0}$	0.13 $We^{0.9}$	$0.08 \cdot K_3 l^{-0.2}$	0.03–0.02	0.04–0.03
Medusae	3.2–8.0	11 $l^{2.7}$	9.91 $We^{0.5}$	$3.13 \cdot K_3 l^{-1.35}$	0.27–0.08	0.42–0.13

weights (energy equivalents – *We*) have been determined for zooplankters of thirty-eight taxonomic groups by the Winkler method. The results of these measurements are given in Figs. 12.4 and 12.5. Having the data on rates of exchange for practically all representatives of pelagic animals in the Pacific area under study, daily *P/B* (production/biomass) coefficients for the most common tropical zooplankters were calculated (Table 12.1). In doing so, based on our data and data from the literature (Sushchenya, 1969; Winberg, 1972; Shushkina, 1972), it was suggested that the K_2 coefficient in constantly replenished populations equals 0.4. By using tentative values of exchange and production rates for all common zooplankters, and placing the latter in definite trophic and dimensional groups (Arashkevich & Timonin, 1970; Petipa *et al.*, 1971) it is possible to construct a scheme of energy flow through the community under consideration (Fig. 12.6).*

* In Fig. 12.6 more elements and relationships are indicated than are incorporated into the model.

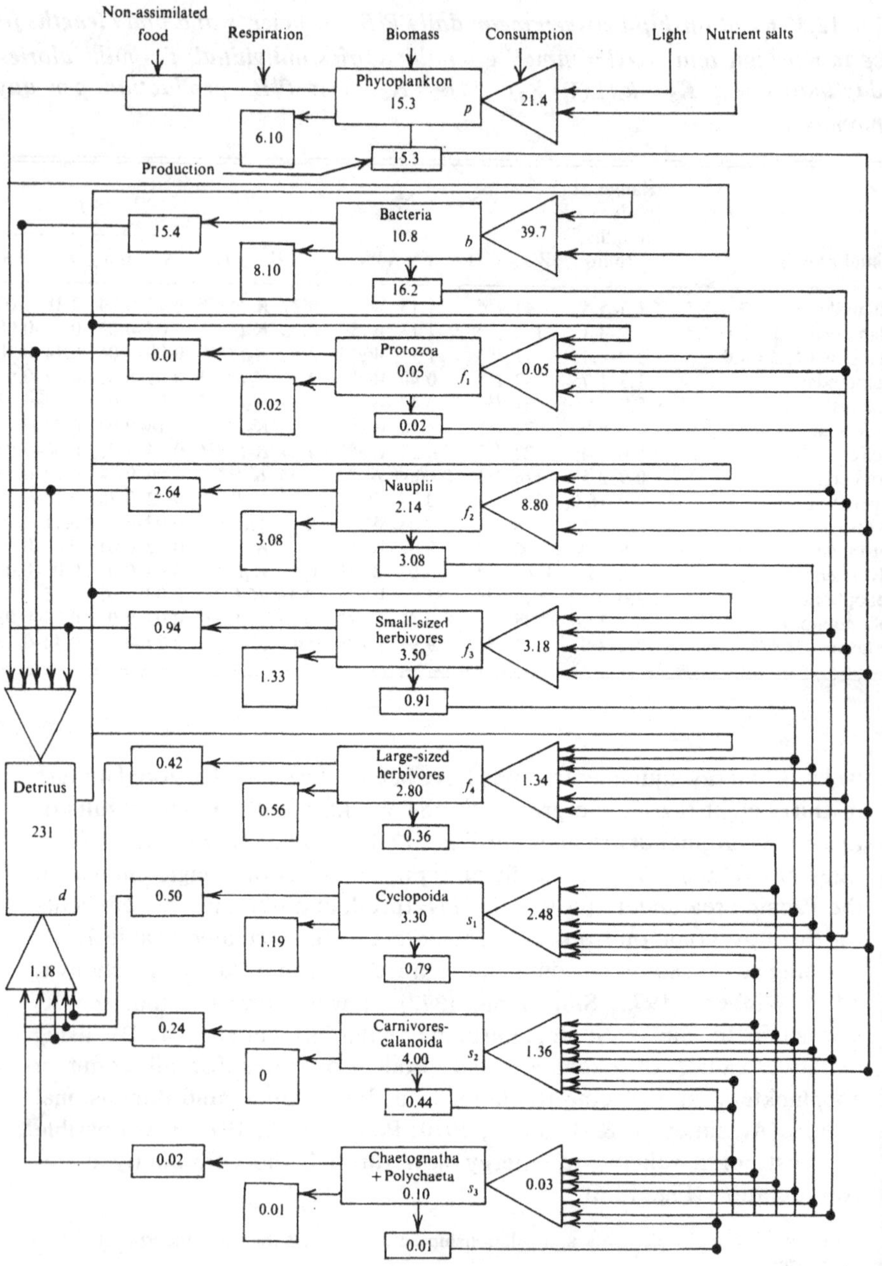

Fig. 12.6. Scheme of energy flow (cal/m^2/day) through the community inhabiting the subsurface 200 m layer in oligotrophic tropical regions of the Pacific Ocean.

Mathematical model of the functioning of a pelagic ecosystem in tropical oceanic waters

On the basis of the scheme mentioned above, and direct determinations of certain other parameters and relationships in the oceanic tropical pelagic ecosystem, a finite-difference model of the functioning of the community was developed (Menshutkin, 1971) and has given interesting results, coinciding well with the real facts (Vinogradov, Menshutkin & Shushkina, 1972). Computations performed on this model gave us the opportunity to follow concentration changes of various ecosystem elements in the upper 200 m layer of tropical oceanic waters as well as the changes in the pattern of vertical distribution of these elements during succession.

The modelling algorithm was written in the ALGOL-60 language and realised on the BESM-3M computer. The scheme of recycling nutrient salts in the ecosystem, proposed by Vinogradov *et al.* (1970) was used as the basis for developing the model. The change in the community with depth is studied (from the surface down to 200 m. at intervals of 10 m) and with time (up to 100 days, at intervals of one day), since the system while developing moves with water flow and changes its position in space. The state of the system at a particular depth is determined from light intensity (e), concentrations of nutrient salts (n)* and detritus (d), biomass of phytoplankton (p), bacteria (b), small-sized herbivores (not exceeding 1.0 mm in body length) (m), large-sized herbivores (f), omnivores (v) and predators (s). Macroplanktonic invertebrates and fishes were not considered in the model.

The operating of the modelling algorithm consists of the transition of the state of the system on a particular day to its state on the following day. The algorithm begins with modelling the light penetration through the water column, only the portion of the spectrum which is used by phytoplankton for photosynthesis being considered. The value of vertical light attenuation is taken to be linear with concentrations of detritus and phytoplankton. Given the intensity of the light flux on the surface and the transparency, the magnitude of illumination is determined in each 10 m layer from 0–200 m.

Vertical transport of nutrient salts between two adjacent layers was simulated with the aid of finite-difference interpretation of the turbulence diffusion equation:

$$W_i = A_z(i, t)\frac{n_{i+1} - n_i}{Z} \tag{2}$$

* Total nitrogen served as a 'nutrient salt' prototype in the model.

where W_i is the flow of nutrient salts between l and $l+1$ layer, Z – vertical distance between layers, A_z – coefficient of turbulence diffusion at the depth of l – layer at 't' instant. A three-layered model was used to describe changes of A_z with time and depth. The position of the thermocline layer with a steep gradient is considered to be removed downwards from the upwelling zone to the convergence area. Moreover, changes in nutrient salt concentration in a particular layer are determined from their uptake by phytoplankton and their excretion by all other community members in the process of metabolism. This excretion is taken to be proportional to energy losses.

Phytoplankton production is limited both by illumination conditions and by the abundance of nutrient salts. The highest permissible value of the P/B coefficient which changes from five to two in the process of succession of the system is also taken into account. It is also suggested in the model that the excretion of dissolved organic matter by phytoplankton constitutes 0.3 of phytoplankton production and that the dissolved organic matter is consumed by bacterioplankton (Sorokin, 1971*b*).

Bacterioplankton production is limited by the amount of food (detritus and dissolved organic matter excreted during phytosynthesis) and by the maximum rate of growth in P/B which is taken as 3 per day (Sorokin, 1971*a*). Bacterioplankton is suggested to consume not more than 10% of the total detritus available per day in a given water layer. The magnitudes of food requirements of the elements of the system are found from equation (1) and are distributed in accordance with food sources (proportional to the biomasses of the latter). The rate of production P is calculated taking into account changes of coefficient $K_2 = P/(P+T)$ with the animal's growth. K_2 is taken to be equal to 0.25 (Shushkina, 1971) in the model but the latest data suggest that the K_2 value may be taken as 0.4 (Shushkina, 1972) for these calculations.

The total sum of all components of highest rations falling within the same food source indicates the pressure by consumers. The real ration of each type of consumer and the grazing by consumers are calculated from the well-known equation by Ivlev (1955).

Diurnal changes in biomass of each living element of the system occur according to the equilibrium equation (Winberg & Anisimov, 1966, 1969):

$$B_i = r_i - T_i - H_i - M_i - \Sigma C_{ij} r_j \qquad (3)$$

where H is non-assimilated food; M is mortality; $C_{ij} r_j$ is the decrease of biomass of the i-element (prey) by grazing; C_{ij} is the share of i-element in the food of the j-element (consumer); r_i is a real ration; T_i is the rate of exchange.

The effect of the vertical migrations of zooplankton on trophic relationships is simulated in the model by assuming that the food requirements of large-sized herbivores, omnivores and carnivores inhabiting the 0–50 m layer are supplemented by a certain portion k of the total food requirements of the same community elements occurring in deeper layers (50–200 m). This coefficient is taken to be dependent on time according to Vinogradov's (1968) data.

The changes in detritus concentration take their origin in its consumption by bacteria and in the supply of dead organic matter in the form of non-digested food and dead individuals. The rate of sinking of detritus was supposed to depend on the water density gradient, while detritus was taken to consist of particles with different diameters and with different rates of sinking. Because of vertical currents, the local ascent of a certain portion of dissolved organic matter is considered possible into the layers with the highest gradient of water density. It is supposed that all elements of the community in its initial state ($t = 0$) are homogeneously distributed in the water column, and that in this initial state of the community, nutrient salt concentration equals 150 mg/m^3, that detritus is absent, and that the biomasses of living community elements range from 0.01 to 0.1 cal/m^3.

Computations show that the biomass of the phytoplankton increases rapidly and reaches its maximum on the fourth to fifth day of the system's existence. Later on, the biomass of the phytoplankton decreases. Small-sized herbivores lag somewhat behind phytoplankton in attaining their maxima. Carnivores prove to be the slowest; their biomass attains its maximum on the thirtieth to fortieth day. Thus the model demonstrates the spatial separation and succession of the biomass maxima of phytoplankton, herbivores and carnivores, as the water moves farther away from the upwelling zone (Vinogradov, Voronina & Sukhanova, 1961; Vinogradov & Voronina, 1964; Timonin, 1969; etc.). Note the prevailing dominance of herbivores in the area adjacent to the upwelling zone; in oligotrophic areas, lying far away from it, carnivores make up about 50% of the total zooplankton biomass. All these findings are in good agreement with the real state of things observed in tropical oceanic regions (Guerdrat, 1971; Timonin, 1971).

The computed curves of vertical distribution in various elements of the system were compared with the actual curves obtained by direct measurement, the latter being obtained by water bottle samples taken mainly in the extreme points of the bioluminescent field and at standard depths (nutrient salts, bacteria, phytoplankton). This method and its results were described by Vinogradov *et al.* (1970). The picture of zooplankton distribution turned out to be more complex. To obtain comparative data, zooplankton was sampled at 10 m vertical intervals as was done in the model. The computed curves proved to differ sharply from those obtained from samples procured from standard layers (0–25, 25–50, 50–100, 100–200) when maxima and minima are averaged (Fig. 12.7). For

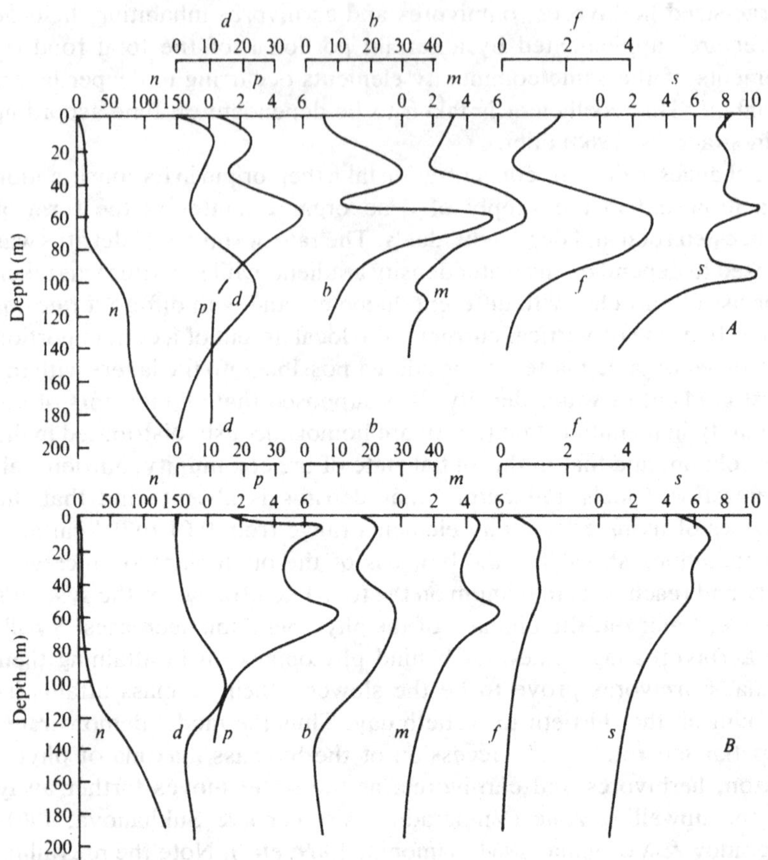

Fig. 12.7. Comparison of vertical distribution of some elements of the system obtained by model computations ($t = 40$ days) and from field observations made at stations with subsurface waters whose 'age' was supposed to be of 30–50 days duration. *A*, Field observations; *B*, computed data; (*n*) nutrient salts (nitrogen); (*d*) detritus; (*p*) phytoplankton; (*b*) bacteria; (*m*) small-sized herbivores; (*f*) large-sized herbivores; (*s*) predatory zooplankters. Nutrient salts are given in mg/m³, all other elements are given in cal/m². From Vinogradov *et al.* (1973).

example as well-marked maximum at the depth of 50–70 m 'disappears' here, since it is combined with the minimum of zooplankton biomass registered at 80–100 m depth. As Timonin has shown, a detailed vertical sampling reveals clear-cut differences in the distribution of various trophic groups of zooplankton. Their vertical distribution is also characterised by well-expressed maxima and minima, while the positions of their maxima do not coincide.

Comparison between simulated data and those obtained by direct measurements demonstrates general agreement in the patterns of vertical distribution of elements of the system, and even in some absolute values

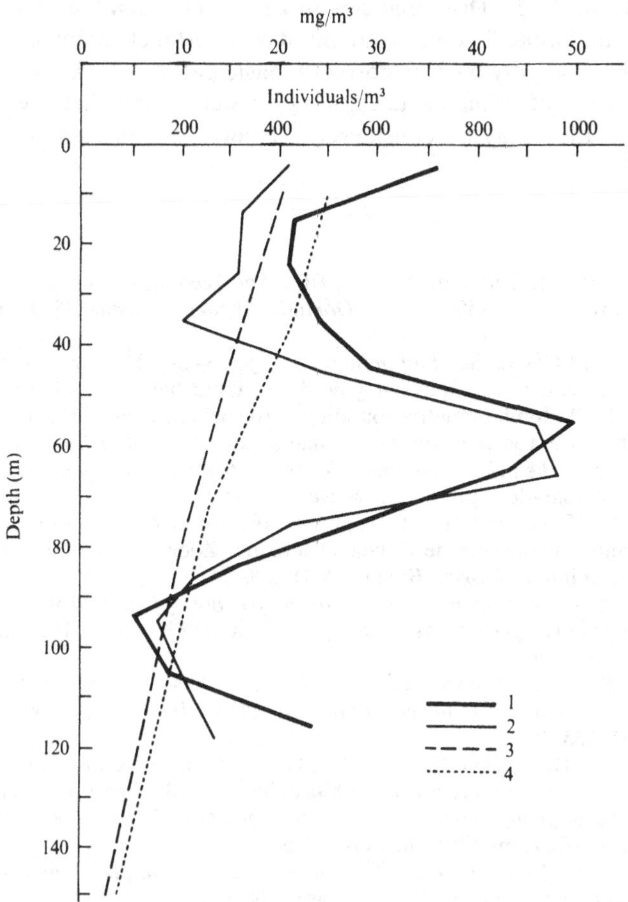

Fig. 12.8. Vertical distribution of zooplankton at station 6469. Nocturnal data obtained through 10 m interval sampling (solid lines) and through sampling at 'standard' layers (0–25; 25–50; 50–100 m) (dashed lines). (1) individuals/m^3; (2) mg/m^3.

Table 12.2. *Biomass (cal/m^2) of individual elements of the system in the 0–200 m layer*

Elements	Simulated data 30 days	Simulated data 40 days	Simulated data 50 days	Field data of the 44th cruise corresponding to 40–50 days of the existence of the system
Phytoplankton	650	520	430	1000
Bacteria	340	330	260	670
Herbivores	820	690	520	770
Predators	690	750	570	630
Zooplankton	1510	1440	1090	1760

(Fig. 12.8, Table 12.2). Discrepancies in details between the model and the 'original' undoubtedly take place, but it is important to emphasise here that the model evidently mirrors correctly basic patterns in the formation and development of complex oceanic ecosystems, and that the mathematical simulation may be considered a helpful tool in studying these systems.

References

Arashkevich, E. G. & Timonin, A. G. (1970). The feeding of copepods in the tropical part of the Pacific Ocean. *Doklady Akademiia Nauk SSSR*, **191**(4), 935–8.

Gitelzon, I. I. (ed.) (1969). *Sea bioluminescence*, pp. 3–181. Moscow: Nauka.

Gitelzon, I. I., Levin, L. A., Shevyrnogov, A. P., Utyushev, R. N. & Artemkin, A. S. (1971). Bathyphotometric sounding of the pelagic zone of the ocean and its possible use for the studies of spatial structure of a biocenosis. In *Functioning of the pelagic communities in the tropical regions of the ocean*, ed. M. E. Vinogradov, pp. 50–64. Moscow: Nauka.

Gueredrat, I. A. (1971). Evolution d'une population de copépodes dans le système des courants equatoriaux de l'Océan Pacifique. Zoogéographie, écologie et diversité specifique. *Marine Biology*, **9**(4), 300–14.

Ivlev, V. S. (1955). *Experimental ecology of the feeding of fishes*. Moscow: Pishchepromizdat. (English translation by D. Scott (1961). New Haven: Yale University Press.)

Jørgensen, C. B. (1962). The food of filter-feeding organisms. *Rapport et Procès-Verbaux des Réunions (Conseil Permanent International pour l'Exploration de la Mer)*, **153**, 99.

Koblentz-Mishke, O. I., Tsvetkova, A. M., Gromov, M. M. & Paramonova, L. I. (1971). Primary production and chlorophyll *a* in the Western Pacific. In *Functioning of pelagic communities in the tropical regions of the ocean*, ed. M. E. Vinogradov, pp. 70–9. Moscow: Nauka.

Menshutkin, V. V. (1971). *Mathematical modelling of populations and aquatic faunal communities*, pp. 3–196. Leningrad: Nauka.

Pavlova, E. V., Petipa, T. S. & Sorokin, Yu. I. (1971). The role of bacterioplankton in the feeding of marine pelagic organisms. In *Functioning of pelagic communities in the tropical regions of the ocean*, ed. M. E. Vinogradov, pp. 142–51. Moscow: Nauka.

Petipa, T. S., Pavlova, E. V. & Mironov, G. N. (1970). The food web structure utilization and transport of energy by trophic levels in the planktonic communities. In *Marine food chains*, ed. J. H. S. Steele, pp. 142–67. Edinburgh: Oliver and Boyd.

Petipa, T. S., Pavlova, E. V. & Sorokin, Yu. I. (1971). Radiocarbon studies of feeding of the mass plankton forms in the tropical Pacific. In *Functioning of pelagic communities in the tropical regions of the ocean*, ed. M. E. Vinogradov, pp. 123–41. Moscow: Nauka.

Ponomareva, L. A., Tsikhon-Lukanina, E. A. & Sorokin, Yu. I. (1971). On phytoplankton and bacteria consumption by tropical euphausiids. In *Functioning of pelagic communities in the tropical regions of the ocean*, ed. M. E. Vinogradov, pp. 152–6. Moscow: Nauka.

Shushkina, E. A. (1968). Production of calculations of copepods on their metabolism regularities and the coefficient of utilisation of the assimilated food for growth. *Okeanologia*, **8**(1), 126–38.

Shushkina, E. A. (1971). Estimates of the production intensities of the tropical zooplankton. In *Functioning of pelagic communities in the tropical regions of the ocean*, ed. M. E. Vinogradov, pp. 157–66. Moscow: Nauka.

Shushkina, E. A. (1972). Intensities of production and the use of assimilated food for growth of the Japan Sea mysids. *Okeanologia*, **12**(2), 326–37.

Shushkina, E. A. & Pavlova, E. V. (1973). Rates of metabolism and production of zooplankton in the Equatorial Pacific. *Okeanologia*, **13**(2), 339–75.

Shushkina, E. A. & Sokolova, I. A. (1972). Calorific equivalents of the body mass of tropical organisms from the pelagic part of the ocean. *Okeanologia*, **12**(5), 860–7.

Sorokin, Yu. I. (1959). On the effect of water stratification on the primary production of photosynthesis in the sea. *Zhurnal Obshchei Biologii*, **20**(6), 455–63.

Sorokin, Yu. I. (1971*a*). On the role of bacterioplankton in the biological productivity of the tropical waters of the Pacific Ocean. In *Functioning of pelagic communities in the tropical regions of the ocean*, ed. M. E. Vinogradov, pp. 92–122. Moscow: Nauka.

Sorokin, Yu. I. (1971*b*). On the role of bacteria in the productivity of tropical oceanic waters. *Internationale Revue der gesamten Hydrobiologie*, **56**(1), 1–48.

Sushchenya, L. M. (1969). Quantitative regularities in crustacean metabolism and transformation of energy and substance. Doctor of Science Thesis, manuscript in the Library of the PP Shirshov Institute of Oceanology, Moscow.

Timonin, A. G. (1969). Structure of the pelagic communities. The quantitative relationship between different trophic groups of plankton in the frontal zones of the tropical ocean. *Okeanologia*, **9**(5), 846–56.

Timonin, A. G. (1971). The structure of plankton communities of the Indian Ocean. *Marine Biology*, **9**(4), 281–9.

Vinogradov, M. E. (1968). *Vertical distribution of the oceanic zooplankton*, pp. 7–320. Moscow: Nauka.

Vinogradov, M. E., Gitelžon, I. I. & Sorokin, Yu. I. (1970). The vertical structure of a pelagic community in the tropical ocean. *Marine Biology*, **6**(3), 187–94.

Vinogradov, M. E., Krapivin, V. F., Menshutkin, V. V., Fleischman, B. S. & Shushkina, E. A. (1973). Mathematical simulation of the functioning of a pelagic ecosystem in the tropical ocean. (Based on the materials of the 50th cruise of RV *Vityaz*.) *Okeanologia*, **13**(5), 852–66.

Vinogradov, M. E., Menshutkin, V. V. & Shushkina, E. A. (1972). On mathematical simulation of a pelagic ecosystem in tropical waters of the ocean. *Marine Biology*, **16**, 261–8.

Vinogradov, M. E. & Voronina, N. M. (1964). Some peculiarities of the plankton distribution in the Pacific and Indian Oceans' equatorial currents. *Okeanologia Issledovaniya*, 10, Section, I.G.Y. Program No. 13, 128–36.

Vinogradov, M. E., Voronina, N. M. & Sukhanova, I. N. (1961). The distribution of the tropical plankton and its relation to some peculiarities of the structure of water in open sea areas. *Okeanologia*, **1**(2), 283–93.

Vishkwartsev, D. I. & Sorokin, Yu. I. (1971). Some results of studying osmotic nutrition in marine invertebrates. Accounts of works on the 50th voyage of *Vityaz*. Manuscript, Library of Oceanology Institute.

Winberg, G. G. (1966). The rate of growth and intensity of metabolism in animals. *Uspechi Sovremennoi Biologii* (Advances in modern biology), **61**(2), 274–93.

Winberg, G. G. (ed.) (1968). *Methods for the estimation of production of aquatic animals.* Minsk: 'High School'.
Winberg, G. G. (1972). Some interim results of Soviet IBP investigations on lakes. In *Productivity problems of freshwaters, proceedings of the IBP-UNESCO symposium on problems of freshwaters.* Warszawa-Krakow.
Winberg, G. G. & Anisimov, S. I. (1966). A mathematical model of an aquatic ecosystem. In *Photosynthesising system of high productivity*, pp. 213–23. Moscow: Nauka.
Winberg, G. G. & Anisimov, S. I. (1969). An attempt to investigate the mathematical model of aquatic ecosystems. *Trudy VNIRO*, **67**(1), 49–76.

13. Soviet investigations of the benthos of the shelves of the marginal seas

A. A. NEYMAN

These investigations of the benthos of the shelves, being a part of IBP, were started in 1964. Their main aim was the study of the distribution of the benthos of the shelves in different geographical zones. This research involved both the collecting of new materials as well as their study, and the interpretation of the data obtained earlier.

The main device used by the Soviet investigators for the quantitative collection of benthos is the 'Ocean' bottom-sampler, covering an area of 0.25 m^2. In most cases, this bottom-sampler provides quite satisfactory samples; it takes a sufficiently thick layer, but it does not work well on compact sands near shore (Neyman, 1965*a*, *b*). V. D. Gordeev's prismatic bottom-sampler (Gordeev, 1945) operates very well on these grounds, where it takes a layer up to 30 cm thick and provides a true sample of compact near-coast sands, remarkable for its very high biomass (Neyman, 1965*a*, *b*). It is, however, very sensitive to the vessel's drift and requires anchoring; for the study of large sea areas the 'Ocean' sampler is used.

Epifauna and large mobile bottom animals are collected by trawls, including commercial otter-trawls, and by dredges of different designs. For the near-shore zone (depths less than 20 m) SCUBA diving methods of collecting and direct observations are becoming more and more commonly used, but these methods and the results obtained by them are not considered here. Only the work performed outside shallow near-shore zones from ships surveying large areas, is discussed here.

In all publications based on materials obtained with bottom-samplers, the total benthos biomass for each station (sample) as well as the biomass of its individual components (separated according to different criteria), are determined in g/m^2. Often the biomass of these components is expressed as the per cent of total biomass at a given station. The quantitative distribution of the benthos and of its components is mapped on the basis of these data.

One of the main methods of interpretation of the data on benthos distribution is the identification of trophic groups in each sample and their distributional dependence on environmental factors. This method was originally applied by Turpajeva (1949, 1953, 1954, 1957) and later became widely adopted. Irrespective of their taxonomical position, Turpajeva divided all non-predatory bottom invertebrates into five trophic groups: (1) swallowing deposit-feeders that inhabit subsurface bottom layers and swallow sediments without discrimination; (2) collecting deposit-feeders that feed on detritus from the surface of the bottom; (3) filter-feeders 'A', feeding on suspended detritus within the thin near-bottom water layer; (4) filter-feeders 'B', feeding on suspended detritus from the uppermost water layer; (5) passive suspension-feeders. Thus, Turpajeva classified benthic non-predatory invertebrates mainly according to their food source, taking into consideration their mode of feeding as well.

Turpajeva has clearly shown that the distribution of trophic groups depends on the type of bottom sediments. She found: (1) that the distribution of bottom communities in which swallowing and collecting deposit-feeders predominate (by weight) is positively correlated with small-sized sediment particles; (2) the distribution of communities in which filter-feeders 'A' prevail is positively correlated with medium-sized particles, while (3) the distribution of communities in which filter-feeders 'B' and passive suspension-feeders predominate is positively correlated with large-sized sediment particles.

Further investigations of benthic trophic groups was most distinctly elaborated by Sokolova (1956, 1960, 1964, 1965, 1969, 1972), who studied the distribution of the deep-sea macrobenthos. She has considerably simplified the classification of trophic groups of non-predatory invertebrates by reducing their number to three: (1) swallowing deposit-feeders feeding on detritus buried within sediments; (2) collecting deposit-feeders feeding on the surface of sediments; and (3) suspension-feeders that consume detritus suspended in water. The main characteristic used by Sokolova for distinguishing trophic groups is the food source of benthic animals, i.e. their ability to obtain food from one of three habitats: subsurface layers of sediments, the surface sediments and the near-bottom water layers.

It was distinctly shown by her that trophic groups in their distribution are closely connected with sedimentation conditions, on which depends whether the major part of the detritus is suspended in the near-bottom water layer, settled on the surface of the bottom or buried within the subsurface layer of the sediments. In establishing these relationships Sokolova was able to elaborate a scheme of spatial distribution of trophic groups for deep-sea benthos of the north-western part of the Pacific Ocean. She also formulated the concepts of the 'trophic zone' and 'trophic zonality'. The trophic zone is a certain area of the bottom occupied by the communities of the same trophic type, i.e. characterised by the predominance by weight of a certain trophic group. Each trophic zone is characterised by similar feeding conditions for bottom invertebrates, i.e. by a similar type of detritus distribution.

The dependence of the distribution of trophic groups on the conditions of sedimentation is closely connected with dependence of sedimentation on bottom topography; on the protrusions and elevations characterised by erosion or absence of sedimentation, zones where suspension-feeders predominate are developed, while in the depressions characterised by the accumulation of sediments, zones where deposit-feeders predominate are developed. Since from the intertidal zone down to the ultra-abyssal (hadal) zone the bottom is repeatedly curved forming successive protrusions and depressions, assemblages of these three trophic zones occur repeatedly.

Most recent investigations by Sokolova reveal the effect of the amount of organic matter in sediments, and of its composition, on the distribution of trophic groups and on the trophic structure of the benthos of the deep-sea, i.e. on weight ratios of trophic groups. Here, too, the application of her classification of trophic groups proved to be successful, as it made it possible to show the relationship between benthic trophic structure and productivity in the euphotic zone of the ocean, as well as the sedimentation rate. Both these factors determine the abundance and the degree of transformation of organic matter in sediments (Bordovsky, 1964), i.e. the abundance and nutritive value of detrital organic matter. This research resulted in the formulation of the concepts of eutrophic and oligotrophic structures of the deep-sea benthos.

The eutrophic structure is developed under conditions of sufficiently abundant supply of sedimented organic matter and a sufficiently high sedimentation rate, which results in burying the organic matter in a biochemically active state. Therefore conditions of intense sediment accumulation are favourable for the development of zones characterised by the predominance of deposit-feeders. Zones in which suspension-feeders prevail are formed only under the conditions of erosion or absence of sedimentation. Thus the predominance of alternate trophic groups depends only on bottom topography.

The oligotrophic structure of the benthos is developed under conditions of poor supply of organic matter and a slow rate of sedimentation, when the organic matter is buried in a considerably transformed state, i.e. when buried, it has no nutritive value for swallowing deposit-feeders. Therefore suspension-feeders prevail here, everywhere, irrespective of the bottom topography.

This approach to the classification of trophic groups made it possible to reveal the relationship of the benthic population to the character of distribution and degree of transformation of the organic matter, i.e. with the biogeochemical processes. It was found that similar types of accumulation and a similar degree of transformation of the organic matter determine similar development and similar weight ratios of trophic groups even under conditions which differ widely in other respects (Vinogradov, 1962, 1963, 1966). This clear dependence was disclosed because the trophic groups were classified strictly according to a single characteristic only, namely the source of food. It is natural that only non-predatory benthic invertebrates depend directly on the distribution of organic matter. The distribution of predatory invertebrates depends, in its turn, on the distribution of non-predatory invertebrates and should be investigated separately.

The achievements in the investigations of the deep-sea benthic fauna (which may appear to be remote from the subject of the present paper) are considered here because it is only in the deep-sea zone (on the ocean floor) that the oligotrophic structure of the benthos may be observed. It is only in the central parts of the oceans under subtropical halistases that conditions are suitable for the development of oligotrophy in its pure form. These conditions result from two types of zonality in the oceanic plane: the latitudinal zonality in the water productivity (in these tropical parts of the oceans the productivity is the lowest); and the circum-continental zonality in the sedimentation rate, the latter being also the lowest in these parts of the oceans due to the great remoteness from continental coasts (Bezrukov, 1964).

Since shelves are confined to the coastal area they all lie within the same circum-continental zone and therefore the oligotrophic structure of the benthos in its full development cannot be observed there, although, as will be shown below, its manifestation is an important feature of benthic populations in some regions.

Now let us consider the influence of the latitudinal zonality in the productivity of waters on the benthos of the shelves. The distributional pattern of trophic groups on the shelves in the sub-Arctic region has been investigated by a number of authors: in the Okhotsk Sea (Savilov, 1961; Vinogradov & Neyman, 1969; Neyman, 1969*a*, *b*, *c*, 1972), in the eastern Kamchatka and in the western portion of the Bering Sea (Kuznetsov, 1963, 1964), in the eastern part of the Bering Sea (Neyman, 1963; Semenov, 1964), in the Gulf of Alaska (Shevtsov, 1964; Semenov, 1965), in the Barents Sea (Kuznetsov, 1970), in the Baltic Sea (Luksenas, 1969), in the Newfoundland–Labrador region (Nesis, 1965); the distributional pattern of trophic groups of sub-Antarctic water structure was investigated on the Uruguay shelf (Semenov, 1969) and on the shelf of eastern New Zealand (Neyman, 1970).

Trophic zonality as observed on the shelves of these regions is basically the same as in the bathyal (200–2000 m) and abyssal (2000–6000 m) zones

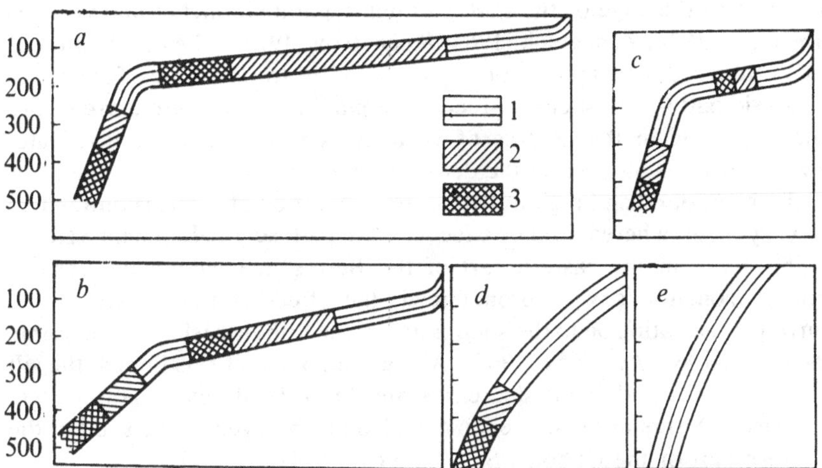

Fig. 13.1. Scheme of trophic zonality on the shelves with different bottom topography. (1) Suspension-feeders. (2) Collecting deposit-feeders. (3) Swallowing deposit-feeders.

of the north-western Pacific; a distinct relationship was established between the bottom topography, the accumulation of sediments and the distribution of trophic groups. Consequently it was possible to elaborate a scheme of dependence of the distribution of trophic zones on the shelf topography, i.e. on the width and steepness of the shelf. On the shelves of the Bering and Okhotsk Seas it was possible to show a direct dependence of the distribution of the benthos on organic carbon in the sediments (Neyman, 1961, 1963, 1972). On the shelves of the North Pacific the vertical trophic zonality is very distinct, caused by a very distinct vertical zonality in the distribution of the sediments (Gershanovich, 1964, 1965*a*, *b*; Lisitzin, 1966). The vertical zonality in sediment distribution in its turn is affected by the fact that in the North Pacific Ocean and in marginal seas the currents flow along the edges of the shelves (Arsenjev, 1967; Moroshkin, 1966). It is thus in the North Pacific Ocean that the relationship of the distribution of the trophic groups to the width and steepness of the shelves is so clear. On broad shelves there is a more or less pronounced zone of accumulation of silty sediments lying either in the middle of the shelf or close to its edge (see Fig. 13.1*a*, *b*). On narrow shelves this zone of accumulation is either very narrow, or it is shifted towards the upper portion of the slope, or it is absent entirely (see Fig. 13.1*c*, *d*, *e*).

According to Derjugin (1928, 1939), Derjugin & Somova (1941), Filatova & Zenkevich (1957), Brotskaja, Zdanova & Semenova (1962), the vertical tropic zonality is clearly developed in the White Sea, in the Kara Sea and in the Sea of Japan. In the Barents Sea, on the other hand, the currents flow across the edge of the shelf forming separate branches and vortices (Derjugin, 1924; Zenkevich, 1963; Kuznetsov, 1970). The distribution of the bottom sediments (and, consequently, of the trophic zones) forms here a mosaic pattern. Suspension-feeders prevail in areas where currents are intense, while in the centres of vortices deposit-feeders predominate. Strong currents also affect the distribution of trophic zones on the shelf of the Newfoundland region, where the proportion of suspension-feeders is very high whereas deposit-feeders are not so predominant (Nesis, 1965). Even in the eastern part of the Bering Sea, where the vertical trophic zonality is very pronounced, the effect of the branch of the current penetrating on to the shelf north of the Pribylov Isles is perceptible (Natarov, 1963; Arsenjev, 1967). Along the course of this branch the silt content of the sediments decreases and the role of suspension-feeders increases. A similar phenomenon can also be observed on the shelf of the western part of the Okhotsk Sea (Savilov, 1961).

Outside the limits of the marine sub-Arctic, it was possible for the author to investigate the distribution of the benthos on the shelves of the East China Sea, of the Great Australian Bight, of the western coasts of Australia and New Zealand, of western India and Burma (Gershanovich & Neyman, 1964; Neyman, 1965*a*, *b*, 1968, 1969*a*, *c*, 1970).

The shelf of the East China Sea is influenced by the waters of the Kuroshio current, i.e. of the waters of the subtropical gyre; the shelves of western and southern Australia and of western New Zealand are influenced by the waters of the subtropical gyres of the Southern Hemisphere, those of the Pacific and Indian Oceans.

The mean biomass of the benthos of these regions is one or two orders smaller than that on the shelves of the sub-Arctic region (Neyman, 1965*a*). This phenomenon can be regarded as similar to the decrease of the biomass of the benthos on the deep-sea floor to the south of latitude 30° N in the Pacific, i.e. eutrophic and oligotrophic regions are also encountered on the shelf. For the solution of this problem it was necessary to compare the distribution of trophic groups of the benthos on shelves situated in different water masses.

For the purpose of elimination of the effect of the relief on the trophic groups distribution in the course of this comparison, regions with most similar shelf topography were chosen, e.g. the continental shelf of the East China Sea and the continental shelf of the eastern part of the Bering Sea. These regions have many similar features: both seas are fully saline marginal seas separated from the ocean by arcs of islands with deep-water straits between them; the continental shelves are very wide, gently sloping, and abruptly changing to the precipitous slope (Gershanovich, 1963; Kotenev, 1963); and the outer parts of both shelves are influenced by currents penetrating from the ocean through the straits of the island arc.

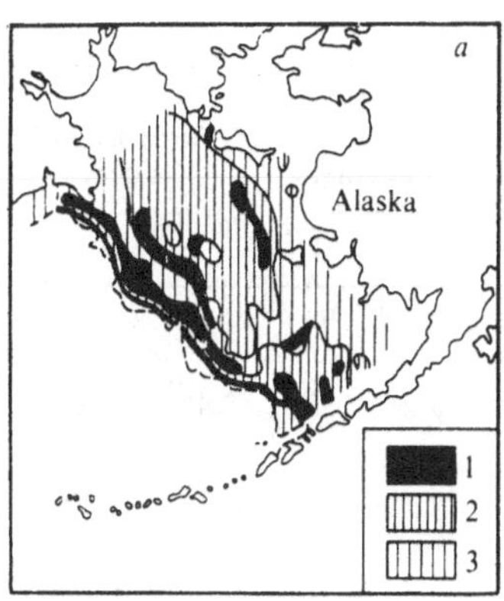

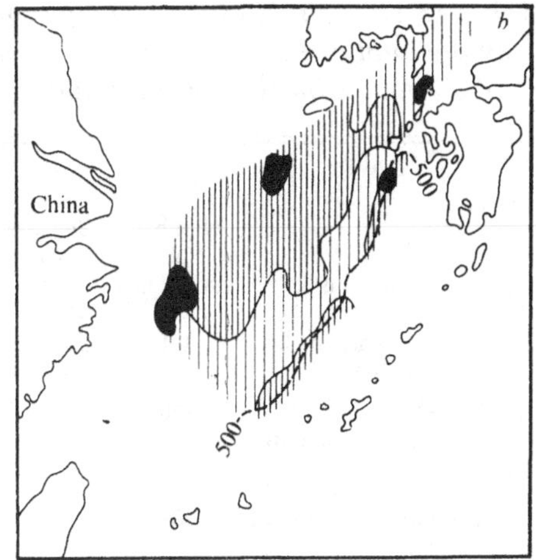

Fig. 13.2. The trophic zonality in the Bering Sea (*a*) and the East China Sea (*b*). (1) Suspension-feeders. (2) Collecting deposit-feeders. (3) Swallowing deposit-feeders.

The shelf of the eastern part of the Bering Sea is influenced by the waters of the sub-Arctic system (Natarov, 1963). The greatest part of the shelf of the East China Sea is influenced by the subtropical waters of the Kuroshio current, while on the boundary with the Yellow Sea there is a zone of mixing of the subtropical waters with those of the Yellow Sea (Gershanovich, 1963; Kharchenko, 1968). The waters of the east part of the Bering Sea are characterised by a high plankton biomass (Meshcheryakova, 1964). In the Kuroshio waters the biomass of the plankton is small, increasing somewhat in the zone of mixing with the waters of the Yellow Sea (Krylov, 1969).

The biomass of the benthos decreases abruptly towards the south of the Tzushima Strait (Neyman, 1965*a*). On the shelves of the Bering, Okhotsk and Japan Seas the biomass of the benthos is measured by hundreds g/m^2 (Zenkevich, 1963), on the shelf of the East China Sea the mean biomass of the benthos is only 15 g/m^2 (Gershanovich & Neyman, 1964).

On most of the East China Sea shelf the biomass of the benthos is small; separate patches of large biomass are formed only by suspension-feeders; collecting deposit-feeders are scanty, while swallowing deposit-feeders are entirely absent on many stations (see Fig. 13.2). It should be emphasised that it is not the prevalence of suspension-feeders that is considered here (and that in this case is not surprising, since the outer part of the shelf is covered by sandy grounds), but the fact that deposit-feeders, particularly swallowing feeders, are practically absent in this zone. On

Table 13.1. *The role of the trophic groups of non-predatory invertebrates in the total benthos. (1) per cent of the total mean biomass; (2) per cent of stations with the given trophic group*

Region[a]	Number of stations	Deposit-feeders: Swallowing 1	Swallowing 2	Collecting 1	Collecting 2	Suspension-feeders 1	Suspension-feeders 2
I. The east part of the Bering Sea	261	22	98	49	97	29	98
II. The shelf of the western Kamtshatka	140	15	98	40	98	45	98
III. The East China Sea	60	13	56	42	92	45	92
IV. The Great Australian Bight	101	4	26	16	68	80	92

[a] Regions I, II are situated in the subpolar waters; regions III, IV in the periphery of the subtropical waters; regions I, III have a large platform shelf; regions II, IV have a narrow rugged shelf.

similar grounds in the zone of predominance of suspension-feeders in the waters of the sub-Arctic region swallowing deposit-feeders are always present in perceptible numbers (Kuznetsov, 1963; Neyman, 1963, 1972; Filatova & Barsanova, 1964).

All these specific features of the benthic composition characterise the benthos of the shelf of the East China Sea as somewhat oligotrophic.

Comparing the Bering Sea with the East China Sea (Fig. 13.2), it is apparent that they resemble each other in the distribution of collecting deposit-feeders. In both regions they form two distinct zones of predominance, one in the middle part of the shelf, another on the upper part of the slope; in the East China Sea there persist the same distribution patterns of collecting deposit-feeders as those inherent in broad gently sloping shelves (see Fig. 13.1) with a conspicuous vertical trophic zonality.

On the outer part of the shelf of the East China Sea suspension-feeders are significantly more abundant than in the eastern part of the Bering Sea, there are differences in the development of the swallowing deposit-feeders. In the eastern part of the Bering Sea they occur everywhere and form a distinct zone of predominance on the outer part of the shelf. In the East China Sea they are absent at many stations and there is no zone in which they predominate (see Table 13.1).

These differences do not in fact allow a decisive conclusion about a real oligotrophy of the benthos on the shelf of the East China Sea, since there persists though with some disturbances, the trophic zonality inherent in wide gently sloping shelves. Evidently this is due to the fact that on the shelf, i.e. in the close vicinity of the continent, the rate of sediment accumulation and the water productivity does not fall so low as to determine the formation of such oligotrophic regions as those encountered in the deep-sea floor (Sokolova, 1964, 1965, 1966, 1969, 1972; Sokolova & Neyman, 1966).

In the Great Australian Bight (Neyman, 1965*a*, *b*, *c*, 1970) the ratio of suspension-feeders is still greater, while the proportion of deposit-feeders, particularly swallowing feeders, is still smaller, than in the East China Sea (see Table 13.1), i.e. the features of oligotrophy are still more conspicuous, as compared to the latter sea. But a considerable part of these differences can be attributed to the differences in bottom topography and the characteristics of the bottom sediment distribution associated with the bottom topography. Therefore data pertaining to the shelves of Kamtchatka and the western part of the Bering Sea, similar in bottom topography, were used for the comparison (Kuznetsov, 1963, 1964; Vinogradov & Neyman, 1969; Neyman, 1972). All these narrow shelves can be assigned to type *c* or *d* (see Fig. 13.1). Although on the narrow shelves of the northern Pacific suspension-feeders play an important role and a great part of the shelves is occupied by the zone of their predominance, still deposit-feeders, including swallowing ones, are distributed all over the shelf, being sometimes fairly abundant. They are obligatory components of each community (Kuznetsov, 1963; Filatova & Barsanova, 1964; Neyman, 1972).

The effect of the water productivity on the distribution of the benthos and its trophic groups is clearly perceptible on the shelves of the northern part of the Indian Ocean. Thus almost all over the shelf of western India influenced by highly productive waters of the Arabian Sea (Vinogradov & Voronina, 1962; Elizarov, 1968), the benthos is fairly rich, the biomass at certain areas exceeding 100 g/m^2. According to Sokolova (1972), the benthos of the deep-sea floor of the Arabian Sea is eutrophic. Belyaev & Vinogradova (1961), as well as Sokolova & Pasternak (1962) have noted a large benthos biomass on the deep-sea floor of the Arabian Sea, quite comparable to that of the deep-sea floor of the highly productive northern Pacific.

The eutrophic nature of the benthos on the shelf of western India persists only at depths less than 70–100 m. At greater depths the trophic structure changes due to the effect of the waters with an oxygen deficit that reach the shelf, at least in certain seasons. This affects swallowing deposit-feeders, since silty sediments with a high content of organic matter aggravate the deficit of oxygen. The trophic structure of the benthos of the greater depths of these shelves proves to be similar to that of poor oligotrophic regions. However, its causes prove to be directly opposite.

Therefore it is proposed to designate the structure of the benthos of the lower part of the western India shelf as hyper-eutrophic.

The effect of the oxygen deficit on the relationship of the deposit-feeders distribution to the content of organic matter in bottom sediments is also apparent in the Baltic Sea. Here the phenomenon is due to the fact that the maximum content of organic matter in the sediments is confined to depressions, where the waters are considerably impoverished in oxygen or even practically devoid of it, and therefore the usual positive correlation of the deposit-feeder distribution to that of the organic matter in the sediments is observed only in shallow waters, where there is no oxygen deficit (Luksenas, 1969).

Distinct features of oligotrophy are observed in the benthos of the western India shelf south of 5–10° N influenced by the poorly productive equatorial waters (as compared to those of the Arabian Sea – Elizarov, 1968). Here the biomass of the benthos does not exceed 5 g/m^2. The trophic structure of this benthos is characterised by features of oligotrophy both on the shelves of the Bay of Bengal and the Andaman Sea. Sokolova & Pasternak (1962, 1963) pointed out the extremely small biomass of the benthos of the deep-sea floor of the Bay of Bengal. On the shelves of the Bay of Bengal and the Andaman Sea the eutrophic benthos structure was detected only in the close vicinity of the estuaries of the Ganges and the Irrawady rivers where this structure is coupled with a fairly conspicuous vertical trophic zonality (Neyman, 1969*a*).

Thus the character of the distribution and the relative role of the benthic trophic groups on the shelves, just as on the deep-sea floor, are determined by the latitudinal zonality, or, more exactly, by the productivity of waters dependent on the latitudinal zonality; the development of all three trophic groups being in accordance with the bottom topography. An increase of the relative abundance of suspension-feeders and a considerable decrease of the relative abundance of the swallowing deposit-feeders are characteristic of the benthos of the poorly productive zones irrespective of bottom topography.

As for the differences between the benthos of the shelf and that of the deep-sea floor situated under waters of equal productivity, they depend on the fact that in the system of circum-continental zonality the shelves are situated in one zone – the coastal zone, whereas the deep-sea floor, according to the distances from the coast, is subdivided into several subzones (Sokolova, 1969). The greatest part of the terrigenous sediments is deposited in the coastal zone, which inevitably affects the conditions of burying of organic matter (Bordovsky, 1964). Therefore on the shelves, even in very poor regions, the oligotrophic benthos structure is never expressed in its pure form, the trophic zonality inherent in the given bottom topography of the shelf in the euphotic regions being always retained to some extent.

In regions with eutrophic benthic structure the influence of different local factors is but slightly perceptible. Thus on the shelf of western India the effect of the discharge of the Indus on the distribution of the benthos is practically imperceptible. On the other hand, in regions with oligotrophic benthos structure the effect of any local factors is conspicuous. For instance, the regions influenced by the discharge of the Ganges and Irrawady can be clearly seen because of their relative richness and eutrophy. Apparently the differences in the degree of oligotrophy of the East China Sea and the Great Australian Bight not only depend on the bottom topography, but are also determined by the influence of the great discharge of the Yangtze.

Thus during recent years data characterising the benthic populations of vast shelf areas have been obtained by Soviet investigators, using the method of classification of trophic groups and of the establishment of the zones of their predominance. In the first approximation the flux of energy through a benthic community is directly proportional to the biomass of the community; the amount of energy flowing through some component of a community being likewise correlated with the proportion of this component in the total biomass of the community. Therefore a conclusion can be drawn that in the highly productive regions the energy flux through a benthic community is distributed among trophic groups depending on bottom topography. In the oligotrophic regions it flows mainly through the group of suspension-feeders. In this case the turn-over must be quick, since the organic matter does not reside for any length of time in the sediments awaiting consumption.

It should be mentioned in conclusion, that the effect of the productivity of waters is the same irrespective of the species composition of the fauna; thus, the eutrophic structure is inherent in the benthos of both the Bering Sea and the Arabian Sea.

The causes of the oligotrophy on the deep-sea floor are more or less clear, but not for the oligotrophy of the shelves. On the shelves, neither the sedimentation rate, nor the water productivity can be as low as on the deep-sea floor of the central parts of the ocean. Therefore the oligotrophy of the benthos of the shelves must have other causes. On the deep-sea floor the oligotrophy is formed at low water temperature. In contrast, on the shelves the oligotrophy was found only at high bottom temperature. So in the Kara Sea, where the water productivity is very low (Zenkevich, 1963) outside of the brown sediment zone the benthos has a clearly eutrophic structure.

It is possible to suppose that at high bottom temperatures and low productivity in the euphotic zone the organic matter on the bottom is rapidly transformed and when buried has a low nutritive value. It is possible also that in such conditions the high rate of terrigenous mineral sedimentation results in a decrease in the concentration of the organic matter in the sediments.

The boundaries between the regions with different benthos structure coincide with zoogeographic boundaries. The boundary between the tropical and Arcto-Boreal regions in the Tsushima Strait coincides with the boundary between eutrophic and oligotrophic regions. The same was found on the shelf of western India – the boundary at 5–10° N between the areas of some species of the western and eastern parts of the Indian Ocean.

The zoogeographical boundaries run along boundaries of water masses (Beklemishev, Neyman, Parin & Semina, 1972). Water masses differ in productivity, which influences the trophic structure of the benthos. This may explain why the zoogeographical boundaries coincide with the boundaries between regions differing in trophic structures.

References

Arsenjev, V. S. (1967). *Currents and water masses of the Bering Sea.* (In Russian). Moscow: Nauka.

Beklemishev, C. W., Neyman, A. A., Parin, N. V. & Semina, H. I. (1972). Le biotope dans le milieu marin, *Marine Biology*, **15**, 57–73.

Belyaev, G. M. & Vinogradova, N. G. (1961). Quantitative distribution of bottom fauna in the northern part of the Indian Ocean. (In Russian). *Doklady Akademiia Nauk SSSR*, **158**(5), 132–6.

Bezrukov, P. L. (1964). Zonality and distribution of sediments in the oceans. (In Russian). In *Problems of modern geography; contributions of Soviet geographers at the 20th international congress of geography* (*London, 1964*), pp. 246–8. Moscow: Nauka.

Bordovsky, O. K. (1964). *Accumulation and transformation of organic matter in marine sediments.* (In Russian). Moscow: Nedra.

Brotskaja, V. A., Zdanova, N. N. & Semenova, N. L. (1962). The bottom fauna of the Velilaja Salma Strait and adjacent parts of the Kandalaksha Bight of the White Sea. (In Russian). *Trudy Kandalakshskogo Gosudarstvennogo Zapovednika, Petrozavodsk*, **5**, 28–42.

Derjugin, K. M. (1924). The Barents Sea along the Kola Meridian. (In Russian). *Trudy Instituta Izuchenija Severa, Leningrad*, **19**, 1–102.

Derjugin, K. M. (1928). The fauna of the White Sea and the conditions of its existence. (In Russian). *Issledovanija Morei SSSR, Leningrad*, **7–8**, 1–203.

Derjugin, K. M. (1939). Zones and communities of the Peter the Great Bight (Japan Sea). (In Russian). In *The scientific works, dedicated to N. M. Knipovich*, pp. 115–43. Moscow: Pishchepromisdat.

Derjugin, K. M. & Somova, M. N. (1941). Materials on the quantitative evaluation of the benthos of the Peter the Great Bight (Japan Sea). (In Russian). *Issledovanija Morei SSSR, Leningrad*, **11**, 13–36.

Elizarov, A. A. (1968). Preliminary results of oceanographic investigations of the western coast of India. (In Russian). *Proceedings, All-Union Research Institute of Marine Fisheries and Oceanography, Moscow*, **64**, 94–101.

Filatova, Z. A. & Barsanova, N. G. (1964). The bottom communities of the western part of the Bering Sea. (In Russian). *Trudy Instituta Okeanologii, Moscow*, **64**, 112–205.

Filatova, Z. A. & Zedkevich, L. A. (1957). The quantitative distribution of the bottom fauna of the Kara Sea. (In Russian). *Trudy Vsesojusnogo Hydrobiologicheskie Obshchestva, Moscow*, **8**, 105–127.

Gershanovich, D. E. (1963). Oceanological studies in the north part of the Pacific Ocean. (In Russian). *Okeanologija*, **3**(6), 1114–17.

Gershanovich, D. E. (1964). Bottom sediments in the central and eastern Bering Sea. (In Russian). *Proceedings, All-Union Research Institute of Marine Fisheries and Oceanography, Moscow*, **53**, 31–82.

Gershanovich, D. E. (1964). New data on the accumulation of organic matter in the recent sediments of the North Pacific. (In Russian). *Okeanologia*, **5**(2), 228–333.

Gershanovich, D. E. (1965). Recent bottom sediment strength and sedimentation rate in the Bering Sea. (In Russian). *Proceedings, All-Union Research Institute of Marine Fisheries and Oceanography, Moscow*, **57**, 261–70.

Gershanovich, D. E. & Neyman, A. A. (1964). Sediments and bottom fauna of the East China Sea. (In Russian). *Okeanologia*, **4**(6), 1064–70.

Gordeev, V. D. (1945). The prismatic bottom sampler. (In Russian). *Investija TNIRO*, **19**, 45–58.

Kharchenko, A. M. (1968). The currents and the water masses of the East China Sea. (In Russian). *Okeanologia*, **8**(1), 38–48.

Kotenev, B. N. (1963). On the bottom geomorphology of the East China Sea. (In Russian). *Vestnik Moskovskogo Universiteta, Geography Series, Moscow*, **5**, 13–22.

Krylov, V. V. (1969). Distribution of plankton in the East China Sea. (In Russian). *Proceedings, All-Union Research Institute of Marine Fisheries and Oceanography, Moscow*, **65**, 198–215.

Kuznetsov, A. P. (1963). *The bottom invertebrate fauna of the Kamtshatka and North Kuril Islands Water of the Pacific Ocean.* (In Russian). Moscow: Isdatelstvo Akademii Nauk SSSR.

Kuznetsov, A. P. (1964). The distribution of the bottom fauna of the western Bering Sea in trophic zones and some general aspects of trophic zonality. (In Russian). *Trudy Instituta Okeanologii, Moscow*, **69**, 6–72.

Kuznetsov, A. P. (1970). The trophic groups and trophic zones in the Barents Sea benthos. (In Russian). *Trudy Instituta Okeanologii, Moscow*, **88**, 5–80.

Lisitzin, A. P. (1966). *Processes of recent sedimentation in the Bering Sea.* (In Russian). Moscow: Nauka.

Luksenas, Ju. K. (1969). Trophic zones and communities of the benthos of the south Baltic Sea. (In Russian). *Okeanologia*, **9**(6), 1069–75.

Meshcheryakova, I. M. (1964). The quantitative distribution of plankton in the southeastern Bering Sea. (In Russian). *Proceedings, All-Union Research Institute of Marine Fisheries and Oceanography, Moscow*, **49**, 141–50.

Moroshkin, K. V. (1966). *Water masses of the Okhotsk Sea.* (In Russian). Moscow: Nauka.

Natarov, V. V. (1963). On the water masses and currents of the Bering Sea. (In Russian). *Proceedings, All-Union Research Institute of Marine Fisheries and Oceanography, Moscow*, **48**, 111–33.

Nesis, K. N. (1960). The changes of the bottom fauna of the Barents Sea under the influence of the changes of the hydrological regime. (In Russian). In *Soviet investigations in the North European Sea*, pp. 129–37. Moscow: Rybnoje Khozaistvo.

Nesis, K. N. (1965). Communities and biomass of the benthos in Newfoundland-Labrador region. (In Russian). *Proceedings, All-Union Research Institute of Marine Fisheries and Oceanography, Moscow*, **57**, 453–89.

Neyman, A. A. (1961). Some regularities of the quantitative distribution of the benthos in the east part of the Bering Sea. (In Russian). *Okeanologia*, **1**(2), 294–304.

Neyman, A. A. (1963). Quantitative distribution of benthos and food supply of demersal fish in the eastern part of the Bering Sea. (In Russian). *Proceedings, All-Union Research Institute of Marine Fisheries and Oceanography, Moscow*, **48**, 145–206.

Neyman, A. A. (1965*a*). Some regularities of the quantitative distribution of the benthos on the North Pacific shelves. (In Russian). *Proceedings, All-Union Research Institute of Marine Fisheries and Oceanography, Moscow*, **57**, 447–52.

Neyman, A. A. (1965*b*). Some data on the quantitative distribution of the benthos on the Australian shelves. (In Russian). *Okeanologia*, **5**(1), 82–5.

Neyman, A. A. (1965*c*). Quantitative distribution of the benthos of the West Kamtchatka shelf and some methods of its investigation. (In Russian). *Okeanologia*, **5**(6), 1052–9.

Neyman, A. A. (1966). La répartition des groupes trophiques du benthos sur le plateau continental. In *Second international oceanography congress (30 May–9 June 1966), abstracts of papers*, pp. 270–1. Moscow: Nauka.

Neyman, A. A. (1968). Characteristics of benthic populations on the shelf of western and southern Australia. (In Russian). *Proceedings, All-Union Research Institute of Marine Fisheries and Oceanography, Moscow*, **64**, 204–10.

Neyman, A. A. (1969*a*). Some data on the benthos of the North Indian Ocean shelves. (In Russian). *Okeanologia*, **9**(6), 1071–7.

Neyman, A. A. (1969*b*). Benthos of the West Kamtchatka shelf. (In Russian). *Proceedings, All-Union Research Institute of Marine Fisheries and Oceanography, Moscow*, **65**, 223–33.

Neyman, A. A. (1969*c*). On the distribution of the benthic trophic groups on the shelf in different geographic zones. (In Russian). *Proceedings, All-Union Research Institute of Marine Fisheries and Oceanography, Moscow*, **65**, 282–96.

Neyman, A. A. (1970). Quantitative distribution of the benthos on the shelves of the Great Australian Bight and New Zealand. (In Russian). *Okeanologia*, **10**(3), 517–20.

Neyman, A. A. (1972). *Species composition of the bottom communities of the shelf of the northeastern part of the Okhotsk Sea*. (In Russian). Moscow: Isdanie Otdela Nauchno-Tekhniceskoi Informacii VNIRO.

Savilov, A. I. (1961). Ecological characteristics of the bottom invertebrate communities of the Okhotsk Sea. (In Russian). *Trudy Instituta Okeanologii, Moscow*, **46**, 3–84.

Semenov, V. N. (1964). Quantitative distribution of the benthos of the south-eastern Bering Sea. (In Russian). *Proceedings, All-Union Research Institute of Marine Fisheries and Oceanography, Moscow*, **53**, 177–86.

Semenov, V. N. (1965). The quantitative distribution of the bottom fauna of the shelf and the upper continental slope in the Gulf of Alaska. (In Russian). *Proceedings, All-Union Research Institute of Marine Fisheries and Oceanography, Moscow*, **58**, 49–78.

Semenov, V. N. (1969). Biomass distribution and trophic zonality of the benthos of the Uruguay shelf. (In Russian). *Trudy Molodykh Uchenykh VNIRO, Moscow*, **1**, 74–83.

Shevtsov, V. V. (1964). Quantitative distribution and trophic groups of benthos in the Gulf of Alaska. (In Russian). *Proceedings, All-Union Research Institute of Marine Fisheries and Oceanography, Moscow*, **53**, 161–76.

Sokolova, M. N. (1956). On the distribution pattern of the deep-sea benthos. (In Russian). *Doklady Akademiia Nauk SSSR*, **110**(4), 692–5.

Sokolova, M. N. (1960). Distribution of the groupings (biocoenoses) of bottom fauna in the deep trenches of the northwest part of the Pacific Ocean. (In Russian). *Trudy Instituta Okeanologii, Moscow*, **34**, 21–59.

Sokolova, M. N. (1964). Some regularities of the distribution of trophic groups of the deep-sea benthos. (In Russian). *Okeanologia*, **4**(6), 1079–88.

Sokolova, M. N. (1965). On varying distribution of deep-sea benthic trophic groups depending on varying accumulation of sediments. (In Russian). *Okeanologia*, **5**(3), 498–506.

Sokolova, M. N. (1969). The distribution of the deep-sea bottom invertebrates depending on their mode of feeding and food conditions. (In Russian). In *The Pacific Ocean*, ed. V. G. Kort, vol. 7 (Biology of the Pacific Ocean) book 2 (The Deep-Sea Bottom Fauna, Pleuston), pp. 182–201. Moscow: Nauka.

Sokolova, M. N. (1972). Trophic structure of the deep-sea macrobenthos. *Marine Biology*, **16**, 1–12.

Sokolova, M. N. & Neyman, A. A. (1966). The trophic groups of the bottom fauna and regularities of their distribution. (In Russian). In *Ekologia Vodnykh Organismov*, pp. 42–9. Moscow: Nauka.

Sokolova, M. N. & Pasternac, F. A. (1962). Quantitative distribution of the bottom fauna in the north part of the Arabian Sea and the Gulf of Bengal. (In Russian). *Doklady Akademiia Nauk SSSR*, **144**(5), 645–8.

Sokolova, M. N. & Pasternac, F. A. (1963). Quantitative distribution and trophic zonality of the bottom fauna in the Gulf of Bengal and Andaman Sea. (In Russian). *Trudy Instituta Okeanologii, Moscow*, **64**, 271–96.

Turpajeva, E. P. (1949). The feeding of some bottom invertebrates in the Barents Sea. (In Russian). *Zoologicheskyi Zhurnal*, **27**(6), 503–12.

Turpajeva, E. P. (1953). Feeding and food grouping of marine bottom invertebrates. (In Russian). *Trudy Instituta Okeanologii, Moscow*, **7**, 259–99.

Turpajeva, E. P. (1954). The types of the sea-bottom biocoenoses and the dependence of their distribution on abiotic factors. (In Russian). *Trudy Instituta Okeanologii*, **2**, 100–141.

Turpajeva, E. P. (1957). The feeding relations between the dominant species in the sea-bottom biocoenoses. (In Russian). *Trudy Instituta Okeanologii*, **20**, 150–168.

Vinogradov, L. G. (1962). Complexes and biocoenoses. (In Russian). In *Voprosy Ekologii* (*Materials of the 4th Ecology Conference*), Vol. 4, pp. 15–17. Kiev: Izdatelstvo Kievskogo Universiteta.

Vinogradov, L. G. (1963). Marine bottom biocoenoses and the application of the data on their distribution in fish scouting operations. (In Russian). *Proceedings, All-Union Research Institute of Marine Fisheries and Oceanography, Moscow*, **48**, 135–44.

Vinogradov, L. G. (1966). Les rapports des groupes trophiques, des complexes zoogéographiques et des biocoenoses benthiques marins. In *Second International Oceanography Congress* (*Moscow, 30 May–9 June 1966*), *Abstracts of Papers*, pp. 381–2. Moscow: Nauka.

Vinogradov, L. G. & Neyman, A. A. (1965). On the distribution of zoogeographical complexes of bottom invertebrates in the eastern Bering Sea. (In Russian). *Proceedings, All-Union Research Institute of Marine Fisheries and Oceanography, Moscow*, **58**, 45–9.

Vinogradov, L. G. & Neyman, A. A. (1969). Bottom fauna of the shelf of the eastern part of Okhotsk Sea and some features of the Kamtshatka King Crab. (In Russian). *Okeanologia*, **9**(2), 329–40.

Vinogradov, M. E. & Voronina, N. M. (1962). Some features of the distribution of the zooplankton of the north part of the Indian Ocean. (In Russian). *Trudy Instituta Okeanologii*, **53**, 44–78.

Zenkevich, L. A. (1963). *The Biology of the Seas of URSS*. (In Russian). Moscow: Izdatelstvo Akademii Nauk SSSR.

14. Studies of trophic relationships in bottom communities in the southern seas of the USSR

E. A. YABLONSKAYA

Introduction

The Azov, Caspian and Aral Seas have long been famous for heavy catches of such valuable species of fish as white sturgeon, stellate sturgeon, bream, carp, roach, etc. The extensive areas of productive spawning grounds on flood plains and in deltas of rivers, together with the rich feeding grounds in the seas ensure the reproduction of abundant stocks of these species. Most valuable species of fish from these water bodies (*Acipenseridae*, *Cyprinidae*) feed on bottom and off-bottom invertebrates. The same food species are used by the *Gobiidae* and by other benthic fishes which although unimportant for fisheries (except for *Neogobius melanostomus* in the Azov Sea), are favourite food species of many predators, for example, *Huso huso*, *Lucioperca lucioperca*, *Acipenser stellatus*, *Aspius aspius* and others.

In recent decades, due to unfavourable climatic conditions and some anthropogenic factors, fish production in the Azov, Aral and Caspian Seas has decreased. It has now become necessary to take certain practical measures aimed at restoring the fish and invertebrate faunas in these bodies of water. At the same time along with studies of the feeding habits of the fish, it is also necessary to investigate the feeding habits and food relations of the benthic organisms associated with them.

The composition of the bottom fauna

The characteristic element of the populations in the Azov, Caspian and Aral Seas is a peculiar relict fauna, the so-called ancient autochthonous complex representing a fragment of the marine tertiary fauna. It is replaced to a considerable extent by euryhaline organisms originating in fresh waters and in the Mediterranean Sea, as well as by immigrants from the Arctic basin (Zenkevich, 1947, 1963). Elements of different origin are unequally represented in faunas of the Caspian, Aral and Azov Seas. Thus according to Mordukhay-Boltovskoy (1960) the Mediterranean elements predominate (76% of species) in the Azov Sea and also in the benthos of the Azov Sea (Vorobyev, 1949).

Since temperature, salinity, depth and bottom-type are not significantly variable throughout the Azov Sea, the species composition of the benthos is also somewhat similar in different parts of it; biocoenoses differ in the quantitative ratios of the same species rather than in the qualitative composition of the populations. According to Vorobyev (1949), the dominant and characteristic species in the typical biocoenosis of the Azov Sea account for only 27% of the total species, but for 99% of the total biomass. These abundant species of bottom invertebrates play a leading role in the trophic dynamics and productivity of the Azov Sea. The benthonic biocoenoses of the Aral Sea are also oligomixic (Zenkevich, 1947; Yablonskaya, 1960).

Organisms originating in the Caspian Sea, molluscs in particular, predominate in the benthos of the Aral Sea and until recently constituted about two-thirds of its biomass (Yablonskaya, 1960). Recently certain mysids from the Caspian Sea have been introduced into the Aral Sea, but only *Paramysis lacustris* and *P. intermedia* have been acclimatized. The second important group of organisms in the Aral Sea includes species which are widely distributed in fresh and brackish inland water bodies; in the bottom fauna these are the larvae of insects and oligochaetes. The latter are of minor importance, their biomass being negligible. Of the insect larvae *Chironomus behningii* is the most abundant.

Species of Mediterranean–Atlantic origin make up the third element of the invertebrate fauna in the Aral Sea. These are widely distributed in the Caspian Sea, in the Azov and Black Seas, the Mediterranean Sea and the Atlantic. Of the bottom invertebrates, two species will be mentioned here: *Cerastoderma lamarcki* and *Bowerbenkia imbricata.* Recently the Mediterranean community of bottom invertebrates in the Aral Sea has been augmented by the introduction of *Palaemon elegans* Rathke, *Abra ovata* Phil., and *Nereis diversicolor* O.F. Muller. Before this introduction the main component of the benthonic communities consisted of a few species of molluscs (*Adacna*, *Dreissena*, *Cerastoderma*, *Caspiohydrobia*) and larvae of Chironomidae. Other insects, and the only representative of higher crustaceans *Dikerogammarus aralensis* were of minor importance (Yablonskaya, 1960).

Autochthonous brackish-water fauna, endemic for the Ponto-Caspian basin, is very well represented in the Caspian Sea. In terms of the number of species, this fauna predominates over the other three faunal groups (Mediterranean, Arctic and freshwater). Whole genera and even families of Amphipoda, Mysidacea, Cumacea and Cardiidae consist entirely or almost entirely of autochthonous species of Caspian origin (Mordukhay-Boltovskoy, 1960).

Until recently (in the first quarter of this century) the ancient relicts were dominant in the benthos of the Caspian Sea in terms of the number of species, constituting also the bulk of the biomass of bottom invertebrates

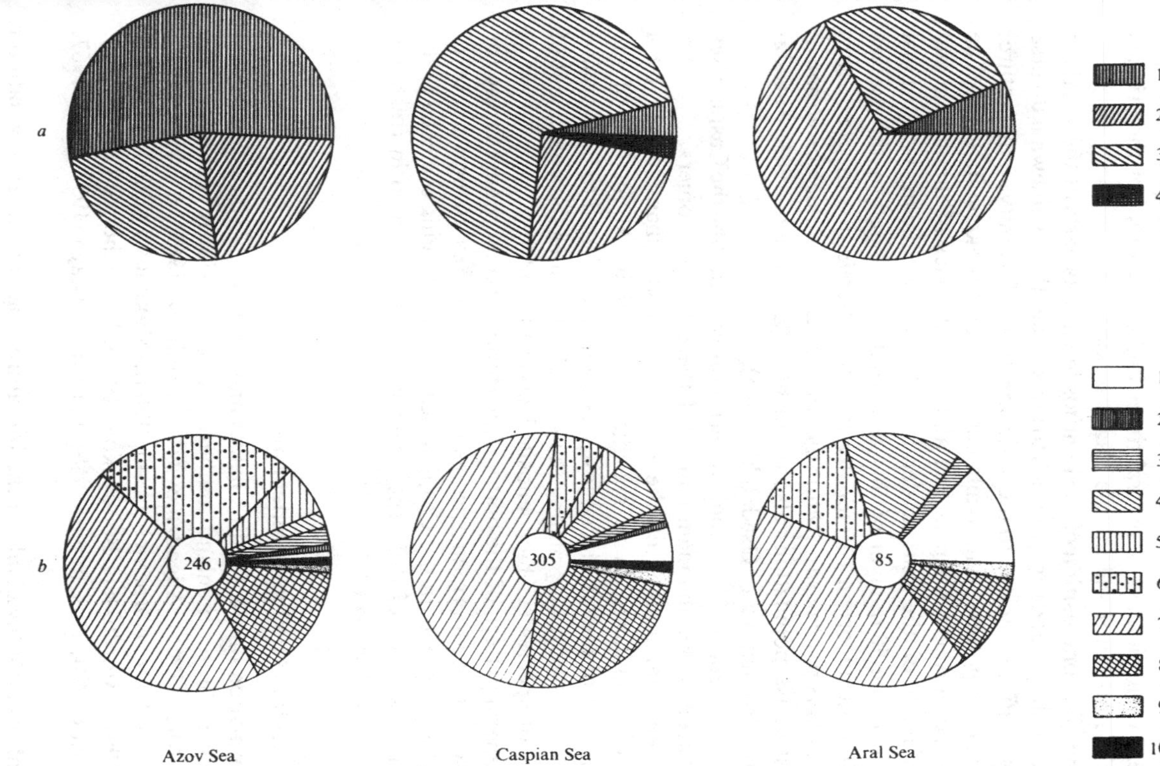

Fig. 14.1. The composition of bottom fauna from the Azov, Caspian and Aral Seas by zoogeographical (*a*) and systematic (*b*) groups. (*a*). 1, The Atlantic–Mediterranean fauna; 2, fauna of freshwater origin; 3, autochthonous fauna in the Caspian Sea; 4, Arctic fauna. (*b*). 1, Protozoa; 2, Porifera; 3, Coelenterata; 4, Platyhelminthes; 5, Nemathelminthes; 6, Annelida; 7, Arthropoda; 8, Mollusca; 9, Bryozoa; 10, others. The number of species is shown in the centres of the circles.

(Zenkevich, 1947; Mordukhay-Boltovskoy, 1960; Zenkevich & Zevina, 1969). In the last three or four decades, due to both incidental penetration and planned introduction, the specific diversity of the Mediterranean complex in the bottom fauna of the Caspian Sea has increased, and furthermore these new species are now widely distributed all over the sea. At present their biomass is several times higher than that of the original fauna in the northern and middle Caspian Sea.

The specific composition of the benthos in various parts of the sea from brackish shallow waters in estuaries to abyssal depths is somewhat diverse and shows differences due to the wide spectrum of salinity, temperature, depth, type of bottom, and availability of food (Zenkevich, 1947). The fauna in the northern Caspian Sea and coastal zones of the middle and southern parts of the sea, at depths of less than 100 m, is the most diverse. It is the 50 m layer of this zone that is inhabited by the species of Mediterranean origin (*Mytilaster lineatus*, *Nereis diversicolor*, *Abra ovata*, *Balanus improvisus*, *Rhithropanopeus harrisii*, *Palaemon elegans*). The most dense populations also occur here (Zenkevich, 1963; Aligadzhiev, 1965; Romanova & Osadchikh, 1965).

Of the 313 species of benthonic invertebrates inhabiting the Caspian Sea only 100 seem to be abundant and 11 of these are newcomers.

The deep communities of the Caspian Sea are characterized by a considerable similarity as to specific composition. About half of the species in five of the six biocoenoses examined are found to be the same (Aligadzhiev, 1965); only among the oligochaetes does the coefficient of similarity of species per biocoenosis decrease to 20%.

The general pattern of the bottom invertebrate fauna changes from the dominance of Mediterrean species in the Azov Sea to prevalence of Ponto-Caspian relicts in the Caspian Sea, the latter being replaced in the Aral Sea by species of different origin which are common in fresh and brackish water bodies (Fig. 14.1).

Feeding habits of bottom invertebrates

The information on the feeding habits and food composition of the benthonic invertebrates in inland southern seas of the USSR comes from the work of Aliev (1965), Aligadzhiev (1965), Babaev (1965), Briskina (1952), Vorobyev (1949), Karpevich (1946, 1960), Maximova (1958), Malinovskaya (1961), Mordukhay-Boltovskoy (1960), Reznichenko (1957, 1958), Romanova (1963), Husainova (1958) and Yablonskaya (1952, 1964, 1967, 1971).

The information available in the literature, together with new data on the feeding habits of the most abundant species of molluscs, crustaceans and worms collected and studied in 1966–70, is presented here.

Table 14.1. (I) *Number of species examined.* (II) *Number of stomachs examined.* (III) *Number of experiments with live animals*

Groups	Azov Sea I	Azov Sea II	Azov Sea III	Caspian Sea I	Caspian Sea II	Caspian Sea III	Aral Sea I	Aral Sea II
Polychaeta	2	28	—	1	629	108	1	10
Cirripedia	1	7	—	—	—	—	—	—
Mysidacea	2	489	21	—	—	—	—	—
Cumacea	1	29	—	—	—	—	—	—
Gammaridae	2	20	—	19	812	193	—	—
Corophiidae	1	19	—	6	530	—	—	—
Decapoda	1	—	8	1	—	46	—	—
Chironomidae	—	—	—	—	—	—	1	37
Bivalvia	4	136	52	9	981	218	2	51
Gastropoda	1	—	44	—	—	—	—	—
Total	15	728	125	36	2952	565	4	98

Materials and methods

The stomach contents of 52 benthic species from the Caspian, Aral and Azov Seas have been analysed. The feeding habits of some species were studied experimentally; different food species were offered to the experimental animals to reveal their food preferences and food consumption rates. The number of species and individuals examined is shown in Table 14.1.

Samples of animals were collected by bottom dredges and preserved in formalin. The treatment in the laboratory included selection of specimens, identification of species, size measurements and analysis of stomach contents. At the same time observations were carried out on live animals kept in the aquarium.

The anterior portions of digestive tracts were analysed. The contents of the intestines taken from a group of animals of the same size were squeezed into a drop of filtered water pipetted into the depression on a slide. The drop was then put into a pycnometer, the suspension was thoroughly stirred and then filtered through membrane filter No. 3. The filter was dried, a segment of the filter (¼) was cut off, dyed with a 3% solution of erythrosine in carbolic water, cleared in a drop of immersion oil on a slide and covered with a cover slip; the preparation was ready for the analysis.

At first the whole preparation was microscopically examined; the major components were recorded and the number of cells in the most representative algal species counted. Then by means of an ocular grid the areas covered by the various components were measured in 10–20 fields of vision. The data obtained were expressed as percentages of the total area

Table 14.2. *Trophic characteristics of some abundant bottom invertebrates from the Azov, Caspian and Aral Seas*

Species	Occurrence			Grouping with regard to		Layer in which food is taken	General chracteristics of contents
	Azov	Caspian	Aral	Substrate	Method of feeding		
Worms							
Nereis diversicolor Müller	+	+	+	Infauna	Collecting feeder	Sediment surface	Microscopic algae and animals, filamentous algae, organic–mineral particles, bacteria
Nereis succinea Leuckart	+	—	—	Infauna	Collecting feeder	Sediment surface	Microscopic algae and animals, filamentous algae, organic–mineral particles, bacteria
Nepthys hombergii Aud. et Edw.	+	—	—	Infauna	Collecting feeder	Sediment surface	Sediment detritus
Harmothoe imbricata (Linné)	+	—	—	Infauna	Collecting feeder	Sediment surface	Microbenthos
Hypaniola kowalewskyi (Grimm)	+	+	—	Infauna	Collecting feeder	Sediment surface	Microscopic algae, organic–mineral aggregations
Hypania invalida (Grube)	+	+	—	Infauna	Collecting feeder	Sediment surface	Microscopic algae, organic–mineral aggregations
Manajunkia caspica Annenkova	—	+	—	Infauna	Filter feeder	Off-bottom layer	Microscopic algae, organic–mineral aggregations
Limnodrillus newaensis Michaelson	+	—	—	Infauna	Burrowing	Sediments	Organic–mineral particles, bacteria
Molluscs							
Dreissena polymorpha (Pall.)	+	+	+	Epifauna	Suspension feeder	Off-bottom layer	Phytoplanktonic algae, fine-grained detritus, mineral particles
Dreissena caspia Eichw.	—	+	+	Epifauna	Suspension feeder	Off-bottom layer	Phytoplanktonic algae, fine-grained detritus, mineral particles
Dreissena rostriformis (Desh.)	—	+	—	Epifauna	Suspension feeder	Off-bottom layer	Planktonic algae, fine-grained detritus

Didacna trigonoides (Pall.)	—	+	—	Epifauna	Suspension feeder	Off-bottom layer	Planktonic algae fine-grained detritus
Didacna longipes (Grimm)	—	+	—	Epifauna	Suspension feeder	Off-bottom layer	Planktonic algae fine-grained detritus
Hypanis angusticostata Logv. et Star.	—	+	—	Infauna	Suspension feeder	Off-bottom layer, sediment surface	Phytoplanktonic algae, detritus, organic–mineral aggregations
Hypanis colorata (Eichw.)	+	+	—	Infauna	Suspension feeder	Off-bottom layer	Phytoplanktonic algae, fine-grained detritus
Hypanis vitrea (Eichw.)	—	+	+	Infauna	Suspension feeder	Off-bottom layer, sediment surface	Planktonic algae, organic–mineral aggregations
Hypanis minima (Ostr.)	—	+	+	Infauna	Suspension feeder	Off-bottom layer, sediment surface	Microscopic planktonic and bottom algae, organic–mineral aggregations
Cerastoderma lamarcki (Reeve)	+	+	+	Infauna	Suspension feeder	Off-bottom layer, sediment surface	Microscopic planktonic and bottom algae, organic–mineral aggregations
Abra ovata (Phil.)	+	+	+	Infauna	Collecting feeder	Sediment surface	Planktonogenous detritus, organic–mineral aggregations
Mytilaster lineatus (Gmel)	+	+	—	Epifauna	Suspension feeder	Off-bottom layer	Phytoplanktonic algae, fine-grained detritus
Corbulomya maeotica Mil.	+	—	—	Epifauna	Suspension feeder	Off-bottom layer	Phytoplanktonic algae, fine-grained detritus
Hydrobia ventrosa (Mont.)	+	—	—	Epifauna	Collecting feeder	Sediment surface	Microscopic algae, detritus, bacteria
Theodoxus pallasi Ldh.	+	+	+	Epifauna	Collecting feeder	Weed bed	—
Crustaceans							
Balanus improvisus Darwin	+	+	—	Epifauna	Suspension feeder	Off-bottom layer	Phyto- and zooplankton, organic–mineral aggregations
Ampelisca diadema A. Costa	+	—	—	Epifauna	Collecting feeder	Sediment surface	Planktonic and bottom algae, bacteria, organic–mineral aggregations
Axelboeckia spinosa (G. O. Sars)	—	+	—	Epifauna	Collecting feeder	Sediment surface	Bottom algae
Amathillina cristata Grimm	+	+	—	Epifauna	Collecting feeder	Sediment surface, weed bed	Bottom plants, animal remains, detritus

Table 14.2.

Species	Occurrence			Grouping with regard to		Layer in which food is taken	General chracteristics of contents
	Azov	Cas-pian	Aral	Substrate	Method of feeding		
Amathillina pusilla G. O. Sars	—	+	—	Epifauna	Collecting feeder	Sediment surface	Algal detritus, planktonic algae
Dikerogammarus haemobaphes (Eichw.)	+	+	—	Epifauna	Collecting feeder	Sediment surface	Bottom algae, detritus, crustaceans
Dikerogammarus caspius (Pallas)	—	+	—	Epifauna	Collecting feeder	Sediment surface weed bed	Bottom filamentous algae
Dikerogammarus aralensis (Uljanin)	+	+	+	Epifauna	Collecting feeder	Sediment surface weed bed	Bottom diatoms, filamentous green algae, animal carcasses
Akerogammarus contiguus (Pjatakova)	—	+	—	Epifauna	Collecting feeder	Sediment surface	Algal detritus, planktonic algae
Pontogammarus robustoides (Grimm)	+	+	—	Epifauna	Collecting and scratching feeder	Sediment surface weed bed	Periphyton algae, plant tissues
Pontogammarus abbreviatus (G. O. Sars)	—	+	—	Epifauna	Collecting and scratching feeder	Sediment surface, weed bed	Bottom algae, plant tissues, detritus
Pontogammarus maeoticus (Sowinsky)	+	+	—	Epifauna	Collecting and filter feeder	Sediment surface, surf zone	Bottom ålgae, vegetable detritus
Stenogammarus compressus (G. O. Sars)	—	+	—	Epifauna	Collecting and scratching feeder	Sediment surface, weed bed	Filamentous algae, periphyton detritus
Stenogammarus similis (G. O. Sars)	+	+	—	Epifauna	Collecting feeder	Sediment surface	Bottom green algae, detritus
Niphargogammarus quadrimanus G. O. Sars	—	+	—	Infauna	Collecting and suspension feeder	Sediment surface, off-bottom layer	Algae of periphyton, fine-grained detritus

Pandorites podoceroides (Grimm)	+	+	—	Infauna	Suspension feeder	Sediment surface, off-bottom layer	Microscopic algae, fine-grained detritus
Pandorites platycheir (G. O. Sars)	—	+	—	Epifauna	Scratching and collecting feeder	Sediment surface, weed bed.	Periphyton, detritus
Gmelinopsis tuberculata G. O. Sars	+	+	—	Epifauna	Collecting deposit feeder	Sediment surface	Small planktonic algae, fine-grained detritus
Gmelina pusilla G. O. Sars	+	+	—	Epifauna	Collecting deposit feeder	Sediment surface	Small planktonic algae, fine-grained detritus
Gmelina brachyura Derzh et Pjat.	—	+	—	Epifauna	Collecting deposit feeder	Sediment surface	Small planktonic algae, fine-grained detritus
Gmelina costata Grimm (G. O. Sars)	+	+	—	Epifauna	Collecting deposit feeder	Sediment surface, weed bed	Filamentous algae, macrophyte tissues
Chaetogammarus placidus Grimm	—	+	—	Epifauna	Collecting deposit feeder	Ground surface	Small planktonic algae, algal fine-grained detritus
Chaetogammarus behningi (Mart.)	—	+	—	Epifauna	Collecting deposit feeder	Ground surface	Small planktonic algae, algal fine-grained detritus
Chaetogammarus ischnus Stebbing	+	+	—	Epifauna	Collecting deposit feeder	Ground surface	Small planktonic algae, algal fine-grained detritus
Chaetogammarus pauxillus Grimm	—	+	—	Epifauna	Collecting deposit feeder	Ground surface	Small planktonic algae, algal fine-grained detritus
Pontoporeia affinis microphthalma Grimm	—	+	—	Epifauna	Collecting deposit feeder	Ground surface	Small planktonic algae, algal fine-grained detritus
Corophium chelicorne G. O. Sars	+	+	—	Infauna	Suspension feeder	Off-bottom layer, sediment surface	Small planktonic algae, algal fine-grained detritus
Corophium robustum G. O. Sars	+	+	—	Infauna	Suspension feeder	Off-bottom layer, sediment surface.	Small planktonic algae, algal fine-grained detritus

Table 14.2.

Species	Occurrence			Grouping with regard to			
	Azov	Caspian	Aral	Substrate	Method of feeding	Layer in which food is taken	General chracteristics of contents
Corophium mucronatum G. O. Sars	—	+	—	Infauna	Suspension feeder	Off-bottom layer	Small planktonic algae, algal fine-grained detritus
Corophium monodon G. O. Sars	—	+	—	Infauna	Suspension feeder	Off-bottom layer	Fine-grained detritus
Corophium volutator (Pallas)	+	—	—	Infauna	Suspension feeder	Off-bottom layer	Microscopic algae, fine-grained detritus
Corophium curvispinum G. O. Sars	-	+	—	Infauna	Suspension feeder	Off-bottom layer, sediment surface	Vegetable detritus, microscopic algae
Corophium nobile G. O. Sars	—	+	—	Infauna	Suspension feeder	Off-bottom layer, sediment surface	Planktonic algae, fine-grained detritus
Schizorhynchus bilamellatus G. O. Sars	—	+	—	Epifauna	Collecting deposit feeder	Sediment surface	Microscopic plankton and bottom algae, organic–mineral aggregations
Pterocuma pectinata G. O. Sars	+	+	—	Epifauna	Collecting deposit feeder	Sediment surface	Microscopic plankton and bottom algae, organic–mineral aggregations
Stenocuma gracilis G. O. Sars	—	+	—	Epifauna	Collecting deposit feeder	Sediment surface	Microscopic plankton and bottom algae, organic–mineral aggregations
Paramysis lacustris (Czerniavsky)	+	+	+	Epifauna	Filter feeder, predator	Off-bottom layer	Microscopic plankton, algae, zooplankton, organic–mineral aggregations

Limnomysis benedeni Czerniavsky	+	+	—	Epifauna	Suspension feeder	Off-bottom layer, sediment surface	Detritus, microscopic algae
Mesopodopsis slabberi	+	—	—	Epifauna	Suspension feeder	Off-bottom layer, and mid-water	Phytoplanktonic algae
Brachynotus sexdentatus	+	—	—	Epifauna	Predator	Ground surface	Molluscs, worms
Rhithropanopeus harrisii (Gould)	+	+	—	Epifauna	Predator	Ground surface	Molluscs, worms crustaceans
Palaemon elegans Rathke	+	+	+	Epifauna	Predator	Plant and weed bed	Algae, chironomid larvae, crustaceans
Insects							
Chironomus behningii Goetgh.	—	—	+	Infauna	Collecting feeder	Sediment surface	Bottom microscopic algae, organic–mineral particles

occupied by the particles. By this procedure the following information was collected: intact single cells of small planktonic algae; large algae, cell chains and filaments; light fine-grained detritus formed due to the decomposition of planktonic algae; dense brown mass of organic-mineral character; mineral suspension.

In some samples wet preparations of the intestinal contents were examined to estimate the frequency of occurrence of various components. A small portion of the contents taken from the anterior part of the intestines was stirred up in a drop of water, covered with a cover slip, and examined microscopically. Remains of microscopic plants and animals were identified and recorded, when possible. Quantitative estimates of the remains were made visually as to the frequency of occurrence in various fields of vision and the area covered by them (Table 14.2).

In some cases (e.g. samples from the Azov Sea), to obtain a general pattern of the intestine content, the percentage of such major components as sand and 'other constituents' were visually estimated. The term 'other constituents' means brown flocculent mass, a lighter fraction of the ground which includes organic detritus, algae and light inorganic constituents of the ground (argillaceous particles with organic substances absorbed) and other material.

In addition to the examination of the gut contents in preserved animals, observations were made on feeding habits, food preferences and the amount of food consumed by amphipods (*Dikerogammarus haemobaphes*, *Pontogammarus maeoticus*, *Amathillina cristata*, *Axelboeckia spinosa*); mysids (*Paramysis lacustris*, *Mesopodopsis slabberi*); decapods (*Rhithropanopeus harrisii*); molluscs (*Abra ovata*, *Corbulomya maeotica*, *Hypanis colorata*, *H. angusticostata*, *Cerastoderma lamarcki*, *Didacna trigonoides*, *D. longipes*, *Mytilaster lineatus*, *Hydrobia ventrosa*); and worms (*Nereis diversicolor*, *Hypaniola kowalewskyi*).

The feeding habits of other abundant bottom invertebrates, such as Protozoa, Coelenterata, Oligochaeta, and Ostracoda, however, remain almost unknown and require further study. Such quantitative aspects as the rate of food consumption, assimilation, and effect of food concentration on the efficiency of feeding, etc., also require further investigation.

This is only the first attempt to generalize the data available and to present the general characteristics of the trophic relations in the benthos of the Caspian, Aral and Azov Seas.

Feeding habits of abundant species of bottom invertebrates

The information available on the frequency of occurrence of abundant species of invertebrates in the benthos of the Caspian, Aral and Azov Seas, on their food habits and on regions of feeding is summarized in Table 14.2

(Vorobyev, 1949; Mordukhay-Boltovskoy, 1960; Birstein *et al.*, 1968; Identification book of the fauna of the Black and Azov Seas, 1969; Aligadzhiev, 1965; Romanova, 1963; Karpevich, 1960; Greze, 1965; Yablonskaya, 1952, and other authors mentioned above). The general characteristics of the gut contents and habitats of animals both on the surface of bottom sediments (epifauna) and in the sediment (infauna) are also shown.

Molluscs

Analysis of the intestinal contents shows that bivalved molluscs which are suspension feeders filter mostly small (up to 100 μm, commonly 15–20 μm) rounded cells of diatoms, dinoflagellates, green and blue-green algae, planktonogenous detritus with bacteria and small mineral particles (Yablonskaya, 1971). Large acerates and diatoms arranged in chains (*Rhizosolenia calcar avis*, *Skeletonema costatum*, *Chaetoceros* spp. and others), filamentous green (*Spirogyra*) and blue-green algae (*Aphanizomenon flos aquae*), although known to be abundant in the plankton, occur rarely in the intestines of molluscs.

The food contents of sedentary suspension feeders without long siphons (*Mytilaster*, *Dreissena*, *Didacna*, *Corbulomya*) differ from those of collecting deposit feeders with long proboscis-like siphons, which burrow into the ground (*Abra ovata*). The food of the former consists of light fine-grained detritus with a large amount of planktonic algae, and is similar to the composition of the suspension; whereas these components constitute an insignificant part of the food content of the latter. The intestines of collecting deposit feeders are filled largely with dense brown aggregations of organic-mineral character, formed in the surface layer of the sediment as a result of sedimentation of detritus and argillaceous mineral particles. A significant amount of large mineral particles with a diameter of 75–78 μm are found in the intestines of *Abra*, whereas in *Dreissena*, *Mytilaster* and *Didacna* the sizes of particles of mineral suspension are 13–27 μm and the bulk of the particles measure only a few. *Cerastoderma*, *Monodacna* and *Adacna* are crawling, burrowing, suspension feeders; they occupy an intermediate position.

The amount of small planktonic algae is considerably less in the gut contents of those molluscs which obtain food particles not only from the water but also filtered from off the sediments, than in those of sedentary suspension feeders. The bulk of the gut contents in these molluscs in the middle Caspian Sea consists of planktonogenous detritus including abundant remains of *Rhizosolenia* (Yablonskaya, 1971).

No zooplanktonic organisms or their eggs or larvae were found in the molluscs examined. All bivalves examined consume phytoplankton and vegetable detritus; thus they are consumers of the first food link (herbivores). Of planktonic algae, *Exuviaella cordata* is the most common food species found in molluscs from the middle Caspian Sea. *Exuviaella cordata*, *Actinocyclus ehrenbergii*, *Thalassiosira variabilis*, *Scenedesmus quadricauda*, *Binuclearia* sp., *Microcystis pulverea*, and various species of *Merismopedia*, occur most frequently in the guts of molluscs from the north Caspian Sea.

In the Sea of Azov, *Coscinodiscus jonesianus*, *Coscinodiscus granii*, *Cyclotella caspia*, *Thalassiosira decipiens*, *Melosira granulata*, *Biddulphia mobiliensis*, *Binuclearia* sp., *Scenedesmus quadricauda*, and various species of *Merismopedia* are found in the stomach contents of molluscs. *Actinocyclus ehrenbergii*, *Chaetoceros wighamii*, and various species of *Merismopedia* and *Navicula* occur frequently in the intestines of Aral Sea bivalves. No intact cells of *Rhizosolenia* are found in molluscs. *Rhizosolenia calcar avis* is consumed as detritus, the largest amount of it being found in *Cerastoderma* of the north Caspian Sea as well as in *Cerastoderma* and *Monodacna* of the middle Caspian Sea. The only gastropod from the Azov Sea studied, *Hydrobia ventrosa*, collects small particles of detritus and microscopic algae from the sediment surface, the preference being for newly-settled detritus enriched with diatoms (Yablonskaya, 1971).

Crustaceans

Among benthonic crustaceans there are herbivorous, omnivorous and carnivorous species. *M. slabberi* in the Azov Sea feeds on planktonic algae. The intestines of mysids of this species are filled with crushed cells of *Coscinodiscus*, together with yellow-green detritus formed from decomposed chromatophores.

In *Paramysis lacustris*, which stays off the bottom in the daytime, the intestines contain, along with benthonic and planktonic algae, a substantial amount of sediment material (amorphous brown flocculent mass and grains of sand) as well as some remains of planktonic crustaceans and rotifers. This is supported by an experiment in which *Paramysis* was kept together with rotifers (*Keratella*, *Brachiounus*, *Asplanchna*), copepods (*Heterocope*) and larvae of cirripedes and copepods. The abundance of planktonic animals was found to be reduced considerably on the following day. Large specimens of *Paramysis lacustris* were observed to be attacking the young *Mesopodopsis* (*Macropsis*) *slabberi* and sucking out the soft parts of their bodies (Yablonskaya, 1971). The carnivorous habit of *Paramysis lacustris* was also emphasized by Mordukhay-Boltovskoy (1960). The observations show that this mysid feeds largely on detritus and phytoplankton and only for lack of such food, turns to animal food.

The intestines of *Balanus* examined from the north Azov Sea contained copepod remains (segments of pleopods, furcae), carapaces of Cladocera, setae of Polychaeta, crushed valves of *Coscinodiscus* and fine-grained flocculent amorphous material. Vorobyev (1949) characterizes *Balanus* as an 'off-bottom layer' feeder.

Typical of herbivorous crustaceans is *Idothea baltica* in the Azov Sea (Reznichenko, 1958) which feeds on soft filamentous algae and detritus of vegetable origin. The shrimp *Palaemon adspersus* consumes both plants (*Cladophora*) and animals (mysids and amphipods). According to Malinovskaya (1961) *Palaemon elegans* of the Aral Sea feeds largely on vegetation (filamentous green and diatomous algae and other plants), whereas animals (larvae of Chironomidae and benthonic crustaceans) constitute 7–34% of the total food by weight. Of the carnivorous bottom invertebrates of the Azov Sea, Vorobyev (1949) mentions that *Brachynotus lucasi* feeds on worms, *Mytilaster* and the 'carcasses of animals'.

The amphipods can be classified, by the type of food consumed, into two large groups.

(i) Consumers of small food particles such as fine-grained detritus, rounded dinoflagellates, small diatoms, green and blue-green algae. Usually the diameter of food particles is not more than 20 μm and rarely reaches 40–50 μm. This group includes all Corophiidae examined, all species of *Chaetogammarus*, *Akerogammarus contiguus*, *Gmelinopsis tuberculata*, *Gmelina brachyura*, *Amathillina pusilla*, *Pontoporeia affinis microphthalma*.

(ii) Consumers of large particles of coarse detritus (up to 600 μ) of vegetable and animal origin, and sometimes of live animals. This group includes larger-sized species of Amphipoda inhabiting mostly the weed bed, such as *Dikerogammarus haemobaphes*, *D. caspius*, *D. aralensis*, *Stenogammarus similis*, *Nipharogammarus robustoides*, *Pontogammarus abbreviatus*, *Amathillina costata* and other dwellers of shallow waters and weed beds. The characteristic of this group is the euryphagous habit of feeding.

The intestines of Cumacea examined were filled with fine-grained detritus supplemented with planktonic algae. Judging from the character of the food content we may assume that bacteria play an important role in the feeding of the species.

The feeding habits of numerous Ostracoda have not been studied. Like freshwater ostracods, they are believed to feed on plants, bacteria and detritus of both vegetable and animal origin (Luferova & Sorokin, 1970).

Worms

The food of Ampharetidae from the Azov and Caspian Seas consists of flocculent organic-mineral particles with some remains of diatoms, blue-green and green algae. *Nereis diversicolor*, an omnivorous species, consumes the surface layer of the sediments along with detritus, living plants and small animals without giving preference to any particular item or displaying hunting habits. Bacteria play an important role among the food species for *N. diversicolor* from the Caspian Sea (Zhukova, 1954, 1957).

In *Nereis diversicolor* (accidentally introduced into the Aral Sea just recently), the intestines contain filaments of *Vaucheria dichotoma* and a mass of small diatoms (Pennatae).

The food habits of oligochaetes from the Azov and Caspian Seas have not been studied. The representatives of the family Tubificidae feed in the sediments (Romanova, 1963). As in other waters, they seem to consume to a large extent bacteria and transformed organic matter (Poddubnaya & Sorokin, 1961).

Chironomid larvae

Konstantinov (1958) believes that larvae of the genus *Chironomus* are omnivorous, but according to his data on the composition of the gut contents examined in *Ch. annularis*, *Ch. obtusidens* and *Ch. plumosus*, the larvae of this genus are largely herbivorous although they consume a small amount of cladocerans. That they feed on vegetable food (diatoms, blue-green and green algae) is supported by Borodich (1956) and Kajak & Warda (1968).

Konstantinov (1958) has ascertained that the larvae of *Chironomus* feed only in the superficial layer of the sediments. The food is taken by various ways: by collection of various particles from the sediment surface; by 'webbing' the sediment surface round the burrow and then swallowing the web together with particles adhering to it; and by similar treatment of the particulate matter adhering to the walls of the burrow itself. Observations show that algae developed near the bottom and on the surface of the sediments (*Navicula* sp., *Pleurosigma*, *Vaucheria* sp., *Spirogyra* sp.), are important food for *Chironomus behningii* (Yablonskaya, 1964). The bulk of the gut contents of their larvae is formed of flocculent organic-mineral particles.

The feeding habits of the larvae of *Procladius skuze* and *Cryptochironomus gr. defectus* have not been studied. One may assume that as in other regions, the Aral representatives of the genera are mainly predators.

The most abundant chironomids of the Caspian Sea, *Chironomus albidus* and *Clunio marinus*, are herbivorous and feed on detritus (Birstein *et al.*, 1968).

The feeding habits of abundant benthonic invertebrates and the general trophic structure of the benthos

Proceeding from the data available and information cited in the literature, an attempt has been made to present the general trophic characteristics of benthos in the Azov, Caspian and Aral Seas, using the biomass values for the most abundant benthonic invertebrates (Table 14.3). The bottom invertebrates may be divided into groups depending on feeding habits and zones of feeding, as adopted in Soviet literature (Turpaeva, 1953; Sokolova & Neiman, 1966; Savilov, 1961; Kuznetsov, 1964; and others). Using these criteria we can distinguish an epifaunal group consisting of sedentary suspension feeders or filtrators (*Mytilaster*, *Dreissena*, *Didacna*, *Balanus*) which settle on various objects on the bottom and filter food particles from the 'off-bottom' layer. The infaunal group is represented by moving suspension feeders or filtrators of infauna, such as most Cardiidae, some Gammaridae (*Pandorites podoceroides*, *Niphargoides quadrimanus*) and Corophiidae. Another group consists of moving collecting feeders, which move over the bottom and collect food from the sediment surface or from various objects available on the bottom. This mode of feeding is typical of many scratching and gnawing gammarids.

We can also distinguish a group of detritus feeders, which get food mainly from the sediment surface near their burrows and tubules in the sediment column (*Abra*, *Nereis* and some Ampharetidae); these constitute the infaunal group of detritophagous animals. Other groups are represented by predators (crabs, shrimps) and organisms boring the ground to take food from the sediment layer (Tubificidae).

It is clear from the above that the benthos of the Azov Sea is characterized not only by its oligomixic type of biocoenoses, but also by the uniformity of its trophic structure. The bulk of the biomass (94–95%) is formed by primary consumers – phytophagous forms feeding on microscopic phytoplanktonic algae and other small particles of the suspension. Omnivorous and carnivorous animals are poorly represented and even in omnivorous animals (e.g *Nereis*) the bulk of the gut contents is formed by various vegetable remains and amorphous flocculent particles of organic–mineral character.

Sometimes the food of some carnivorous forms (e.g. crabs) includes a considerable amount of vegetable remains (Reznichenko, 1964). The major portion of benthic biomass (71–81%) in the Azov Sea is formed by animals living in the sediments which filter food out of the demersal layer (filtrators 'A' according to Turpaeva, 1953). Most animals either collect detritus and amorphous organic–mineral aggregations on the sediment surface (*Abra*, *Hydrobia*, *Hypaniola*, some Gammaridae and others) or swallow the sediment (*Nereis*) with the whole complex of its components; larger particles, which cannot be consumed due to their size, being thrown

Table 14.3. *Food structure of benthos in the Azov, Aral and Caspian Seas (percentages of the biomass)*

	Sea of Azov		Aral Sea		Caspian Sea					
					Northern		Middle		Southern (East)	
Feeding groups Years...	1934–5	1956–7	1954–7	1968–9	1935	1962	1934–5	1962	1935	1962
Feeding type										
Herbivorous	93.8	94.6	97.8	83.9	94.2	91.8	98.4	95.9	98.0	89.8
Omnivorous	6.1	4.2	0.8	16.1	0.8	4.3	0.4	2.4	0.2	1.4
Carnivorous	0.1	0.2	1.2	— }	5.0	3.9	1.2	1.7	1.8	4.1
Unidentified	—	1.0	0.2	— }						
The zone of feeding										
'Off-bottom' layer	71.3	81.3	61.9	50.2	88.6	69.3	97.6	72.6	97.6	44.4
Sediment surface	28.3	18.2	36.5	49.8	6.7	24.2	0.9	24.7	0.4	51.2
Sediment column }	0.4	0.3	—	—	0.8	3.6	0.2	1.0	0.2	0.3
Unidentified }		0.2	1.6	—	3.9	2.9	1.3	1.7	1.8	4.1
The mode of feeding										
Suspensions feeders	71.3	81.3	61.9	50.2	88.6	69.3	97.6	72.6	97.6	44.4
epifauna	15.0	5.9	36.7	20.8	58.4	61.6	86.2	70.7	90.7	44.4
infauna	56.3	75.4	25.2	29.4	30.2	7.7	11.4	1.9	6.9	—
Collecting deposit feeders	28.3	17.0	36.5	49.7	6.7	24.2	0.9	24.7	0.4	46.5
Predators	—	0.2	—	—	—	—	—	—	—	4.7
Sediment-boring animals }	0.4	0.3	—	—	0.8	3.6	0.2	1.0	0.2	0.3
Unidentified }		1.2	1.6	0.1	3.9	2.9	1.3	1.7	1.8	4.1
Total biomass (g/m^2)	363.3	348.3	27.7	12.1	40.2	71.6	243.4	207.2	50.8	36.3

away. Carnivorous forms (crabs), feeding largely in the mollusc biocoenoses, and animals feeding in the sediment itself (oligochaetes) constitute less than 1% of the benthic biomass.

The feeding habits of most benthonic animals in the Azov Sea are associated with phytoplankton and detritus, the latter being formed largely from decayed planktonic algae, since according to Datsko (1959) more than 96% of primary organic matter is produced by phytoplanktonic algae. Microscopical analysis indicated a large amount of bacterial cells in the amorphous flocculent mass found in the intestines of many benthic animals, and these cells also serve as a food component (Zhukova, 1954). The fact that most of the animals under consideration either take food from the 'off-bottom' (hyperbenthic) layer (suspension feeders: *Corbulomya, Cerastoderma, Monodacna, Hypaniola, Mesopodopsis*) or collect it from the sediment surface (*Nereis, Abra, Hydrobia*) leads us to assume that most benthic organisms are dependent upon the supply of fresh organic matter.

Macrophytobenthic organisms (benthonic algae, remains of higher plants) are unimportant or only locally important as a source of food for invertebrates in the Azov Sea. Fragmentary plant remains have been revealed in the alimentary tracts of some crustaceans (*Paramysis lacustris, Pontogammarus robustoides, Amathillina cristata, Dikerogammarus villosus*) inhabiting the Don River estuary rich in bottom vegetation. *Idothea baltica, Palaemon adspersus*, and to some extent *Pontogammarus maeoticus* feed on delicate bottom algae and vegetable detritus. None of them uses coarse vegetation such as *Ceramium* and *Zostera*, which are abundant on the rocky coastal area of the Azov Sea (Reznichenko, 1958).

Among the Aral Sea benthonic invertebrates herbivorous animals distinctly predominate. They either filter food from the hyperbenthic layer or collect it from the sediment surface. During the 'aboriginal' period of the Aral fauna, omnivorous and carnivorous organisms constituted a negligible part of the benthic biomass. Recently the role of omnivorous animals has increased because the biomass of the Caspian relicts and of the species of freshwater origin has been reduced, and marine polychaetes (*Nereis diversicolor*) and decapods (*Palaemon elegans*) of Mediterranean origin have been introduced. A marked prevalence of herbivorous animals feeding in the hyperbenthic layer or on the surface layer of the sediments has been recorded for all parts of the Caspian Sea. The sedentary suspension feeders (*Dreissena, Mytilaster*) and other suspension feeders of the epifauna (*Didacna*) are predominant. Next are the suspension feeders of the infauna (*Cerastoderma, Monodacna, Adacna*, some Gammaridae and Corophiidae) and epifaunal collecting feeders, which get food from the hyperbenthic layer and from the surface of the sea floor. Various scratching and gnawing gammarids obtain food from the surface of objects on the bottom. On the other hand semi-sedentary collecting deposit feeders of the infauna (*Nereis*, *Abra*, Ampharetidae and Chironomidae) feed largely at the sediment surface. The forms which obtain food from within the sediments (Tubificidae exclusively), and predators, are poorly represented in the Caspian Sea.

The poor development of predatory bottom invertebrates which reduces the useful production of benthos in the high seas is one of the most typical features of the bottom population in the Caspian, Aral and Azov Seas. Herbivorous animals consuming largely phytoplankton and planktogenous detritus prevail in the benthic biomass. The benthic organisms also use benthic microscopic algae (in the Aral Sea) and filamentous green algae (in the Aral and north Caspian Seas). Plants and their remains which occur in abundance in the pre-delta zone of the north Caspian Sea and in the coastal zone of the Aral Sea are of limited food value. The observations made (Romanova, 1963; Reznichenko, 1958) show that various kinds of periphyton developed on the plants are eaten rather than the plant tissues themselves. Bacteria constitute about 10% of the food of worms, crustaceans and molluscs.

The suspended matter brought down in rivers is of very limited value as food because of the low content of organic matter; moreover the organic content consists largely of cellulose which is hard to digest (Gorshkova, 1951). Marine bottom invertebrates are very likely to utilize this allochthonous organic matter through protozoans and bacteria which are intensively developed in the zones of suspended matter deposition and decomposition of detritus (Zhukova, 1955). All the data available lead to the conclusion that the feeding of invertebrates in the Azov, Caspian and Aral Seas is associated largely with organic matter produced in the sea water itself.

The uniformity of the food of the benthos and the evident prevalence of herbivorous species are obviously associated with the availability of the food, which is in agreement with emphasis on high phytoplankton production in the inland southern seas (the Caspian and Azov Seas in particular) due to the abundant runoff rich in biogenic substances derived from the drainage area. Because the Azov and Aral Seas, and the areas of intensive development of benthos in the Caspian Sea, are relatively shallow, the organic matter of phytoplankton produced in the photic zone is only slightly transformed when it reaches the biotopes inhabited by benthic organisms. Therefore suspension feeders and collecting deposit feeders are dominant.

Although the specific composition of benthic communities in the seas is fairly uniform, there are differences in terms of the dominance of one or another trophic group of invertebrates. Suspension feeders of the infauna are dominant in the Azov Sea, and in the Aral Sea infaunal suspension feeders and collecting deposit feeders predominate; in the Caspian Sea sedentary suspension feeders are predominant. The differences result from the peculiarities of the distribution and sedimentation of small suspended particles, which in turn depend on the character of the bottom relief and the dynamics of the water. In the shallow saucer-like Azov Sea intensive sedimentation of fine mineral and organic suspensions is known to occur, and the entire bed of the Azov Sea is filled with silt rich in organic matter (Gorshkova, 1955; Pakhomova, 1960). These conditions are most favourable for infaunal organisms which filter out food particles settling on the bottom or collect them from the sediment surface.

Although the sedimentation of organic particles in the Aral Sea is not so intensive as in the Azov Sea, the rich diatomous flora on the surface of the sediments (Yablonskaya, 1964) also favours the development of deposit and suspension feeders.

In the north Caspian and in the peripheral zones of the middle and south Caspian Sea the transport of fine-grained particles predominates over deposition because of the bottom relief and the prevailing currents (Gorshkova, 1951; Bordovsky, 1964; Yablonskaya, 1969). Accumulation of fine-grained silty sediments occurs largely in the middle and south Caspian Sea hollows, where the organic matter loses, to a large extent, its nutrient value (Pakhomova, 1961; Bordovsky, 1964; Yablonskaya, 1969). Under conditions of continuous movement and transport of food particles the infaunal suspension feeders are at an advantage and are dominant in the benthonic biomass of the Caspian Sea.

Stability of the benthic trophic structure in the Azov, Aral and Caspian Seas

Figs. 14.2–14.4 illustrate long-term changes in the composition of the trophic groups in the total benthic biomass of the Azov, Aral and north Caspian Seas. The changes that occurred in the total biomass of the bottom invertebrates and in the biomasses of certain geographical complexes are also shown. Similar data for the middle Caspian Sea are presented in Fig. 14.5. The data indicate a fairly stable trophic structure except in the years characterized by a lower rate of primary production, when the stable trophic structure of benthos in the Azov Sea was drastically affected; the abundance of species feeding on fresh detritus from the sediment surface decreased (Fig. 14.2). During the years 1957–66 the primary production and the total content of biogenic elements in the sea increased as a result of increased Don River runoff. It was in this period (from 1956–7 onward) that the disturbed benthic trophic structure began gradually to be restored and the role played by collecting deposit feeders in the production of the benthic biomass increased.

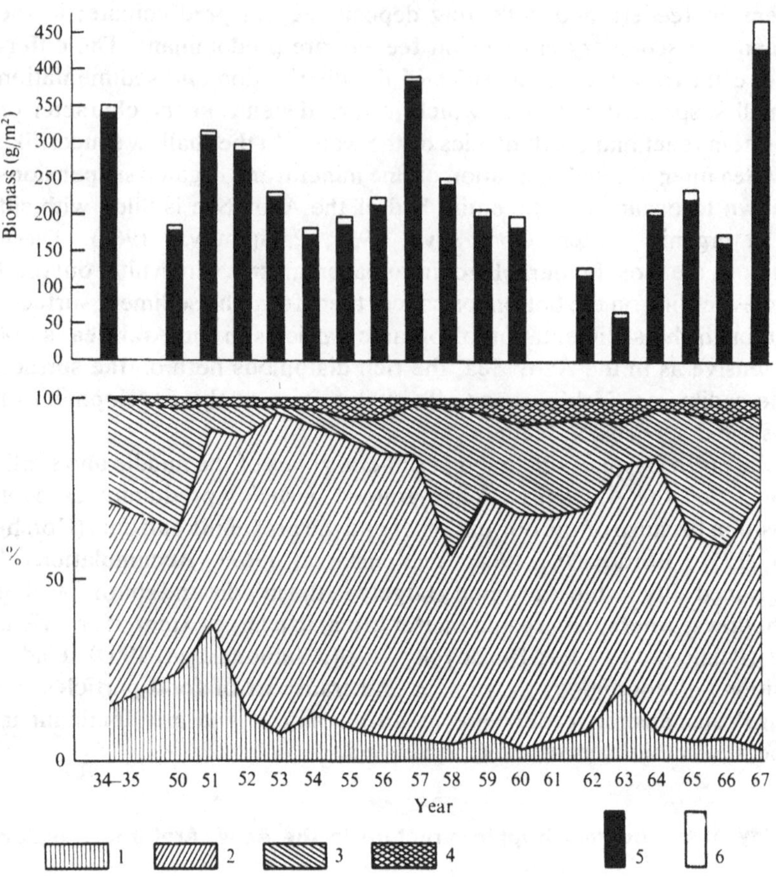

Fig. 14.2. Long-term changes in the trophic structure of benthos in the Azov Sea, 1934–7. 1, Suspension feeders of epifauna (%); 2, suspension feeders of infauna (%); 3, collecting deposit feeders (%); 4, other invertebrates (%); 5, the biomass of invertebrates of Mediterranean origin (g/m²); 6, the biomass of invertebrates of Caspian origin (g/m²).

The benthos in the Aral Sea is also characterized by a highly stable trophic structure, which is almost constant on a long-term basis, although certain changes are observed in the biomass value and faunistic composition of the bottom populations (Fig. 14.3).

The same holds true for the Caspian Sea (Figs. 14.4 and 14.5). In the middle Caspian Sea the organisms of Mediterranean origin had displaced, to a certain extent, the aboriginal benthic forms by 1962. This is illustrated by the way in which *Mytilaster lineatus* has displaced the heavy populations of *Dreissena caspia* and *D. elata* on the rocky and shell grounds in the middle and south Caspian Sea as well as on the border between the north and middle Caspian Sea (Logvinenko, 1965). In the course of the

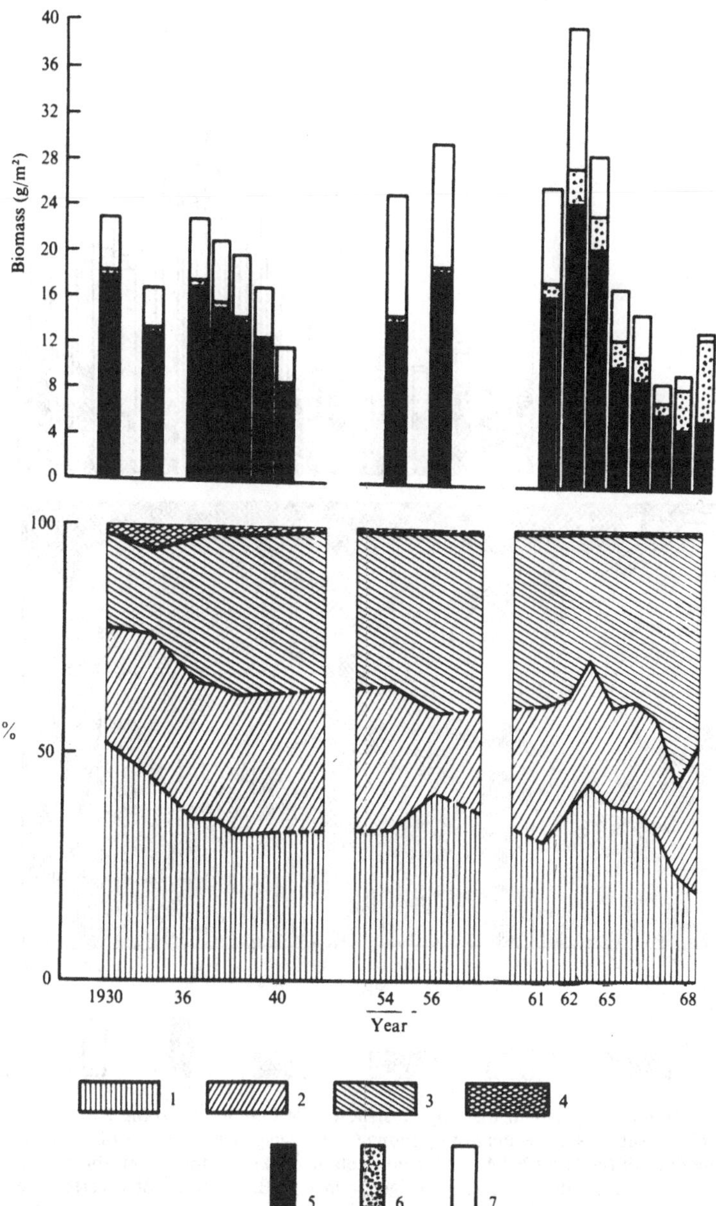

Fig. 14.3. Long-term changes in the trophic structure of benthos in the Aral Sea, 1930–68. 1, Suspension feeders of epifauna (%); 2, suspension feeders of infauna (%); 3, collecting deposit feeders (%); 4, other invertebrates (%); 5, the biomass of invertebrates of Caspian origin (g/m²); 6, the biomass of invertebrates of Mediterranean origin (g/m²); 7, the biomass of other invertebrates (g/m²).

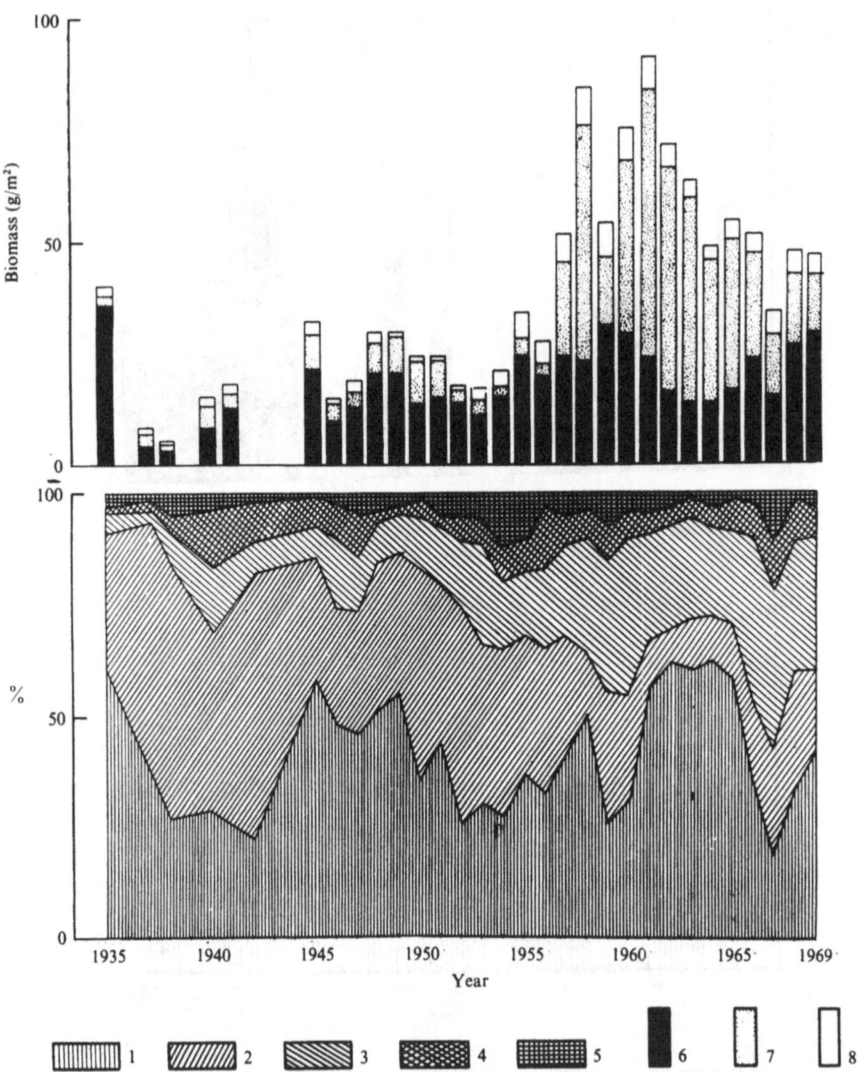

Fig. 14.4. Long-term changes in the trophic structure of benthos from the North Caspian Sea, 1935–69. 1, Suspension feeders of epifauna (%); 2, suspension feeders of infauna (%); 3, collecting deposit feeders (%); 4, boring invertebrates (%); 5, other invertebrates (%); 6, the biomass of invertebrates of Caspian origin (g/m^2); 7, the biomass of invertebrates of Mediterranean origin (g/m^2); 8, the biomass of other invertebrates (g/m^2).

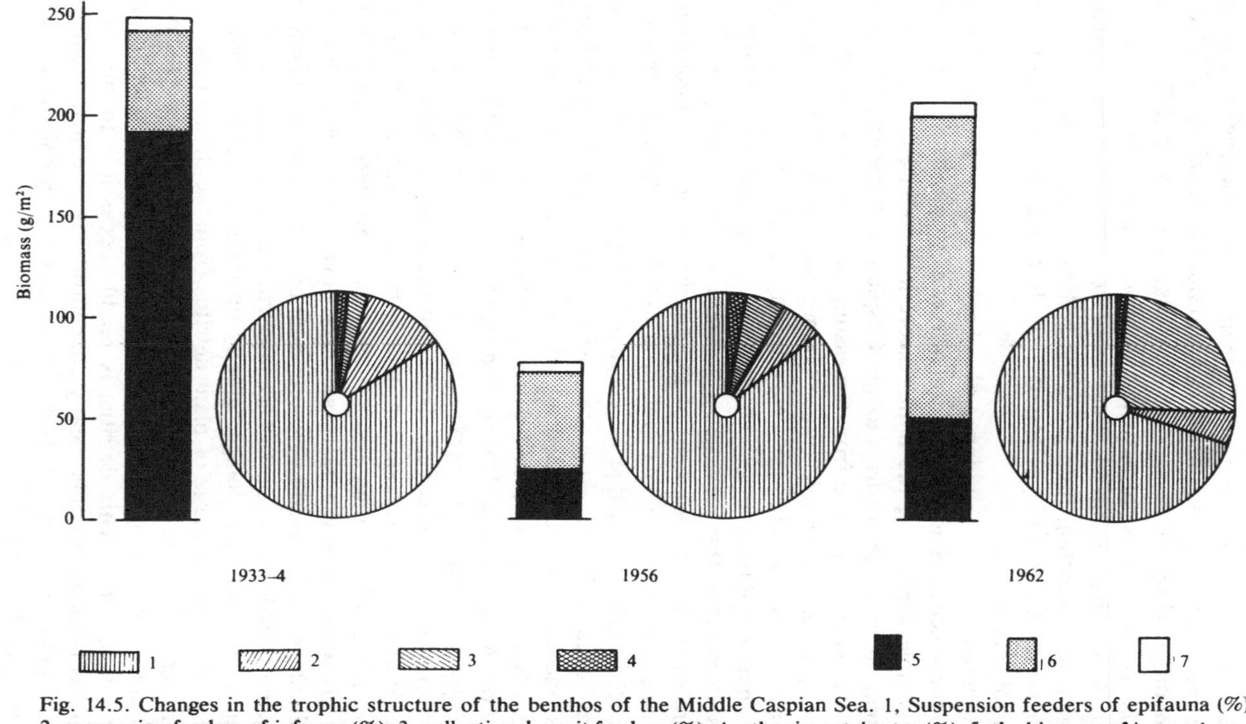

Fig. 14.5. Changes in the trophic structure of the benthos of the Middle Caspian Sea. 1, Suspension feeders of epifauna (%); 2, suspension feeders of infauna (%); 3, collecting deposit feeders (%); 4, other invertebrates (%); 5, the biomass of invertebrates of Caspian origin (g/m^2); 6, the biomass of invertebrates of Mediterranean origin (g/m^2); 7, the biomasss of other invertebrates (g/m^2).

displacement process the epifauna of suspension feeders preserved their dominant position (Fig. 14.5). For the most recent decade recorded here the benthic biomass in the north Caspian Sea has increased on account of the intensive development of the introduced organisms of Mediterranean origin (Fig. 14.4). By 1962, in comparison with 1934–5, some decrease in the biomass of crawling infaunal suspension feeders (mainly *Cerastoderma*) had occurred along with the increase in the abundance of organisms collecting food from the sediment surface (*Nereis*, *Abra*, Gammaridae, Ampharetidae) and of animals which feed within the sediments (oligochaetes).

The increase in the abundance of collecting deposit feeders is largely due to the successful acclimatization of *Nereis* and *Abra*. It is an illustrative example supporting the initial theoretical premises concerning the introduction of these forms into the Caspian Sea suggested by several scientists. They proceeded from the assumption that the Caspian bottom fauna is poor in forms able to use the organic matter of the sediments. The increase in the biomass of collecting deposit feeders and other representatives of the aboriginal Caspian fauna (gammarids, ampharetids) seems to indicate that new general conditions in the Caspian Sea are favourable to the development of this trophic group. Of importance, in this respect, is the intensive development of detritis-forming algae of the genus *Rhizosolenia* in saline waters and *Spirogyra* in freshened areas in the past decade. These algae are, owing to their morphological peculiarities, inadequately consumed by pelagic suspension feeders, but they are readily used by collecting detritus eaters after they have died and settled on the sea bottom. Olive-green layers formed by *Rhizosolenia* are frequently observed in the surface sediments in the middle Caspian Sea (Yablonskaya, 1969). It is very likely that the feeding habits of molluscs taking food from the suspension, whose biomass had decreased in the middle and south Caspian Sea by 1962 as compared to 1933–5, were adversely affected by the decrease in the development of small diatoms and dinoflagellates caused by the introduction of *Rhizosolenia* (Zenkevich & Zevina, 1969). In the north Caspian Sea, where the feeding habits of benthic organisms have changed greatly since the regulation of the Volga flow in 1956, the discharge of plant detritus from the delta of the Volga, which is growing more and more silty and overgrown with weeds, is very important as a source of food for collecting deposit feeders (Gershanovich & Grunduls, 1969). This detritus is readily accessible to such mobile feeders as large gammarids, heavy concentrations of which occur in the coastal shallow waters; their biomass is more than double what it was in the period of 1935–55.

It is very likely that after the introduction and intensive development of the crab *Rhithropanopeus harrisii* the role of carnivorous forms among the Caspian benthos has increased somewhat, but this cannot yet be substantiated because of inadequate estimates of the abundance of the crab stocks.

These long-term changes in the benthos of the southern seas support the view that the trophic pattern of the benthonic populations alters much more slowly than their specific composition. Since the trophic structure is formed under the influence of such relatively stable physical–geographical factors as the morphology of the water body and the dynamics of the water, the peculiarities of the trophic structure of the benthos in the Azov, Caspian and Aral Seas should be taken into consideration in any attempts to solve problems pertaining to the reconstruction of the bottom fauna. In view of the local specific features and the stable character of the trophic structure of benthos in the Azov, Caspian and Aral Seas it may be concluded that a substantial functional stability is inherent in these communities and that they show considerable adaptive powers toward optimum utilization of food.

Conclusion

The seas under consideration are epi-continental land-locked (the Caspian and Aral Seas) or semi-land-locked (the Azov Sea) water bodies, the most striking peculiarity of which is to accumulate river runoff from very large drainage basins. A large amount of mineral and organic compounds of biogenic elements are discharged into them, and the biological use of these substances is favoured by the fact that the seas are shallow and the water column is intensively warmed up during the vegetative period. The brackish-water character of the seas resulting from the continental runoff, which entails qualitative impoverishment in marine fauna, is nevertheless responsible for the high biological productivity. The mean annual values of primary production of the seas are close to those of eutrophic lakes, and range from 1390 Cal/m^2 (the Caspian Sea) to 2600 Cal/m^2 (the Azov Sea).

The organic matter produced by phytoplankton is rapidly transferred from the pelagic zone into the benthic layer and maintains the abundant bottom fauna. Herbivorous animals, consumers of planktonic algae and detritus, are dominant in the benthos; their feeding habits are associated with the hyperbenthic layer and the surface of the sediments, and they are mainly deposit feeders. Predatory invertebrates are poorly represented.

The most abundant bivalves of these seas either filter out the small rounded cells of the planktonic algae and detritus from the demersal layer (*Mytilaster*, *Dreissena*, *Didacna*, *Corbulomya*) or suck in food particles settled on the bottom (many of the Cardiidae) or collect them from the surface of the sediments (*Abra*). Most small ampharetids also either collect particles of vegetable detritus from the sediment surface or filter them from the demersal layer. Larger polychaetes – *Nereis* and *Nephthys*

– are omnivorous; they consume the surface layer of sediments together with vegetable remains and small animals. The bulk of the benthonic crustaceans (Amphipoda, Cumacea, Mysidae) are herbivorous, consuming either planktonic algae or small grain planktonogenous detritus (most Corophiidae, some species of the genus of *Chaetogammarus*, *Gmelinopsis tuberculata*, *Gmelina brachyura*, *Mesopodopsis slabberi* and others). Larger amphipods inhabiting the estuarine coastal areas (*Pontogammarus maeoticus*, *P. robustoides*, *Dikerogammarus caspius*, *D. haemobaphes*, etc.), consume tissues of plants and benthonic filamentous algae. Some species are recorded as consuming animal food as well (crabs and shrimps).

The main source of food for the bottom invertebrates of the Caspian and Azov Seas is organic matter produced by phytoplanktonic algae in the sea proper. Sinking into the demersal layer they are consumed either directly by numerous benthic species of suspension feeders or, after having settled on the bottom and undergone some morphological and biochemical change, by various detritus-eating species collecting food from the bottom sediments. Of local importance, largely for mobile collecting deposit feeders (Amphipoda) and burrowing species (Tubificidae), are phytobenthos and detritus brought in with river runoff.

There are four main sources of food for herbivorous bottom invertebrates in the Aral Sea: phytoplankton; plants and bottom macroscopic algae; microscopic bottom algae developed in the surface sediments; and detritus brought in with river runoff. Direct consumption of macrophytobenthos by bottom animals is limited at least by two principal factors; firstly, macrobenthos is intensively developed only within certain limited areas of the sea (bays and bights); secondly, the bulk of macrophytobenthos consists of plants (*Zostera*, *Chara*) the outer tissues of which are coarse and stiff and thus almost inaccessible for invertebrates. Of macrobenthic algae the bottom invertebrates consume *Vaucheria* and *Spirogyra* (Yablonskaya, 1964). Macrophytobenthos serves also as organic fertilizer: dead plants, and detritus formed as a result of decomposition, are discharged from bays into the open sea and enrich the bottom sediments with organic matter. The bulk of suspended substances delivered by the rivers of the Aral basin settle in deltas and upper estuaries; organic particles form part of this suspension and are buried in brown silty sediments sparsely inhabited by animals. It is only fine dispersed particles that are carried away by currents and accumulated in the sea (off-shore sediments).

The temperature stratification prevents nutrients (formed in the process of mineralization of detritus on the bottom) from coming to the surface layers. This factor along with satisfactory illumination of the off-bottom layer in the Aral Sea favours the development of abundant bottom and off-bottom microscopic flora and is responsible for the specific vertical distribution of algae in the Aral Sea, the maximum being close to the bottom (Yablonskaya, 1964; Pichkily, 1970).

The thin surface layer of suspended silt covering the bottom deposits of the Aral Sea contains a mass of live diatoms together with empty valves. The biomass of microscopic algae on the bottom is rather heavy, ranging from 5.8 to 17.2 g/m^2 in the samples collected (Yablonskaya, 1964). These algae are a very important source of food for bottom invertebrates, larvae of Chirinomids in particular, and require further study.

The principal role of allochthonous detritus and detritus formed *in situ* as a result of bottom vegetation decomposition is to produce a fertilization effect and to arrange a certain reserve buffer mechanism, as was emphasized by Darnell (1967). Bottom invertebrates probably use this material through bacteria and protozoa, a process which is associated with some loss of energy.

Dying phytoplanktonic algae are also subject to certain morphological and biochemical transformations before zoobenthic organisms start to feed on them. The degree of biological and chemical destruction of detritus brought into bottom biocoenoses determines, to a large extent, their trophic structure. The benthic structure of the Caspian, Aral and Azov Seas is composed under conditions in which the organic matter of the phytoplankton formed in the photosynthetic zone attains the bottom biotopes only slightly modified, at high rates of sedimentation.

Summary

The characteristics of the food habits of 55 species of bottom invertebrates from the Azov, Caspian and Aral Seas are described with reference to information available in the literature.

Herbivorous animals consuming microscopic algae and detritus are predominant in the benthos of these seas. Their feeding habits are associated with the hyperbenthic layer and the sediment surface, collecting deposit feeders and suspension-eaters are predominant, whereas predators are poorly represented. The main source of food for bottom invertebrates from the Azov and Caspian Seas is organic matter produced by phytoplanktonic algae. Phytobenthos and allochthonous detritus is locally important largely for mobile collecting deposit feeders. In the Aral Sea the bottom invertebrates use microscopic bottom algae in addition to planktonic algae. The role of allochthonous detritus and macrobenthos is largely the production of fertilizers.

The trophic structure of the benthos is determined by local specific features and stability in time. It is formed under conditions in which the high production of primary organic matter reaches the bottom biotopes only slightly biologically and chemically modified.

References

Aligadzhiev, G. A. (1965). Reconstruction of bottom fauna in the Dagestan area of the Caspian Sea in view of the intensive development of newly introduced species from the Azov and Black Seas. In *Izmeneniya biologicheskikh complexov Caspiiskogo morya poslednie desyatiletiya*, pp. 166–99. Moscow: Izdatelstvo Nauka.

Aliev, A. D. (1965). On the biology of *Brachyodontes (Mytilaster) lineatus* from the Caspian Sea. In *Hydrobiologicheskie i ichthyologicheskie issledovaniya v Yuzhnom Caspii i vnutrennikh vodoemakh Azerbaijana*, pp. 36–41. Azerbaydzhan SSR, Baku: Izdatelstvo AN.

Babaev, G. B. (1965). The role of phytoplankton in the feeding of some bottom invertebrates of the South Caspian. In *Hydrobiologicheskie i ichthyologicheskie issledovaniya na Yuzhnom Caspii i vnutrennikh vodoemakh Azerbaijana*, pp. 54–63. Azerbaydzhan SSR, Baku: Izdatelstvo AN.

Birstein, J. A., Vinogradov, L. G., Kondakov, N. N., Kun, M. S., Astakhova, T. V. & Romanova, N. N. (eds.) (1968). *Atlas of invertebrates of the Caspian Sea*. Moscow: Pishchevaya Promysh.

Bordovsky, O. N. (1964). Accumulation and transformation of organic matter in marine sediments, pp. 1–125. Moscow: Izdatelstvo Nauka.

Borodich, N. D. (1956). On the food habits of larvae of *Chirinomus f.l. plumosus* and their wintering in sediments of drained rearing ponds. *Trudy Vsesoyusnogo Hydrobiologicheskogo Obshchestva, Moscow*, **7**, 123–47.

Briskina, M. M. (1952). The food composition of bottom invertebrates in the North Caspian Sea. *Trudy VNIRO*, **28**, 121–6.

Darnell, R. M. (1967). Organic detritus in relation to the estuarine ecosystem. In *Estuaries*, publ. no. 83, pp. 376–82. American Association for the Advancement of Science.

Datsko, V. G. (1959). *Organic matter in the southern seas of the USSR*. USSR: Izdatelstvo AN.

Gershanovich, D. E. & Grunduls, Z. S. (1969). Suspended substances in the North Caspian Sea. *Trudy VNIRO*, **65**, 57–84.

Gorbunov, K. V. (1976). *The influence of regulated flow of the Volga River on the biological processes in its delta and biological runoff*. Moscow.

Gorshkova, T. I. (1951). Investigations of detritus in water and sediments of the North Caspian Sea. In *Sbornik pamyati academika VD Arkhangelskogo, Moskva*, pp. 568–82. USSR: Izdatelstvo AN.

Gorshkova, T. I. (1955). Organic matter from the sediments of the Azov Sea and the Bay of Taganrog. *Trudy VNIRO*, **31**, 95–122.

Greze, I. I. (1965). On the biology of the amphipod *Ampelisca diadema* A. Costa in the Black Sea. In *Benthos*, pp. 3–8. Kiev.

Husainova, N. Z. (1958). *Biological peculiarities of some abundant bottom food invertebrates from the Aral Sea*, pp. 1–115. Izdatelstvo Kazakhskogo Universiteta, Alma-Ata.

Identification book of the fauna of the Black and Azov Seas. I, II, III (1968, 1969, 1972), pp. 1–437, 1–536, 1–340. Kiev: Izdatelstvo Naukova dumka.

Kajak, Z. & Warda, J. (1968). Feeding of benthic non-predatory Chironomidae in lakes. *Ann. Zool. Fennici*, **5**, 57–64.

Karpevich, A. F. (1946). Food habits of *Pontogammarus maeoticus* in the Caspian Sea. *Zoologicheskyi zhurnal*, **25**, 517–22.

Karpevich, A. F. (1960). Bioecological characteristic features of the mollusc *Monodacna colorata* (Eichwald). *Trudy VNIRO*, **43**, 244–56.

Konstantinov, A. S. (1958). The biology of chironomids and their rearing. *Trudy Saratovskogo otdeleniya VNIRO*, **5**, 3–358.

Kuznetsov, A. P. (1964). The distribution of the bottom fauna of the West Bering Sea and some general problems of trophic zones. *Trudy Instituta Oceanologii AN USSR*, **69**, 98–177.

Logvinenko, B. M. (1965). On changes in the fauna of molluscs of the genus *Dreissena* after the introduction of *Mytilaster lineatus* (Gmel.). *Nauchnyi Doklady Vysshei Shkoly. Biologicheskie nauki*, **4**, 14–19.

Luferova, L. A. & Sorokin, Yu. I. (1970). The role of ostracods in the trophic chains of the Rybinsk Reservoir. In *Biologicheskie processy v morskikh i kontinentalnykh vodoemakh*, pp. 227–8. Kishiniev, Moldavskoi SSR: Redakcionno-izdatelstvo otdel Akademii nauk.

Maximova, L. P. (1958). The feeding habits of invertebrates introduced into the Tsymlansk reservoir. *Izvestiya VNIORKH*, **45**, 317–26.

Maximova, L. P. (1968). The feeding habits of *Pontogammarus robustoides* and *Limnomysis benedeni* in the Volga delta. In *Materialy nauchnoi sessii, posvyashchennoi 50-letiyu Astrachanskogo Gosudarstvennogo zapovednika*, pp. 99–101. Astrakhan: Astrachanskyi Gosudarstvennyi zapovednik.

Malinovskaya, A. S. (1961). On the biology of shrimp introduced into the Aral Sea. In *Sbornik rabot po ichthyologii i hydrobiologii*, vyp. 3, pp. 113–23. Alma-Ata, Kasakh SSR: Izdatelstvo AN.

Mordukhay-Boltovskoy, F. D. (1960). *The Caspian fauna in the Azov and Black Seas*. Moscow–Leningrad, USSR: Izdatelstvo AN.

Osadchikh, V. F. (1968). Changes in the benthic biomass in the North Caspian Sea in the past five years. *Trudy KaspNIRH*, **24**, 100–12.

Pakhomova, A. S. (1959). On the chemical composition of suspended matter in the sediments of the Volga delta and North Caspian Sea. *Trudy GOIN*, **45**, 117–44.

Pakhomova, A. S. (1960). Organic matter in the bottom sediments of the Caspian Sea. *Trudy GOIN*, **59**, 58–84.

Pichkily, L. O. (1970). The population dynamics and biomass of phytoplankton in the Aral Sea. *Hydrobiologichesky zhurnal*, **6**, 31–6.

Poddubnaya, T. L. & Sorokin, Yu. I. (1961). The optimum layer depth of feeding of tubificids in view of their movement in the sediments. *Bull. Instituta Biologii vodokhranilishch AN USSR*, **10**, 14–17.

Reznichenko, O. G. (1957). Ecology and food significance of *Pontogammarus maeoticus* of the Azov Sea. *Zoologicheskyi zhurnal*, **36**, 1312–22.

Reznichenko, O. G. (1958). The feeding habits of certain bottom crustaceans in the Azov Sea. In *Annotacii k rabotam, vypolnennym VNIRO* v. 1956. *Sbornik* 1, *Moscow*, pp. 20–4. Moskva: VNIRO.

Reznichenko, O. G. (1964). Transoceanic acclimatization of *Rhithropanopeus harrisii* (Crustacea, Brachyura). *Trudy Instituta Okeanologii AN USSR*, **85**, 136–77.

Romanova, N. N. (1963). The modes of feeding and food groups of bottom invertebrates of the North Caspian Sea. *Trudy Vsesoyuznogo Gidrobiologicheskogo obshchestva*, **3**, 9–177.

Romanova, N. N. & Osadchikh, V. G. (1965). The present status of zoobenthos in the Caspian Sea. Changes in the biological complexes of the Caspian Sea in recent decades. In *Izmeneniya biologicheskikh kompeksov Kaspiiskogo morya za poslednie desyatiletiya*, pp. 138–65. Moskva: Izdatelsvo Nauka.

Savilov, A. I. (1961). The ecological characteristics of bottom invertebrate biocoenoses in the Okhotsk Sea. *Trudy Instituta Okeanologii AN USSR*, **46**, 3–84.

Sokolova, M. N. (1964). Some patterns in the distribution of food groups of deep-sea benthos. *Okeanologiya*, **4**, 1079–88.

Sokolova, M. N. & Neiman, A. A. (1966). Trophic groups of bottom fauna and their distribution patterns in the ocean. In *Ecologiya vodnykh organizmov*, pp. 42–50. Moskva: Izdatelstvo Nauka.

Turpaeva, E. P. (1953). The food habits and food groups of marine bottom invertebrates. *Trudy Instituta Okeanologii AN USSR*, **7**, 259–99.

Vorobyev, V. P. (1949). *Benthos of the Azov Sea*, pp. 9–192. Simferopol, *Krymizdat*.

Yablonskaya, E. A. (1952). The feeding of *Nereis succinea* in the Caspian Sea. In *Materialy k poznaniyu fauny i flory, isdsvaemye MOYP, Novaya Seriya, Otdel Zoologicheskyi*, vyp. 33, pp. 285–351.

Yablonskaya, E. A. (1960). The present status of benthos in the Aral Sea. *Trudy VNIRO*, **43**, 115–49.

Yablonskaya, E. A. (1964). On the problem of the role of phytoplankton and phytobenthos in the food web of organisms in the Aral Sea. In *Zapasy morskikh rastenii i ikh ispolzovanie*, pp. 71–91. Moskva: Izdatelstvo Nauka.

Yablonskaya, E. A. (1967). On the food habits of Caspian molluscs. *Informatsionny Bull. Instituta Biologii Vnutrennikh Vod*. **1**, 47–50.

Yablonskaya, E. A. (1969). Suspension as food for bottom organisms in the Caspian Sea. *Trudy VNIRO*, **65**, 85–146.

Yablonskaya, E. A. (1971). The food habits of bottom invertebrates and the trophic structure of benthos of the Caspian, Azov and Aral Seas. *Izdatelstvo otdela nauchnotekhnicheskoi informacii VNIRO, Moscow*, pp. 3–146. Moscow: VNIRO.

Zenkevich, L. A. (1947). *Fauna and biological production of the seas*, pp. 1–538. Moscow: Izdatelstvo Sovetskaya Nauka.

Zenkevich, L. A. (1963). *Biology of seas of USSR*, Moscow.

Zenkevich, L. A. & Zevina, G. B. (1969). Fauna and Flora, Kaspiiskoe more, pp. 229–55. Izdatelstvo Moskovskogo Universiteta.

Zhukova, A. I. (1954). The role of microorganisms in the feeding of *Nereis succinea* in the Caspian Sea. *Microbiologia*, **28**, 46–8.

Zhukova, A. I. (1955). The biomass of microorganisms in the sediments of the North Caspian Sea. *Microbiologia*, **24**, 321–4.

Zhukova, A. I. (1957). The role of microorganisms as food for fish. *Voprosy ichthyologii*, **9**, 152–68.

15. Studies of the pattern of biotic distribution in the upper zones of the shelf in the seas of the USSR

A. N. GOLIKOV & O. A. SCARLATO

The upper zones of the shelf prove to possess the most diverse composition of inhabitants, to be the richest in life and to be the most productive in the system of vertical zones in the Ocean. As the main aim of IBP was the study of the organic resources of the biosphere and their exploitation for the welfare of mankind, hydrobiological investigations of this zone are of paramount importance. Detailed investigation of marine shallow regions, with relatively precise quantitative analysis of their populations, became possible only after the development of diving techniques in scientific research.

For several years, and in accordance with IBP objectives, the Laboratory of Marine Researches of the USSR Academy of Sciences has been studying the composition, structure and patterns of distribution of benthic neritic biocoenoses of the upper zones of the shelf, from the littoral down to a depth of 30–40 m in a number of regions of the northern and far eastern seas of the USSR, using SCUBA equipment. In 1962, 1965 and 1966, in different seasons of the year, the biocoenoses of Possjet Bay and waters washing the coasts of south Primorje, were studied (Golikov & Scarlato, 1965, 1967*a*, *b*, 1971). In 1963 investigations were carried out near the coasts of south Sakhalin, which included Tartar Strait in the region of Kholmk-Antonovo, Aniva Bay in the region of Starodubsk (Golikov, 1965, 1966; Golikov & Scarlato, 1968). In 1964, 1967 and 1968 in different seasons of the year the biocoenoses of Chupa Bay in the White Sea were studied. In 1969 investigations were carried out on the biocoenoses of the Kuril Islands: near the southern extremity of Kunashir Island, near the western and eastern coasts of Iturup Island and near Paramushir Island (Golikov & Scarlato, 1970*a*). In autumn 1970 hydrobiological work was carried out in the upper zones of the shelf of Franz-Josef Land (Golikov & Averincev, 1971).

The investigations were performed by the quantitative diving method (Scarlato, Golikov & Grusov, 1964; Golikov & Scarlato, 1965, 1971). Within the limits of each biocoenosis distinguished by the hydrobiologists during reconnoitring diving investigations, organisms were counted irrespective of size, character and density of distribution in areas from 20 mm^2 to 100 m^2. Large, sparsely distributed organisms and patches of underwater plants were counted along ropes stretched over areas of 50 or 100 m^2 of sea bottom. Densities of populations and biomasses of large massed organisms (for example aggregations of oysters or mussels) were determined by repeated collections from 1-m^2 frames. Smaller and more frequently occurring organisms, including mobile organisms, were counted in several areas from 0.05 m^2 to 0.1 m^2 by means of special bottom-grabs. The density of populations and biomass of underwater vegetation and animals associated with it were determined in several areas from 0.1 m^2 to 1 m^2, taking into consideration the size of the plants, by means of a special quantitative diving-bell. The quantity of the young of the macrobenthos and the meiobenthos was estimated by means of soil glasses and frames with areas of 20–100 mm^2.

When it was necessary to estimate the quantity of organisms on large areas of sea bottom of any region exhibiting the zonal type of biocoenosis distribution typical of temperate and cold water, the principle of the so called 'bionomical concordance' was used. The distribution pattern and density of populations of dominant, subdominant, and associated species were determined in biocoenoses situated in areas bionomically similar but distant from one another, and the general character of the sequence in vertical distribution was recorded. After a corresponding diving reconnaissance and analysis of physical and chemical conditions in the intervening regions, taking into consideration the relative constancy of density of populations of species in similar biocoenoses, the total number of individuals (or biomass) of each significant species was estimated in the region under study.

Plankton was taken by nets or by pump, the volume of the filtered water being measured. Collecting of plankton close to the bottom was done by a diver who used a plankton net over the biocoenosis under investigation. Simultaneously with the biological research, physical and chemical data were obtained.

Investigations showed that the bottom type determines only the presence or absence of a given species, which may or may not be adapted morphologically and functionally to life on the different substrates. Of great influence upon diversity and abundance of species is the presence or absence of highly dominant species, which may afford refuge or food for many other organisms (Scarlato *et al.*, 1967; Golikov & Scarlato, 1970*a*). On aggregations of oysters, or in biocoenoses with prevailing clusters of *Mytilus* or *Modiolus*, or in regions of thick growth of marine grasses, *Laminaria* or *Sargassum*, the number of species of macrobenthos often exceeds one hundred and the biomass is often more than several tens of kilograms per square metre, while on similar grounds and at similar

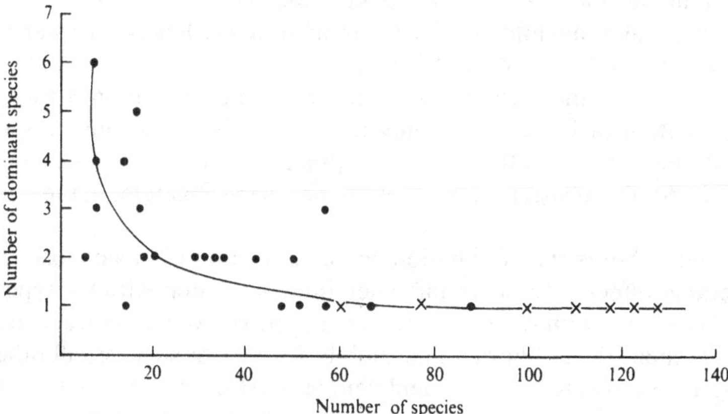

Fig. 15.1. Dependence of the total number of species in biocoenoses upon the number of dominant species forming the background in Possjet Bay. × – biocoenoses in which single dominant species form the background.

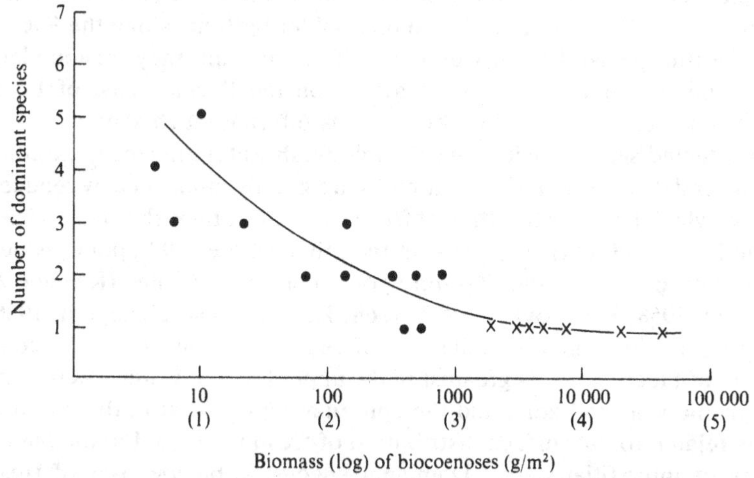

Fig. 15.2. Dependence of the total biomass of biocoenoses upon the number of dominant species. × – biocoenoses in which single dominant species form the background.

depths, but which lack these background forming organisms, the number of species is not more than a few dozen and the biomass at most several hundred grams per square metre. There is thus a tendency toward increasing species diversity and total biomass as the number of dominant species decreases (Figs. 15.1 and 15.2).

The species which form the background create additional ecological niches for existence and rich development of many species which would not exist in the biotope if these background forms were absent. It is necessary to take this into consideration in studying biocoenoses and in estimating energy levels within them.

It is evident that the bottom type determines the composition, structure and distribution of biocoenoses only in areas with similar physical and chemical conditions and that it is of no great importance in determining changes in dominant species caused by change in the character of the water masses.

Near open shores the distribution and composition of biocoenoses are considerably affected by wave influence in combination with the type of bottom. In places with sandy, gravel and pebble or boulder bottoms only firmly attached, deeply buried and mobile forms can survive; all others are cast ashore by storm waves and cannot exist over large areas under such conditions. They may occur in small numbers where special refuges exist. On rough surfaces rich with refuges (for example rocky bottoms) the diversity and abundance of life is far higher and organisms which are absent under the conditions described above can be very numerous. For example, echinoids of the family Strongylocentrotidae, sea stars and some gastropod species are seldom found on boulder bottoms along the Pacific shores of Iturup Island down to depths of 20–30 m, while they are abundant under similar conditions of wave action on the Pacific coast of Paramushir, on rough stony and rocky bottoms offering much shelter.

In protected sections with considerable fresh water influence, the composition and distribution of communities are greatly influenced by reduced salinity, which limits distribution of fresh water and brackish water species on one hand, and marine species on the other, at the 5–9‰ point, which agrees with data from the literature for other water bodies (Remane & Schlieper, 1958; Khlebovich, 1962, 1968; Hartog, 1964; Hedgpeth, 1966; and others). The degree of influence of reduced salinity upon the composition of biocoenoses is greatest in the upper layer of water, decreasing through the seaweed zone and the epifauna, and is least in the infauna. This is related to the surface distribution of fresh water and to the higher salinity of interstitial water. Dominant species in biocoenoses of fresh waters are often found in brackish waters, such as *Corbicula japonica*, *Assiminea lutea*, *Ovassiminea possietica*, *Fluviocingula nipponica*, *Anisogammarus kugi* and others, in the sections of the Far East investigated, and *Gammarus duebeni*, *Gammarus zaddachi* and others, in the northern seas.

The greatest influence upon the distribution, composition and biogeographical structure of biocoenoses is exerted by the temperature regimes of the different water masses. It should be noted that averages and fluctuations in temperature in different layers of one water mass are sharply distinct from averages and fluctuations in bionomically different parts of even one geographical region.

In the protected parts of southern Primorje, southern Sakhalin and southern Kuril Islands which have been studied, the biocoenoses contain many more warm-water elements than are found at similar depths in adjoining unprotected parts of the same regions. For example, contained and summer-warmed bays and lagoons near the coasts of southern Primorje and southern Sakhalin are more similar in type of biocoenosis and in their biogeographical composition than are the open and unprotected areas within the limits of each region (Golikov, 1965, 1966; Golikov & Scarlato, 1968). The protected parts of both regions have a considerable number of species of subtropical origin (such as *Crassostrea gigas*, *Venerupis japonica*, *Musculista senhousia*, *Batillaria cumingii*, *Stichopus japonicus* and others), which form a large biomass and become dominant in the communities. The character of the fauna and flora of bays and lagoons in the regions studied is close to that of subtropical regions. In Ismena Bay near the southern extremity of Kunashir Island the fauna is also more warm-water in affinity than in the adjoining open parts off this Island.

Along relatively open coasts of the regions studied, the fauna and flora are of lower temperature type and here low boreal and pan-boreal species prevail. It is interesting to note here, that in spite of the fact that the open coastal waters of southern Sakhalin and the southern Kuril Islands are under the influence of branches of the warm Tsuchima Current, whereas the waters of southern Primorje are under the influence of the cold Primorsky Current, the biota of the latter region, in accordance with its latitude, is of much warmer-water affinity than that of the former. Thus some warm-water background forming species, dominant in a number of biocoenoses of south Primorje (e.g. *Crenomytilis grayanus*, some *Sargassum* algae, etc.) are of second importance in the biocoenoses, or have disappeared completely, near the shores of southern Sakhalin and the southern Kuril Islands.

Waters of the areas studied cool down to a negative temperature in winter time and are covered with ice over very large areas; and organisms, even those which originated in warm water, as was shown by seasonal observations, tolerate winter lowering of temperature and even freezing into the ice. This shows that the distribution of organisms of the regions studied is primarily determined by summer water temperatures which limit reproduction and survival of species in accordance with the law of Hutchins (1947). The primary results of investigations in the White Sea show that in that region the protected areas have more warm-water elements in the fauna and flora than do open areas. The penetration of warm-water organisms from one protected summer-warmed area into another could apparently happen during periods of climatic warming owing to the heating of the surface layer in the intermediate regions.

The published literature shows that in waters in which summer heating and winter cooling of the surface layer override the thermal influence of the prevailing currents, the dominant factor determining the distribution of organisms throughout the shallow waters is not the thermal balance of the currents but the thermal regime of the coastal waters, which is determined by heat exchange between sea and atmosphere. Take, for example, the European boreal waters and the subtropical waters of Japan. Warm-water species carried by the warm waters of the North Atlantic Drift along the shores of Europe, and by the Kuro-Sio Drift along the Japanese Islands, penetrate far to the North along the open coasts. Half-enclosed areas on the coasts of western and northern Europe, (Grieg, 1914; Soot-Ryen, 1924; Crisp, 1964*a*, *b*; Cushing, 1966; and others) and near Japan (Habe & Horikoshi, 1966), are characterized by much more colder-water type fauna than are the neighbouring open areas. This is apparently correlated with poor tolerance of lowered temperatures in winter shown by subtropical European and tropical Japanese species (Blegvad, 1929; Crisp, 1964*a*, *b*; and others). Along the open shores of Europe and Japan the influence of powerful warm currents reduces seasonal fluctuation in water temperature and counteracts the disasterous influence of low temperatures in extremely cold winters upon populations of shallow waters.

Comparison of investigations in the Sea of Japan, the Okhotsk Sea and adjoining parts of the Pacific Ocean, with data on the distribution of organisms near the shores of northern Europe and Japan shows that there is no one law controlling the distribution of organisms on the upper parts of the shelf of the World's Oceans. In waters in which summer heating of the surface layer exceeds the thermal influence of the constant currents, the degree of thermophily in populations from the upper parts of the shelf increases with their remoteness from the open sea; and exchange of heat between the sea and atmosphere becomes the dominant factor in the distribution of coastal organisms. On the other hand, in coastal waters which are under the influence of powerful warm currents, overriding the temperature influence of the heat exchange between sea and atmosphere upon the surface water, the warm-water species live off the open coasts and are replaced by cold water forms in the protected areas isolated from the warm currents.

In the communities of the upper parts of the shelf in the high Arctic near Franz-Josef Land, the mean number of individuals in species populations is 12–15 times smaller and the biomass 3–5 times smaller, than those of even relatively poor communities in temperate waters, the numbers being on the average 700 individuals/m^2 and (biomass) 400 gm/m^2 in depths of 0–20 m (Golikov & Averincev, 1971). The low abundance and biomass in shallow water regions in the high Arctic is related to the absence or impoverishment of populations in the littoral zone and the scarcity, in most regions, of algae and animals associated with them; and even where algae are present, the numbers of associated animals and epiphytes, etc., are small. The main factors which limit development of life in the high latitudes of the Arctic are lack of light, constant or prolonged ice cover and the mechanical effects of ice, especially of icefloes which plough into the bottom in some places to a depth of 20 m. In the investigated regions of the Arctic (as in temperate and subtropical waters) differences in the character of the distribution of life in bionomically different sections are observed. In the open sections with strong water currents, where the water is periodically free of ice, macrophyte populations develop on the upper parts of the shelf, together with considerable populations of animals (for instance near Victoria Island and Heis Island). On the sections which are covered by ice throughout the year, macrophytes are almost absent, but cryopelagic biocoenoses develop on the lower surface of the pack ice (Golikov & Averincev, 1971), which supply the benthonic populations with organic matter (e.g. in Teplitz Bay on Rudolf Island, and Polar Bay, Alexandra Land).

The vertical distribution of biocoenoses and of the species composing them, is very closely connected with the temperature regime of the water layers and the position of the thermocline at different times of the year. The greatest species diversity of fauna and flora and the maximum quantity of warm-water species in Possjet and Aniva Bays has been observed at a depth of 0–5 m; below that depth sharp changes in the composition of the populations take place, with a decrease in the number of warm water species (Golikov & Scarlato, 1968; and others). This zone is determined by the position of the upper layer of the Japan Sea surface water near the shores of south Primorje and by the position of the upper layer of Okhotsk sea surface water in Aniva Bay. In summer, the water of this zone is characterized by high temperature, which reaches 16–20 °C and to a large extent depends upon the time of day and on weather conditions. Near the Pacific shores of the Kuril Islands the upper zone of the sublittoral region is located at depths from 0 to 10 m in accordance with the position of the upper layer of the Pacific surface water.

The second distinct boundary in the vertical distribution, as can be seen, for example, from the analysis of the distribution of molluscs in Possjet Bay (Golikov & Scarlato, 1966, 1967*b*), is the 25-m depth, at which the subtropical fauna disappears, to be replaced by boreo-arctic species, and at which the interface between the surface water and the spring thermocline occurs. In the White Sea, which seems to be a two-layered water body, the boundary between the boreal and the arctic fauna is at approximately the same depth.

The third zone, which is distinguished in Possjet Bay, for example, on the basis of the distribution of molluscs, is located at depths between 25–30 m and 50–60 m, is characterized by a peculiar composition of fauna and corresponds to the position of the intermediate layer of the Japan Sea upper-water mass. The lower border of this zone, whose waters undergo noticeable seasonal changes in hydrological conditions, is defined by the position of the sharpest summer thermocline.

The fourth zone extends downward from the 55–60 m depth and its fauna is characterized by a sharp increase in the number of boreo-arctic species. The water of this zone is the lower layer of the Japan Sea surface-water mass and possesses comparatively homogeneous hydrological properties.

The data show the distinct dependency of the composition of the benthos upon the physico-chemical properties of the layers of water, and show the necessity of evaluating the distribution of life in terms of the position of water layers when establishing the vertical zonality of the sea.

These principles for the establishment of the vertical zones in the sea have, from our point of view, a number of advantages over other methods, since they do not depend on the transparency of the water or on the degree of penetration of light (the photic principle of Pérès and many other authors), or on the presence or absence of plants (the biotic principle), or on the nature of the substrate and topography of the bottom (the bathymetric principle). The separation of vertical zones according to the general character of the populations and in relation to the physico-chemical properties of the water column of a given layer in the water mass, permits a comparison of different areas and a uniform definition of marine vertical zones.

In the high latitudes of the Arctic the vertical zones of the distribution of biocoenoses, which have been observed in the upper zones of the shelf, may depend upon the duration and position of a layer whose temperature is slightly raised in summer (to −0.3 to −0.6 °C); upon the type of ice cover and upon the mechanical effects of icefloes.

The study of the patterns of distribution of organic resources in the sea at the present time is important in connection with the ever-increasing intensity of commercial activity and the sharp decrease in many commercial species, including invertebrates. There is also a great need for the study of the production possibilities of species of practical importance and of ways of increasing their production.

Investigations in 1965 and 1966 of the phenology of the sea bottom, of succession in biocoenoses and of the ecology and ontogeny of the commoner species in Possjet Bay, made it possible to work out a theoretical recommendation for the creation in southern Primorje of submarine farms for the breeding and cultivation of bivalve molluscs: *Mizuchopecten yessoensis*, *Spisula sachalinensis*, *Crassostrea gigas* and some other animals (Golikov & Scarlato, 1969). It has been established that in April and May, when the temperature of the water rises from 2–3 °C to 10–11 °C, quick maturation of the gonads takes place in all three species, and in many other low boreal species. At the end of May, when the water temperature rises to 12–13 °C, mass spawning of these animals begins. The subtropical species *Crassostrea gigas* and *Stichopus japonicus* start their spawning when the water temperature is 18–19 °C at the end of June and the beginning of July. By the middle of June the veliger larvae of low boreal species of bivalves appear in the plankton in large numbers and by the end of June they are among the dominant forms in neritic waters. The conditions in the plankton in this period prove to be favourable for the normal development and growth of the pelagic stages of these molluscs; the temperature is right, the salinity varies very little, and there is abundant food and relatively few predators. The development continues during July and August in somewhat less favourable conditions, in that rains cause some freshening of the surface water. In shallow enclosed bays which are heated more quickly and where summer temperature is higher, the maturation of sexual products, spawning and settling of young take place 6–10 days earlier than in cooler open regions.

The study of the ecology of the settling of the young of these organisms, taking into consideration the known data on chemotaxis (Allee, 1931; and others) and colour orientation (Thorson, 1964; and others) of the larvae, in settling, made it possible to apply the knowledge of this important stage of ontogenesis to the cultivation of commercial species. It turned out that the abundant larvae of *Mizuchopecten* settle in July on algae, mainly *Sargassum* situated in a fairly narrow belt along the coast; the young of *Crenomytilus grayanus* and *Modiolus difficilis* are concentrated on the byssus of adult individuals of the same species; the young of *Mytilus edulis* on various submerged objects and on submarine plants; and the young of *Spisula sachalinensis* settle on sandy bottoms. In August the larvae of oysters settle on firm substrates of light colour, and the young of *Stichopus* settle on small algae. The high fecundity of these species and the success of survival of their larvae in the plankton, led us to consider the lack of substratum for settling and further development of the young to be one of the most important factors limiting their numbers. This circumstance, reinforced by data contained in the literature, was used in solving the problem of artificial cultivation of commercial species, by means of the use of artificial substrates as collectors, replacing algae and marine grasses by unravelled sisal ropes, pieces of rope, and cotton nets for settling *Mizuchopecten* larvae; birch brooms and bristles for the settling of *Mytilus edulis* larvae; empty shells of masses of sea scallops, oysters and mussels for settling *Crassostrea gigas* larvae. On these collectors used in July (for low boreal species) and August (for subtropical species) a hundred times as many young molluscs settled as on corresponding units of surface under natural conditions. On suspended artificial substrates the rate of growth of young molluscs is considerably higher than on the bottom, the supply of food and dissolved oxygen being better. The further growth of the attached young of *Mytilus edulis* and of oysters is limited only by the size of the artificial substrates.

The young of *Mizuchopecten yessoensis*, during growth, move along the substrate, whether natural or artificial, into deeper water and then become detached, changing to the benthonic mode of life. Observations in nature, and experiments on the transference of artificial substrata with young animals to the bottom, have shown that in the first post-larval stages of development the numbers of these edible molluscs are to a great extent limited by predators, primarily by starfishes, which consume large quantities of young individuals. As a protection against these predators, suspended methods of culture are suggested, together with the periodical removal of predators.

The numbers of *Spisula sachalinensis*, the young of which settle in shallow water with sandy bottom, are limited in the first post-larval stages not only by predators but to a considerably higher degree by the influence of waves, which throw up on the land young molluscs which are unable to dig themselves in deep enough. For the cultivation of these and other similar edible bivalves, it is recommended that the young thrown up on the land be collected and transferred to areas of similar conditions but protected from the surf.

An experimental submarine farm has now been organized in Possjet Bay.

To measure the production rate of populations of abundant commercial species, or of entire biocoenoses, it is necessary to analyse the growth and reproductive characteristics of species which are significant in the energy cycle of the ecosystem. Marine ecosystems are open, and constant control observations of species populations often prove to be impossible, so that the use of production calculation techniques by the method of Boysen-Jensen (1919) or its modification (Winberg, 1968) under remote sea conditions is not feasible. These difficulties, however, are to a considerable extent overcome by using the 'static-dynamic' principle of production identification. This principle is based upon the analysis of the size–weight structure of the population as a result of the annual production process (Golikov, 1970*b*). The population structure itself, as shown by the frequency distribution of individuals of different size or weight for species with interrupted reproduction, can be the basis for establishing the nature of growth and duration of life of species under given conditions.

According to this method, the annual growth of a stationary population represents the increase in weight of all individuals present at a given time and is expressed by the formula:

$$P_g = \sum_{i=1}^{n} N_i(W_i^{t+1} - W_i^{t}) \tag{15.1}$$

Where N_i is the number of individuals of the given generation at the moment of observation; $W_i^{t+1} - W_i^{t}$ is their weight increase per year. The average annual production represents the sum of annual productions, determined at the different moments of observation (seasons) divided by the number of observations.

$$\left(\frac{P_g^1\ P_g^2 \ldots P_g^i}{n}\right) \tag{15.2}$$

Verification of the suggested method using a mathematical model of a population of gastropod molluscs, realized on an electronic computer (where the production was estimated with absolute accuracy, which is impossible in the natural environment) has shown that the growth determined by formulae 15.1 and 15.2 turns out to be a measure of net production in its accepted meaning (Golikov & Menshutkin, 1971). The observation of population growth by means of four samples taken in different seasons lets us determine production with an accuracy differing from the model by less than 5%. Determining the mean production by one sample gives the best results if the sample is taken approximately in the middle of the period of appearance of the young (for most temperate–water species this takes place in summer or the beginning of autumn). To determine the quantity of matter formed during the year and remaining in the population, the idea of 'supporting production' has been suggested. This production is the biomass of individuals under one year old (B_o), plus the weight increase of older age groups per year at the moment of observation ($W_i^{t} - W_i^{t-1}$) and is approximately expressed by the equation:

$$P_s = B_o \sum_{i=1}^{n} N_i(W_i^{t} - W_i^{t-1}) \tag{15.3}$$

The mean annual supporting production is determined by the same method as is used for the mean annual growth production. Functionally, the supporting production maintains the dynamic balance of the population number at a certain necessary level, given the known present life duration of the organisms. The rate of turnover of matter of the supporting production (P_s/B coefficient) is related by linear inverse dependence to the age of individuals prevailing in the population and can be calculated by the straight-line equation.

By this method the life duration, character of growth, seasonal dynamics, and production properties of the commercial trade bivalves *Mizuchopecten yessoensis* (Jay) and *Spisula sachalinensis* (Schrenck) (Golikov & Scarlato, 1970*c*) and of some species of gastropod molluscs (Golikov, 1970; Menshutkin & Golikov, 1971) near the coasts of South Primorje, have been studied. It appears that *M. yessoensis* and *S. sachalinensis*, which are of great economic importance, have life durations established by the analysis of the size-weight structure of the populations in Possjet Bay, of ten and eight years respectively. Maximum mortality was observed in the young at the age of one year; less than 4% of *M. yessoensis* and about 40% of *S. sachalinensis* survive until the age of about one year. Those young individuals which have survived disperse benthonically. They begin to form aggregations on favourable biotopes only at more than three years of age, after reaching maturity. The maximum number, owing to the mass appearance of the young in the populations and the rate of turnover of organic matter (*P*/*B* coefficient), of *Mizuchopecten yessoensis* has been observed at the end of July, and of *S. sachalinensis* in October. The mean annual growth production of *M. yessoensis* (the averaged biomass being 130 gm/m^2), was 130 gm/m^2/year and the supporting production was 32 gm/m^2/year. The corresponding values for *S. sachalinensis* (the averaged biomass being 820 gm/m^2) were 451 gm/m^2/year and 329 gm/m^2/year (Golikov & Scarlato, 1970).

The specific populations investigated are in balance with the external environment; that is, in dynamic balance assuming fluctuations of abiotic factors to be of the same order from year to year. Such stationary populations, in the absence of sharp changes in the normal abiotic conditions, and of excessive fishing or predation, can exist for an indefinite period of time. It is clear that norms of fishery, and the prognosis of possible increase in the number of useful species, should be established on the basis of knowledge of the production potential of their populations.

It is obvious that the annual growth production of a stationary population is equal to the amount of matter lost from the population during a year. Therefore only by means of protection measures, and by maximum increase of the survival of the young can the present biomass of a population be increased. Further increase of biomass, and of the proportion which may be subjected to fishery without catastrophic consequences for the population, is possible only by artificially increasing its size by means of marine-culture.

From the point of view of production theory and the scientific regulation of fisheries, the catching of specific populations with different age structures and numbers is feasible, and naturally it is necessary to know and take into consideration the auto-regulation peculiarities of the populations and of their environment. Further study of production processes in natural specific populations and in entire ecosystems such as those carried on in accordance with the International Biological Programme and its continuation 'Man and the Biosphere' promises a real contribution to the scientific working out of methods of regulation of marine processes and to the increase of the yield of our marine biological resources.

References

Allee, W. (1931). *Animal aggregations*, pp. 1–431. Chicago: University Press.

Blegvad, H. (1929). Mortality among animals of the littoral region in winter ice. *Report of the Danish Biological Station*, **35**, 51–62.

Boysen-Jensen, P. (1919). Valuation of the limfjord. I. Studies on the fish-food in the Limfjord 1909–1917. *Report of the Danish Biological Station*, **26**, 1–44.

Chlebovitsch, V. V. (1962). Peculiarities of the aquatic fauna composition in relation to the salinity of the medium. (In Russian). *Zhurnal Obshchei Biologii*, **23**(2), 90–7.

Crisp, D. J. (1964*a*). The effects of the severe winter of 1962–63 on marine life in Britain. *Journal of Animal Ecology*, **33**, 165–210.

Crisp, D. J. (1964*b*). The effects of the winter of 1962/1963 on the British marine fauna. *Helgoländer wiss. Meeresunters.*, **10**, 313–27.

Cushing, D. H. (1966). Biological and hydrographic changes in British seas during the last thirty years. *Biological Reviews*, **41**, 221–58.

Golikov, A. N. (1965). Comparative ecological analysis of certain marine bottom biocoenoses in South Primorje and South Sakhalin and their exploitation. (In Russian). *The first conference of the All-Union Hydrobiological Society*, pp. 94–5. Nauka.

Golikov, A. N. (1966). Ecological peculiarities of coastal marine bottom biocoenoses of South Primorje and South Sakhalin in connection with hydrological conditions. (In Russian). In *International oceanographic congress*, II, pp. 136–7. Nauka.

Golikov, A. N. (1970). The method of determining productive properties of populations according to size and quantity. (In Russian). *Doklady Akademiia Nauk USSR*, **193**(3), 730–3.

Golikov, A. N. (1970*a*). Seasonal dynamics of the number and production properties of some biocoenoses of the upper part of the shelf near the shores of South Primorje. (In Russian). In *Biological processes in marine and continental waterbodies*. The second conference of the All-Union Hydrobiological Society, Kishenev, pp. 82–3. Moldavskoy SSR: Akad, Nauk.

Golikov, A. N. (1970*b*). The method of determining the production properties of populations by the size structure and number. *Doklady Akademiia Nauk USSR*, **193**(3), 730–3.

Golikov, A. N. & Averincev, V. G. (1971). Some regularities in the life distribution in the upper regions of the shelf of archipelago of Franz-Josef Jand. (In Russian). *Account session Zool. Inst. A.N. USSR. Summary report session on the results of works of 1970*, pp. 11–12. Leningrad: Izd. Nauka.

Golikov, A. N. & Menshutkin, V. V. (1971). Model study of production processes of populations of *Epheria turrita* (A. Adams). *Doklady Academiia Nauk USSR*, **197**(4), 944–7.

Golikov, A. N. & Scarlato, O. A. (1965). Hydrobiological explorations in Possjet Bay by means of the SCUBA diving technique. (In Russian). Fauna of the seas of the North-western Pacific. *Issledovani Fauny Morei.* (Investigation of the fauna of the seas), **3**(11), 5–29.

Golikov, A. N. & Scarlato, O. A. (1966). Ecology and distribution of molluscs in Possjet Gulf (Japan Sea) in connection with the physical-chemical regime and principles of zonation. *Abstracts of papers water science Proceedings, 11th Pacific Science Congress, Tokyo*, **7**, 13–14.

Golikov, A. N. & Scarlato, O. A. (1967*a*). Ecology of bottom biocoenoses in the Possjet Bay (the Sea of Japan) and the peculiarities of their distribution in connection with physical and chemical conditions of the habitat. *Helgoländer Wissenschaftliche Meeresuntersuchungen*, **15**, 193–201.

Golikov, A. N. & Scarlato, O. A. (1967*b*). Molluscs of the Possjet Bay (the sea of Japan). Molluscs and their role in biocoenoses and formation of fauna. (In Russian). *Proceedings of the Zoological Institute, Academy of Science USSR*, **42**, 5–154.

Golikov, A. N. & Scarlato, O. A. (1968). Vertical and horizontal distribution of biocoenoses in the upper zones of the Japan and Okhotsk seas and their dependence on the hydrological system. *Sarsia*, **34**, 109–16.

Golikov, A. N. & Scarlato, O. A. (1969). Scientific basis of the organization of directed submarine farms worked out by means of light diving equipment. (In Russian). In *Marine diving investigations*, pp. 60–6. Submarine Studies Oceanographic Commission AN, SSSR. Publ. Moscow: Nauka.

Golikov, A. N. & Scarlato, O. A. (1970*a*). Regularities of distribution of biocoenoses in the upper parts of the shelf of temperate waters depending upon character and structure of water masses. Biological processes in marine and continental water bodies. *Thesis of reports of the second All-Union Hydrobiological Society Kishinev*, 83–4.

Golikov, A. N. & Scarlato, O. A. (1970*b*). Hydrobiological exploration in cold and temperate waters of the USSR with light diving equipment (SCUBA). *International Revue Gesamten Hydrobiologie*, **55**(3), 305–15.

Golikov, A. N. & Scarlato, O. A. (1970*c*). Abundance, dynamics and production properties of populations of the edible bivalves *Mizuchopecten yessoensis* and *Spisula sachalinensis* related to the problem of organization of controllable submarine farms on the western shores of the sea of Japan. *Helgoländer wissenschaftliche Meeresuntersunchungen*, **20**, 498–513.

Golikov, A. N. & Scarlato, O. A. (1971). Some results of hydrobiological diving investigations in Possjet Bay (Sea of Japan). (In Russian). *Hydrobiologii Zhurnal*, **7**(5), 32–7.

Grieg, J. A. (1914). Bidrag til kundskaben om Hardangerfjordens fauna. *Bergens Museums Årbog*, **1**, 1–147.

Habe, T., Ino, I. & Horikoshi, M. (1966). Marine parks and animals in the Japanese waters. In *Marine Parks of Japan*, pp. 21–9. Nature Conservation Society of Japan.

Hartog, C. (1964). Typologie des Brackwassers. *Helgoländer Wissenschaftliche Meeresuntersunchungen*, **10**, 377–90.

Hedgpeth, J. W. (1966). Aspects of the estuarine ecosystem. A Symposium on Estuarine Fisheries. *American Fisheries Society Special Publ.*, N 3, 3–11.

Hutchins, L. W. (1947). The basis for temperature zonation in geographical distribution. *Ecological Monographs*, **17**, 325–35.

Khlebovich, V. V. (1962). Peculiarities of composition of the water environment depending upon salinity. *Zhurnal Obshchei Biologie*, **23**(2), 90–7.

Khlebovich, V. V. (1968). Some peculiar features of the hydrochemical regime and the fauna of mesohaline waters. *Marine Biology*, **2**(1), 47–9.

Menshutkin, V. V. & Golikov, A. N. (1971). Simulation of the population of a gastropod with the aid of the digital computer. *Oceanologia*, **11**(4), 695–9.

Remane, A. & Schlieper, C. (1958). Die Biologie des Brackwassers. Binnengewässer, **22**, 1–348.

Scarlato, O. A., Golikov, A. N. & Gruzov, E. N. (1964). The SCUBA diving technique for hydrobiological investigations. (In Russian). *Oceanologia*, **4**, 707–19.

Scarlato, O. A., Golikov, A. N., Vassilenko, S. V., Tzvetkova, N. L., Gruzov, E. N. & Nesis, K. N. (1967). Composition, structure and distribution of bottom biocoenoses in the coastal waters of the Possjet Bay (the Sea of Japan). (In Russian). *Explorations of the fauna of the seas*, **5**(13), 5–61.

Soot-Ryen, T. (1924). Faunistische Untersuchungen im Ramfjorde. *Tromsö Museums Årbog*, **45**(6), 1–106.

Thorson, G. (1964). Light as an ecological factor in the dispersal and settlement of larvae of marine bottom invertebrates. *Ophelia*, **1**(1), 167–208.

Winberg, G. G. (1968). *Methods for the estimation of production of aquatic animals*. (In Russian). Handbook and papers, ed. G. G. Winberg, pp. 5–169. Minsk: Izdatelstvo Vyssh. Shkota.

Index

Note: references in italics are to figures

Abra, acclimatization of, 310
Abra ovata, 296–7
abyssal zones, 272
Acanthophora spicifera, 53; preparation of, 230
Acartia clausi, 246
Acartia natalensis, 116
Achnanthes, 166
Acropora assimilis, 53
Acropora conferta, 53
Acropora erythraea, 52
Acropora hyacinthus, 52
Acropora indica, 52–3
Acropora squamosa, 53
Acropora surculosa, 52
Acropora syringodes, 53
Agulhas Bank, sediment distribution on, 114
Agulhas Current: copepod diversity, 101; neritic water fauna in, 102; species diversity, 94; warm, 94; zooplankton biomasses, 101
air temperature, phytoplankton bloom and, 23
Akkeshi Bay, *72*, 82–4; nutrients, 82–3; primary production, 82; respiration loss, 82; water temperature, 82; zooplankton biomasses, 82
algae: chromatic adaptation, 42; density in range of coral reefs, 52; *see also under individual species*
allochthonous organic matter, bacteria utilization of, 255
Amathillina cristata, 296
amberfish, grazing food chain, 79
Amblyops abbreviata, 159
Amblyops kempii, 159
amino-nitrogen, nitrate reduction to, 205
Ampharetidae, 300
Anchovy, *see Engraulis japonica*
Anguilla japonica, *76*
antarctic, *176*; trophogenic layer, 188
aphotic zone, 95
Arabian Sea: eutrophic benthos, 277
Aral Sea, 285; benthonic biocoenoses, 286; benthos food structure, 301–2; bottom fauna, 285–8; bottom population, 304; trophic changes in benthos, 305–*7*; trophic characteristics of invertebrates, 290–5
Ar-arucip, *see Caulerpa racemosa*
Arenicola marina, 207
Arctic: limitation on life development, 323; vertical distribution of biocoenoses, 324
Aspius aspius, 285
Assiminea lutea, 320
Asterias amurensis, in Sendai Bay, 80
Asterionella formosa, 213
Asterionella japonica, 214; photosynthetic rate, *42*
Asterionella kariana, 214
Atlantic Ocean, 184–7; density gradient main pycnocline, 184; maxima cell numbers main pycnocline, 186, 189; nutrient concentrations, 184; stations, *185*; trophogenic layer, 187; vertical distribution phytoplankton, 184
atmospheric pressure, current variations relationships to, 94
atomic ratio measurements, 199; *see also individual regions*
Austroglossus pectoralis: nepheloid layer and, 114; trawling of, 114
authochthonous brackish-water fauna, 286
autochthonous organic matter, bacterial utilization of, 255
Axelboeckia spinosa, 296
Azov Sea, 285; benthos food structure, 301-2; benthic structure, 305–6; biocoenosis, 286; bottom fauna, 285–8; bottom population, 304; trophic characteristics bottom invertebrates, 290–5

Backwaters and Estuaries, 38–50
bacteria: role in bethnic community, 144; biomass, 265; in tropical communities, 244
bacterial: interactions and detritus, 143–4; proteolysis, 207
bacterioplankton: in oligotrophic tropical waters, 254; tropical community production, 255
Balbalulang, *see Hydroclathrus clathratus*
Baltic Sea, 161
bathyal zones, 272
bathymetric samples, non-fixed, 235
bathyplankton: ecosystems energy transfer, 247; food web, *243*
Bay of Naples, phytoplankton in organic matter, 220
Benguela Current, 90–3; structural features, 90–*1*; potential production, 90
benthic: fauna, 104–9; crustacea, 118; *see also under individual regions and species*
bethonic invertebrates, 288; feeding habits of, 286–96, 301–5
benthos: general trophic structure, 301–5; quantitative collection, 269
Bering Sea: continental shelf, 274; trophic zonality, *275*; vertical distribution of phytoplankton, *178*
Biddulphia aurita, 213–14
Biddulphia regia, 214
Biddulphia sinensis, 43; photosynthetic rate, *42*
biocoenoses, *319*; temperature and, 320–1; *see also under individual regions*
black mangrove, *see Avicennia officinalis*
black mussels, toxicity levels in, 98
Black Sea: bathyplankton, 236, 238, 244; commercial fishing, 161; epiplankton, 236–7, *239*, 243–4; planktonic ecological groups, 242; production of phytoplankton, 157

Index

Bombay harbour bay: chlorophyll *a*, 47; primary production rate, 47
Boergesenia forbesii, 53
Boreomysis arctica, 160
Boreomysis nobilis, 160
Boreomysis tridens, 160
bothids, *76*
Bray–Curtis coefficient, 109
Bregmaceros japonicus, *76*
Bullia melanoides, 59; density, 60

Cabot Strait, oxygen concentration, 162–*3*
Calanoidea, 236; rate of exchange, *258*
Calanoides carinatus, 72, 100
Calanus finmarchicus, 101
Calanus helgolandicus, 244, 246
Calanus marshallae, 141, 143
Calanus pacificus, 137–8
Calanus plumchrus, 137–8, 140–1, 143; mortality of, 144
Calanus tonsus, 92
callionymids, 76
Calliurichthys japonicus, *76*
Canot-canot, *see Eucheuma muricatum*
Caocaoyan, *see Gracilaria calicornia, Gracilaria coronopifolia and Gracilaria verrucosa*
Cape hake, *see Merluccius capensis*
Cape Peninsula: production ratio, 96; standing crop, 96
carbon uptake measurements, 10, 145; *see also* primary production
Caspian Sea, 285; benthos food structure, 301–2; bottom fauna, 285–8; bottom population, 304; phytoplankton production, 157; trophic changes in benthos, 305–9; trophic characteristics bottom invertebrates, 290–5
Caulerpa racemosa, preparation of, 230
Central equatorial region, *176*; nutrient concentrations, 182; pycnocline density, 182–3; surface layer, 183; trophogenic layer, 187; vertical distribution of phytoplankton, *183*
centriceae, spring phytoplankton bloom and, 27
Centropages brachiatus, 100
Cerastoderma lamarcki, 296
Ceratium furca, photosynthetic rate, *42*
Chaetoceros lorenzianus, photosynthetic rate, *42*
Chaetoceros, 166–7
Chaetoceros radians, 213–14
chaetognatha, as biological indicators of water movement, 100
Chaetomorpha littorea, 53
Chironomus albidus, 300
Chironomus larvae, 300
chlorophyll, 17–19; see also under individual regions
chlorophyll *a*, *12*; cycle, 17; nitrate related, *18*; phaeophytin and depth to, 189; phosphate related, *18*; in sea-ice, 24, *26*; *see also under individual regions*
Cladophora fascicularis, 53
Clupea pallasii, Fraser River numbers of, *142*
Coastal water: Wadden Sea, 197–200; West Coast of India, 32–5
Cochin Backwater(s), 39; ^{14}C measurement, 39; euphotic zone, 39; microplankton, 40; nannoplankton, 40; nitrate-nitrogen, 39; phosphate-phosphorus, 39; primary production, 40–*1*; zooplankton, 43–4
Cochin beach: ^{14}C uptake, 60; chlorophyll *a*, 60; seasonal changes in feeding types to total biomass of, *61*
Codium intricatum, preparation of, 230
Codium papillatum, preparation of, 230
Colpomenia sinuosa, 53
community respiration, seasonal changes in, *41*
Continental shelf, fauna and species distribution across, *108–9*
copepoda, 236
copper, 212
Coral reefs, 50–6; genera numbers in, 51; primary production, 51
Corbicula japonica, 320
Corbulomya maetica, 296
Coscinodiscus radiatus, photosynthetic rate, *42*
Crassostrea gigas, cultivation of, 325
Crassostrea margaritacea, reproduction study of, 107
crustacea: benthonic, 298–300; herbivorous, 299
ctenophora, rate of exchange in, *258*
cutlassfish, *see Trichiurus lepturus*
culot, *see Gelidiella acerosa, Laurencia papillosa, Laurencia okamura and Acanthophora spicifera*
cycloida, rate of exchange in, *258*
Cyclosalpa pinnata, 101
Cyclotrichium meunier, red water and, 97
Cymodocea isoetifolia, 53, 58–9
Cymodocea serrulata, 59

dab, *see Limanda yokohamae*
Danish Fjords, microflora production on tidal-flats of, 223
Danish Wadden Sea, microflora production on tidal-flats of, 223
detritus, 143–4; bathyplanktonic consumption of, 244; epiplanktonic consumption of, 244; in food chains, 80–*1*; kinetics of decay, 146–7; nutritive value, 43; sinking rate, 263
Diastylopsis dawsoni, 83
Diatoma elongata, 213
diatoms, 20; in seasonal discontinuity layer, 181; silica on, 199; *see also individual species*
Dichtyocha sp., photosynthetic rate, *42*
Dictyosphaeria cavernosa, 53
Dikerogammarus haemobaphes, 296
dinoflagellates, 20
Dinophysis miles, photosynthetic rate, *42*
Diplanthera uninervis, 59
Diplocastrea heliopora, 53
dissolved organic carbon, seas of different climatic zones amounts, 77
Ditylum brightwellii, 214
Donax, population production assessment of, 62
Donax incarnatus, 59–60, 62
Donax serra, 109
Donax spiculum, 59–60

East China Sea: continental shelf, 274; shelf of, 274; trophic zonality, *275*; vertical trophic zonality of, 276; Yangtze discharge and, 279

East Coast of India, 35–44; ^{14}C measurements, 35; primary rates of production, 36–7; shelf region, 36
Eastern equatorial region, 176; 181–2; compensation point, 191; main pycnocline density, 181; maxima cell numbers in main pycnocline, 182; nutrient concentrations, 181; surface layer, 182; trophogenic layer, 187; vertical distribution of phytoplankton, *182*
Eastern Wadden Sea, data on, 206–7
Emerita holthuisi, 59; numbers of, 60
Ems estuary: chlorophyll values of, 216–17; organic carbon concentrations of, 218
energy transfer, food chain, 234
Engraulis capensis, cool upwelled coastal water and, 113
Engraulis japonica, 75–6, 112
Enteromorpha compressa L., preparation of, 230
Enteromorpha plumosa, preparation of, 230
Enteromorpha prolifera, 53
epinephelids, *76*
Epiplanktonic system, energy transfer efficiency of, 246–7
Erythrops abyssorum, 159
Erythrops microps, 159
Etrumeus micropus, *76*
Eucalanus spp., 100
Eucampia zodiacus, 214
Eucheuma muricatum, preparation of, 231
Euphausia lucens, 100, 102
Euphausia nana, respiration rate of, 75
Euphausia pacifica, 143
Euphausia similis, 74, 100
Euphausiacae, rate of exchange in, *258*
euphotic layer: nitrate replenishment of, 16; vertical microdistribution of plankton, 254
euphotic zone, 36; particulate organic matter values, 77; urea in, 78; *see also individual regions*
euryhaline organisms, 285
eutrophic structure 271–3

False Bay, microflora production on tidal-flats at, 223
Far Northern seas, phytoplankton production of, 15
Favia pallida, 52–3
Favites abdita, 52
filter-feeders, 236
Fosliella lejolissii, 53
Fraser River: plume, 140, 143; salinity, 135; silicate content, 135; vitamin B_{12} content, 137; *see also* Strait of Georgia
Frobisher Bay, 9–30; chlorophyll *a*, 11–*12*, 17; euphotic zone, 15; nutrients, 15, 17; photosynthetic activity, 20; phytoplankton bloom in, 22–4; phytoplankton production, 11, *13–14*; sea-ice, 24–8

gamet, *see Porphyra crispata*
gamgamet, *see Ulva lactuca* L.
Gaspé current, 155; primary production, 157
Gaspé passage, oxygen concentration profile of, *163*
Gayong-gayong, *see Halymenia durvillei*, 231
Gelidiella acerosa, eating preparation of, 230
Georgia saltmarsh, microflora production on tidal-flats of, 223
ghost crab, *see Ocypode ceratophthalam*
Glycera alba, 59, 62
gobioids, *76*
Goniastrea retiformis, 52–3
Gonyaulax catenella, 18
Gonyaulax grindleyi, white and black mussel poisoning by, 97
Gonyaulax polygramma, 97
Gonyaulax tamerensis, 99
Gracilaria calicornia, preparation of, 231
Gracilaria coronopifolia, preparation of, 231
Gracilaria verrucosa, preparation of, 231
grasping predators, 236
Great Australian Bight: suspension feeders in, 277; Yangtze discharge and, 279
Gulf of St Lawrence, 151–71; carbon fixation rates, *156*–7; circulation, 155; cold layer, 153; commercial fishing of, 161; estuary, 154, 156–7; horizonal surface water circulation of, 154–5; mean flushing time of, 155; pollution, 154; primary production, 157–8; salinity, 153; sea-ice, 164–8; secondary production, 158–60; temperatures, 153; tidal forces of, 153; upwelling, 153, 156
Gymnodinium galatheanum, 97
Gymnodinium mikimotoii, in Hiuchi-Bingo Nada, 84

Halimeda incrassata, 53
Haliotis midae, 109
halocline (Northern Pacific), 175
halophiles, salt tolerance in estuarian environment, 144
Halophyla ovalis, *59*
Halophyla stipulacea, 59
Halymenia durvillei, preparation of, 231
herbivores: biomass, 265; bottom invertebrates (Aral Sea), 312
herring larvae, *see Clupea pallasii*
Hiuchi-Bingo Nada, 84–6; carbon content, 84–6; nitrogen, 84–*5*; red tide of, 84; phytoplankton of, 84; water temperature, 84–5
homothermal water column, 197
horse mackerel, *see Trachurus japonicus*
Huso huso, 285
Hydroclathrus clathratus, preparation of, 230
Hymenopenaeus triarthrus, 106
Hymenosoma orbiculare, 118
hyperiidea, rate of exchange of, *258*
Hypnea cervicornis, 53
Hypnea charoides, preparation of, 231

ice: chlorophyll concentrations in, *166*; biota, 163–8; coloured sea-, 164; melting time of, 27; nitrate concentration in, *167*; phosphate concentration in, 166; salinity, 166; samples, 10, *164*
India, primary production rates in, 31, 37
internal subsurface waves, 155–6
interstitial water, metal exchange and, 212–13
isohalines, 197

Japanese barracuda, *see Sphyraena japonica*

Index

Japanese bluefish, *see Scombrops boobs*
Japanese codler, *see Bregmaceros japonicus*
Japanese dragonet, *see Calliurichthys japonicus*
Japanese eel, *see Anguilla japonica*
Japanese mackerel, *see Scomber japonicus*
Japanese stargazer, *see Uranoscopus japonicus*
Jasus lalandii, 104
Jasus tristani, 104
Johnius hololepidotus, 118

Kavaratti Atoll, *54;* ^{14}C uptake, 55; chlorophyll *a*, 55; zooplankton biomass, 55–6
Kavaratti bed: production rate of, 58; respiration in, 58
Kerala Backwater, *38*

Labrador current, 157
Lake IJssel, *198*; ammonia concentration, 207; freshwater discharge, 197; nitrate concentration, 207; nitrite concentration, 207; nutrient concentration, 199–200
lanternfish, *76*
Laurencia okamuria, preparation of, 230
Laurencia papillosa, 53; preparation of, 230
leiognathids, *76*
leptocephalus, *76*
Leveillea jungermannioides, 53
light: intensity range, 21; limiting phytoplankton production, 217; penetration measurement, 10; photosynthetic rate and, 19, *20*; phytoplankton bloom and, 23
Limanda herzenteini, 80
Limanda yokohamae, 80
littoral zone, 323
Lobophyllia corymbosa, 53
Long Island Sound, phytoplankton in organic matter of, 220
Lucifer, rate of exchange of, *258*
Lucioperca lucioperca, 285
Lumbriconereis latreilli, 62
Lumot, *see Enteromorpha plumosa and Enteremorpha compressa* L.

macrobenthos, quantity estimation of, 318
macrophytobenthic organisms, 303, 312
Mactra olorina, 59
Madapam bed, 58–9
manatee grass, *see Cymodocea isoetifolia*
Mandovi–Zuari estuarine complex, 44–7; hydrography, 44–5; nutrients, 45–6; primary production, 46; salinity, 44–5
Mangrove swamps: ^{14}C assimilation in, 58; chlorophyll *a*, 57; production rates, 57; seasonal changes, 56–7
Margalef's concept of succession (pelagic community), 253
Medusae, rate of exchange in, *258*
Meganyctiphanes norvegica, annual biomass of, 158
meiobenthos, quantity estimation of, 318
Merluccius capensis, 112
mesopelagic layers, 74
Mesopodopsis slabberi, 296, 298
metal: balance process, 212–13; removal from dissolved state, 211–12
Metapenaeus monoceros, 106; energy utilization of, 43
Metridia lucens, 100
Meuse River, 197
Michaelis–Menten equation for bacterial growth, 42–3
mineralization: aerobic respiration and, 210; sulphate reduction and, 210–11
Mizuchopecten, cultivation of, 324–5
Mizuchopecten yessoensis, 324–5, 328
Mohr's method, chlorinity determination by, 211
molluscs, 297–8; *see also individual species*
Monodactylus argenteus, 119
monsoon system, 37–8
Montipora joliosa, 52
mussels: chlorophyll filtration by, 216; toxicity distribution, 98; *see also individual species*
Mysidacea, rate of exchange in, *258*
Mysidetes farrani, 159
Mysis litoralis, 160
Mysis mixta, 160
Mytilus edulis, 207, 326

Nannocalanus minor, 100
Natal lobster, *see Panulirus homarus*
natural radioactive nuclide in plankton, 103
Neogobius melanostomus, 285
Nephrops andamanicus, 109
Nereis, acclimatization of, 310
Nereis diversicolor, 303, 300
nitrate-nitrogen; annual cycles, 11; phytoplankton productivity chlorophyll to, *22*
Nitzschia, 166
Nitzschia closterium, photosynthetic rate of, *42*
Noctiluca scintillans, 85
non-halophiles, salt tolerance in, 144
non-predatory invertebrates: in benthos, 276; relationships, 233; trophic grouping, 270
North Pacific, shelf vertical trophic zonality, 273
north-eastern Pacific, chlorophyll maximum, 191
North Sea: commercial fishing in, 161; nutrient concentrations, 199; phosphorus loss, 208; phytoplankton production, 157
Northern sub-tropical region, *176*; trophogenic layer of, 187
Nova Scotia, dissolved O_2 values in, 162
nutrition process, methodology of characteristics in, 235
Nyctiphanes capensis, 100, 102

Ocypode ceratophthalma, numbers of, 60
Oithona similis, 244
oligotrophic structure, deep-sea benthos, 271–2
Oliva gibbosa, 59
O_2 depletion, marine fauna mortality and, 97
Ostracoda, rate of exchange in, *258*
Ostrea algoenis, reproduction study of, 107
Ostrea atherstonei, reproduction study of, 107
Ovassiminea possietica, 320

Pacific Coast of Paramushir, 320
Palaemon elegans, 303
Panulirus homarus, factors affecting reproduction of, 105–6
Paracalanus parvus, 75

paralepidids, *76*
Paramysis lacustris, 296
Parathemisto abyssorum, 158
Parerythrops obesa, 159
particulate organic carbon (POC), in sea-waters of different climate, 77
Pegea confoederata, 101
pelagic animals, production/biomass and body length in, 259
pelagic ecosystem: mathematical model, 261–6; trophic interrelationship, 233
pelagic fish: environmental factors affecting recruitment of, 113; stock assessment, 110
Penaeid prawns, production environment of, 106
Penaeus indicus, 106
Penaeus marginatus, 106
Peridinium triquetrum, 97
Phaeocystis poucheti, 213
Philidium, 137–*8*
phosphate-phosphorus, 32; annual cycles, 11; phytoplankton productivity chlorophyll and, *22*
phosphate: chlorophyll *a* cycle and, 17; in sea-ice, 24–*5*
phosphorus compounds, 208–10
photosynthesis: minimum light for (in ice), 28; salinity on rate of, 41–*2*; urea uptake in, 78
phyllosoma larvae of *Jasus lalandii*, 102–3
phytoplankton: chlorophyll *a* to production, *19*, 21; low light adaptation, 21; nitrogen growth inhibition on, 199; in oligotrophic tropical waters, 254; *see also individual regions etc.*
pigment excretion, 74–5
Pinnixa rathbuni, annual production of, 81
Plagiogramma brockmannii, 214
planktonic larvae, distribution of, 102–3
Planktoniella sol, 93; rate of photosynthesis, *42*
Platygyra lamellina, 53
Pleurobachia, 137–*8*
Pocillopora damicornis, 52–3
Pocpoclo, *see Codium papillatum and Codium intricatum*
Pontogammarus maeoticus, 296
Porites spp., 52
Porphyra crispata, preparation of, 230
predator–prey relationship, 233; trophic relationship to, 145
primary production: *ATP* and organic carbon measurement of, 218–21; functional chlorophyll method estimation of, 215–17; measurements of, 10, 32, 138, 140, 257; plankton succession and, 43
Prorocentrum micans, 97
Protomedeia kryer, 83
Pseudocalanus elongatus, 244
Pseudocalanus minutus, 137–*8*, 143
Pseudodiaptomus charteri, *115*–16
Pseudodiaptomus hessei, salinity tolerance of, 116
Pseudodiaptomus nudas, 101
Pseudomma affine, 159
Pseudomma roseum, 160
Pteropoda, exchange rate in, 258
pycnocline ('main'), 175; aggregations in, 190; determination of, 174

red mangrove, *see Rhizophora mucronata*
red tide, 84
red water: bloom, 97; marine fauna mortality, 96–7; mussel poisoning, 98–9
regression curves: pelagic animals, *257*; planktonic crustacea, *256*
Rhine: metal load of, 212; nutrient concentration, 199–200; silt discharge, 201
Rhithropanopeus harrisii, 296, 310
Rhizophora mucronata, 56
rock lobster, environmental factors affecting, 104–5
round herring, *see Etrumeus micropus*

Sagami Bay, 71–8; chlorophyll *a* concentrations, 73; euphotic zone, 71; megalobenthos biomass, 75; particulate organic carbon values, 77; phosphate-phosphorus concentration, 71; primary production rate, 73; salinity, 71
Sagitta elegans, 137–*8*
Sagitta enflata, 102
Sagitta friderica, 100
Sagitta minima, 100
Salpa fusiformis, 101
salps, 101
Sandy beaches, 50–63; trophic ecology, 63; *see also* Cochin *and* Shertallai
Sardinops ocellata, 110; nursery stock and displacement, 113
Saurida elongata, *76*
Saurida undosquamis, *76*
Scheldt river, freshwater discharge from, 197
Scombrops boobs, *76*
Scomber japonicus, *76*
Scylla serrata, 118
sea conger, *see Anago anago*
seagrass beds, *see* Kavaratti bed *and* Mandapam bed
sea-ice, 24–8
seaweed utilization in the Philippines, 229–32
sedimentation, pre-monsoon and post-monsoon on, 39
Sendai Bay, *72*, 78–84; chlorophyll *a*, 78; ecosystem dynamics in, 80; hydrography, 78; nutrients, 78; primary production, 78
sedentary suspension feeders, 297
Shertallai beach: ^{14}C uptake, 60; chlorophyll *a*, 60; biomass, 62; seasonal changes in feeding types, *61*
Shimoda Bay: biomass, 72; *Ecklonia* community, 72; *Sargassum* of, 72
Sillago sihama, *76*
silver whiting, *see Sillago sihama*
Siphonophora, exchange rate in, *258*
Skeletonema costatum, 84, 213–14
snakefish, *see Trachinocephalus myops*
Sonneratia apetala, 56
Southern New England, microflora production on tidal-flats, 223
Southern sub-tropical region, *176*; trophogenic layer, 187
south-west Indian Ocean: organic production distribution, *94*; production rates, 95; zooplankton standing crop, *102*
Sphyraena japonica, 76
spotted-tail grinner, *see Saurida undosquamis*
Spyridia filamentosa, 53

Index

standing stock: chlorophyll *a* measurement of, 135; primary production measurement, 145
'static-dynamic' principle of production, 326
Strait of Georgia, 133–49; hydrographic conditions, 133–5; freshwater runoff, 134–5; primary production, *136*; 138–41; salinity, 135; secondary production, 140–1; standing stock, 135; tertiary production, 140–1; tidal cycle, 134; zooplankton, *138*
sub-Antarctic, trophogenic layer of, 188
sub-Arctic: isohalinity, 177; isothermy, 177; phytoplankton aggregations, 179; pycnocline, *178*; trophogenic layer, 187–8; vertical distribution of phytoplankton, 175–9
sub-tropical regions, 179–81; compensation point, 191; nutrients, 179; main pycnocline, 179–80; vertical distribution of phytoplankton, 179–*80*
Suruga Bay, data on, 71–8; chlorophyll *a*, 73; food web, *76*; pelagic community food web of, *74*; phytoplankton biomass, 73; primary production, 73

Tecticeps japonicus, 83
teleost fish, adaptation mechanism to salinity in, 119
Thalassiosira, 167
Thalassia hemprichii, 58
Thallasiothrix longissima, 93
Thalia democratica, 101
Thalia longicauda, 101
thermocline (tropical zone), 175
thiobacilli, 211
Thysanoessa inermis, annual biomass of, 158
Thysanoessa raschii, annual biomass of, 158
tidal-flats, 221–4; allochthonous matter in, 221–2; phaeopigment–chlorophyll ratio of, 221; sediment, 222
Timoclea imbricata, 59
Trachurus japonicus, *76*
Trachurus trachurus, 110
Trachypenaeus curvirostris, 106
Trade Wind Drift, phytoplankton and, 93
trawling: Agulhas, 114; South Africa, 109–10, *111*–12; South West Africa, 112; *see also individual regions*
Triceratium favus, *42*
Trichodesmium erythraeum, 53
Trichiurus lepturus, *76*
Triglids, *76*
trophic: chains, 234; models, 144–5; series of animals and plants, 27
trophic webs: in neritic ecosystems, 235–43; in oceanic ecosystem, 235–43
tropical ecosystems: detritus consumption, 244; energy transfer efficiency, 247
turbidimetry, sulphate determination, 211
Turbo sarmaticus, 109
turtle grass, *see Thalassia hemprichii*

Ulva lactuca, 53; eating preparation of, 230
unialgal cultures, photosynthetic rate in, 41–2
Upogebia africana, 117–18
Upwelling, 90–3; copepoda in cold, 100–1; geological influence upon, 91–2; induction of, 91; species in warmer, 100
Uranoscopus japonica, *76*

Vellar–Coleroon estuarine system, 47–50; chlorophyll *a*, *b* and *c* values, 47; nutrients, 49; primary production, 49–50; water temperature, 48
vertical distribution of: biocoenoses, 323–4; phytoplankton, 173–5, 187–93; zooplankton, *265*
vertical mixing, 197

Wadden Sea: hydrography, 197–200; material transport, 197–200; mineralization, 209–10; nitrate, 202; nutrient concentration, 199, 201; particulate organic matter, 218; pennate benthic diatoms, 214; phosphate-phosphorus, 202; phytoplankton composition, 213–14; phytoplankton concentrations, 219–20; primary production, 213–25; silica, 201; sediment, 200, 212; tidal-flats, 221–4; tidal watershed, 212
West of Anticosti, oxygen concentration profile of, *163*
West Coast of India, primary production rates, 33, 35
Western equatorial region, *176*, 184; main pycnocline, 182, 184; nutrient concentration, 184
Western India: eutrophic benthos, 277; shelf oligotrophy of benthos, 278
Western Wadden Sea, 205–6; ammonia concentration, 207; chlorophyll values, 216–17; microflora production, 223; nitrate concentration, 206–7; nitrite concentration, 207; organic phosphorus balance in, 210

Ythan estuary, microflora production, 223

zinc, 212
zoogeographic boundaries, 280
zooplankton: biomass, 265; primary productivity and biomass of herbivores, 135; on phytoplankton production, 144; water temperature on concentration of, 100–1; *see also individual species and regions*